HAZARDOUS MATERIALS INCIDENTS

3RD EDITION

HAZARDOUS MATERIALS INCIDENTS

3RD EDITION

CHRIS HAWLEY

DELMAR
CENGAGE Learning

Australia • Brazil • Japan • Korea • Mexico • Singapore • Spain • United Kingdom • United States

Hazardous Materials Incidents, Third Edition

Chris Hawley

Vice President, Technology and Trades ABU: **David Garza**

Director of Learning Solutions: **Sandy Clark**

Managing Editor: **Larry Main**

Product Development Manager: **Janet Maker**

Senior Product Manager: **Jennifer A. Starr**

Marketing Director: **Deborah S. Yarnell**

Marketing Manager: **Erin Coffin**

Marketing Coordinator: **Patti Garrison**

Director of Production: **Patty Stephan**

Production Manager: **Andrew Crouth**

Content Project Manager: **Cheri Plasse**

Art Director: **Bethany Casey**

Technology Director: **Joe Pliss**

Editorial Assistant: **Maria Conto**

Cover Image: **Computer Sciences Corporation (CSC)**

ISBN-13: 978-1-4283-1796-3

ISBN-10: 1-4283-1796-1

Delmar Cengage Learning
5 Maxwell Drive
Clifton Park, NY 12065-2919
USA

Cengage Learning products are represented in Canada by Nelson Education, Ltd.

For your lifelong learning solutions, visit **delmar.cengage.com**

Visit our corporate website at **www.cengage.com**

Notice to the Reader

Publisher does not warrant or guarantee any of the products described herein or perform any independent analysis in connection with any of the product information contained herein. Publisher does not assume, and expressly disclaims, any obligation to obtain and include
information other than that provided to it by the manufacturer. The reader is expressly warned to consider and adopt all safety precautions that might be indicated by the activities described herein and to avoid all potential hazards. By following the instructions contained herein, the reader willingly assumes all risks in connection with such instructions. The publisher makes no representations or warranties of any kind, including but not limited to, the warranties of fitness for particular purpose or merchantability, nor are any such representations implied with respect to the material set forth herein, and the publisher takes no responsibility with respect to such material. The publisher shall not be liable for any special, consequential, or exemplary damages resulting, in whole or part, from the readers' use of, or reliance upon, this material.

Printed in Canada
2 3 4 5 6 12 11 10 9 8

This text is dedicated to the memory of Chief John Eversole, a man who will be missed by the emergency response community.

CONTENTS

Introduction x
Foreword xi
Preface xii
Acknowledgments xv
Reviewers xvi
About the Author xvii
Introduction to Hazardous Materials xviii
NFPA 472 Correlation Guide xix

Chapter 1
Hazardous Materials:
Laws, Regulations, and Standards 1

Objectives 3
Introduction 3
Laws, Regulations, and Standards 4
   Development Process 4
Emergency Planning 5
   State and Local Emergency Response Committees 5
   Local Emergency Response Plans 5
   Chemical Inventory Reporting 6
OSHA HAZWOPER Regulation 8
   Paragraph q 10
   Medical Monitoring 10
Standards 11
   NFPA 472 11
   NFPA 473 12
Standard of Care 12
Additional Laws, Regulations, and Standards 12
   Hazard Communication 13
   Superfund Act 13
   Clean Air Act 13
   Respiratory Protection 13
   Firefighter Safety 13
   NFPA Chemical Protective Clothing 14
Summary 14
Key Terms 14
Review Questions 15
Suggested Readings 15

Chapter 2
Recognition and Identification 17

Objectives 19
Introduction 19
Location and Occupancy 19
Placards, Labels, and Markings 20
   Placards 20
   Problems with the Placarding System 31
   Labeling and Marking Specifics 32
Other Identification Systems 32

NFPA 704 System 32
Hazardous Materials Information System 33
Military Warning System 34
Pipeline Markings 34
Container Markings 36
Pesticide Container Markings 36
Radiation Source Labeling 37
Containers 37
   General Containers 37
   Totes and Bulk Tanks 40
   Pipelines 41
   Radioactive Material Containers 42
   Highway Transportation Containers 43
   Response to Potential BLEVE Situations 50
   Specialized Tank Trucks 51
   Intermodal Tanks 56
   Rail Transportation 58
   Bulk Storage Tanks 62
Senses 67
Chemical and Physical Properties 67
   States of Matter 68
   Vapor Pressure 69
   Vapor Density 70
   Specific Gravity 72
   Corrosivity 73
   Chemical Reactivity 74
   Flash Point 74
   Auto-Ignition Temperature 74
   Flammable Range 74
   Containers and Properties 75
Radiation 76
   Types of Radiation 77
   Toxic Products of Combustion 78
Summary 78
Key Terms 79
Review Questions 80
Endnotes 81
Suggested Readings 81

Chapter 3
Information Resources 83

Objectives 85
Introduction 85
Emergency Response Guidebook 85
Material Safety Data Sheets 100
Shipping Papers 103
   Mode of Transportation 105
Facility Documents 106
Computer Resources 106
Chemtrec 107
Reference and Information Texts 107

Industrial Technical Assistance 108
Summary 109
Key Terms 109
Review Questions 109
Suggested Readings 110
Additional Resources 110

Chapter 4
Protection 111
Objectives 113
Introduction 113
Health Hazards 113
Toxicology 114
Types of Exposures 114
Types of Hazards 114
Categories of Health Hazards 115
Radiation Hazards 116
Radiation Protection 119
Routes of Exposure 119
Exposure Levels 121
Types of PPE 124
Self-Contained Breathing Apparatus 124
Chemical Protective Clothing 126
Level A Ensemble 126
Level B Ensemble 127
Level C Ensemble 129
Level D Suits 130
High Temperature Clothing 130
Cold Temperature Clothing 131
Limitations of PPE 131
Heat Stress 131
Mobility 133
Visibility 133
Communications 133
Summary 134
Key Terms 134
Review Questions 135
Endnote 135
Suggested Readings 135

Chapter 5
Protective Actions 137
Objectives 139
Introduction 139
Incident Management Systems 139
Hazardous Materials Management Processes . . . 142
Isolation and Protection 142
Rescue 143
Site Management 146
Establishment of Zones 148
Evacuations and Sheltering in Place 150
Common Incidents 153
Types of Releases 153
Explosives 155

Gases 157
Flammable and Combustible Liquids 161
Flammable Solids/Water Reactives/Spontaneously
   Combustible 161
Oxidizers and Organic Peroxides 167
Poisons 168
Radioactive Materials 169
Corrosives 170
Other Incidents 171
Decontamination 173
Types of Decontamination 173
Emergency Decontamination 173
Technical Decontamination 174
Fine Decontamination 175
Mass Decontamination 175
Decontamination Process 180
Methods of Decontamination 183
Absorption 183
Adsorption 184
Chemical Degradation 184
Dilution 184
Evaporization 184
Isolation and Disposal 184
Neutralization 184
Sterilization 184
Solidification 184
Vacuuming 184
Washing 185
Summary 185
Key Terms 185
Review Questions 186
Endnotes 186
Suggested Readings 186

Chapter 6
Product Control and Air Monitoring 189
Objectives 191
Introduction 191
Defensive Operations 191
Absorption 191
Diking and Damming 192
Diverting 194
Retention 196
Dilution 196
Vapor Dispersion 196
Vapor Suppression 197
Remote Shutoffs 197
Air Monitoring at the First Responder
   Operations Level 197
Regulations and Standards 198
Air Monitor Configurations 198
Meter Terminology 199
Bump Test 199
Calibration 199
Reaction Time 200
Recovery Time 200
Relative Response 200

Oxygen Monitors 200
Oxygen Monitor Limitations 202
Flammable Gas Indicators 202
Flammable Gas Sensor Types 203
Toxic Gas Monitors 203
Tactical Use of Multi-Gas Detection Devices 204
Other Detectors 205
Carbon Monoxide Incidents 207
Summary 209
Key Terms 209
Review Questions 210
Suggested Readings 210

Chapter 7
Terrorism Awareness 211

Objectives 213
Introduction 213
Types of Terrorism 220
Potential Targets 222
Indicators of Terrorism 223
Incident Actions 225
General Groupings of Warfare Agents 227
Nerve Agents 227
Incendiary Agents 228
Blister Agents (Vesicants) 228
Blood and Choking Agents 229
Irritants (Riot Control) 229
Biological Agents and Toxins 230
Radioactives 232
Other Terrorism Agents 234
Detection of Terrorism Agents 234
Test Strips 234
Colorimetric Sampling 235
Handheld Electronic Detection Devices 236
Biological Agent Detection 236
Federal Assistance 237
Basic Incident Priorities 238
Summary 239
Key Terms 239

Review Questions 240
Endnotes 240
Suggested Readings 240

Chapter 8
HAZMAT and Law Enforcement 243

Objectives 245
Introduction 245
Common HAZMAT Incidents for
Law Enforcement 245
Blood-Borne Pathogens 246
Drugs and Chemical Effects 246
Clandestine Labs 247
Drug Labs 247
Explosives Labs 249
Terrorism Agent Labs 249
Explosives Incidents 251
Bomb Squad and HAZMAT Interface 251
General Rules 251
Special Situations 251
Swat Operations 253
Intel 253
Special Considerations 253
Setup 254
Emergency Conditions 256
Evidence Collection 257
Considerations 257
Sampling and Evidence Collection 258
Operations and Entry 259
Summary 260
Key Terms 260
Review Questions 260
Suggested Readings 260

Glossary 263
Abbreviations 269
Index 271

INTRODUCTION

This text is dedicated to the memory of John Eversole. John Eversole, who wrote the forward for this text, was an incredible human being. He died May 20, 2007 with his family whom he loved and was so proud of. He died in the city that he loved and had dedicated his life's work protecting. In the fire service he was the Division Chief of Special Functions for the Chicago Fire Department, a department he served for more than 30 years. He started on the busiest companies in Chicago, and later was promoted to Lieutenant. As a Lieutenant he started the HAZMAT response program for the city of Chicago when many places had not thought about hazardous materials. He served on the NFPA Hazardous Materials 472 Committee, and was the chairman for 9 years overseeing some of the best changes to those standards. He also served as chairman of the IAFC Hazardous Materials Committee for 15 years, and was the internationally known point man on hazardous materials. With his education, experience, and knowledge Chief Eversole promoted firefighter and emergency responder safety across the globe. He was tireless in his efforts and he spent a considerable amount of time making sure that the responders on the street had what they needed to do the job safely. It really didn't matter who was in Chief Eversole's audience; he provided his opinion without hesitation, and usually with emphasis. The audience could have included other chief officers, manufacturers, high level government officials, and even congressmen. It really didn't matter. He wanted them all to hear his message. He conducted or had people conduct exhausting research before engaging in battle and was always well prepared. He knew the HAZMAT and WMD response business better than most and was a visionary many years ahead of others. Chief Eversole was a leader who had considerable followers, myself included. He knew how to tap into other person's expertise and use it to benefit the cause. He had the ability to encourage many other people to do some great things. He was quick to give young folks a chance, and was a mentor to a large number of people. He truly cared about the people who worked for him, or who were in the emergency services. It didn't matter where you were from, or from how large a department you came; he was truly interested in your opinion, and if he could help you out, he would not hesitate. Chief Eversole's impact will be felt for many years and will be missed by his extended family. If it wasn't going to work on the street benefiting emergency responders it wasn't worth his time. Hopefully you had the chance to meet Chief Eversole as he truly was a unique man, one who will be missed.

Chris Hawley

FOREWORD

Emergency responders have been going to hazardous materials incidents long before the federal government even invented the term *hazardous materials*. In most of these early incidents, which we used to call chemical fires, we rushed into the spill, rescued people, and usually put water on the incident until, somehow, it went away.

In those early days, emergency responders paid a heavy price for their heroic actions. Little concern was given to special tactics. The science of personal protective clothing was archaic and most of the responders truly believed that "real men" did not need masks. This lack of technical information and macho ideals proved disastrous for the American fire service.

We now have established standards that require a greatly improved level of knowledge, which will allow us to do a better and safer job.

Hazardous Materials Incidents is intended to be a textbook used by first responder operations classes.

These are generally the first emergency responders on the scene. The actions of these first responders are absolutely critical to a successful outcome. Their defensive actions will protect the community, minimize the damage, and ensure a safe working atmosphere.

This book meets and exceeds NFPA 472 at the operations level. If emergency responders understand and follow this book, everyone will benefit. What makes this text so important is that it gives more than the basic technical information. It shows responders how to do risk-based response. By understanding the management principles of what risks we take and what gains and losses will result, we need no longer charge forward simply out of a sense of duty.

I encourage every reader to study this text and learn well. Your life could depend on it!

John M. Eversole

PREFACE

Intent of This Book

This text is designed for all emergency responders, including fire, EMS, law enforcement, and private industrial responders. This text is adaptable to a variety of disciplines. For example, as a result of growing concerns over terrorism, many responders who were not involved in hazardous materials response are now involved. To function at a potential terrorism event, one needs to have hazardous materials training. One unique feature is that this text incorporates terrorism training.

Why I Wrote This Book

What I have tried to do in this text is to take what I have learned over the years and incorporate this "street knowledge" into a readable format. Many good training programs exist, but most instructors use a variety of programs in their delivery. This text is a compilation of these various approaches. This text is designed to meet and exceed the objectives outlined in 2008 Edition of NFPA 472 *Competence of Responders to Hazardous Materials/Weapons of Mass Destruction Incidents Standard*. Following this text students will meet both the hazardous materials awareness level and the hazardous materials operations level. The NFPA 472 Hazardous Materials Committee has created mission-specific competencies (MSC), which are skill sets that used to be at the technician level. The use of detection devices that are now common are part of the operations level mission-specific competencies. The material in this text meets most of these MSCs. Some people refer to these competencies as OPS-plus. The division of a responder operating at the OPS level is often blurred. Many emergency response organizations have first responders trained to the operations level, but they are assigned to handle various other tasks. This text addresses that gap in training by covering those types of responses at the OPS-plus level.

How to Use This Book

The text is built on a logical progression with each chapter building on the next:

Chapter 1 covers the laws, regulations, and standards associated with hazardous materials response. It provides important background information that is essential to effective and safe response to incidents.

Chapter 2 provides the necessary tools for recognizing and identifying various chemicals and other hazardous materials by following established systems such as placards, labels, and markings on containers. It also covers various other factors, such as location and occupancy, to consider in identification of hazardous materials.

Chapter 3 outlines various sources of information on chemicals and other hazardous materials. Important references such as the *Emergency Response Guidebook*, Material Safety Data Sheets, and other facility documents, as well as computer resources, provide guidance on emergency response and are thoroughly explained in this chapter.

Chapter 4 details critical information on personal protective equipment. The chapter identifies various types of hazards, exposure levels, and the necessary equipment required for protection in each situation. Different types of PPE are explored, including types of SCBA and Level A, B, C, and D suit ensembles.

Chapter 5 introduces protective actions necessary to the successful execution of a response to a hazardous materials incident. Procedures such as the Incident Management System, Hazardous Materials Management Process, and Site Management are explained, as well as types and methods of decontamination.

Chapter 6 is a follow-up to the information presented in the preceding chapter, and provides an explanation of different types of defensive operations as well as an overview on air monitoring and equipment.

Chapter 7 covers the threat of terrorism and terrorist activities. This chapter identifies types of terrorism and potential targets, as well as factors involved in the response to such events. It is a must-read for the current-day responder.

Chapter 8 explains how law enforcement personnel are involved in various hazardous materials response situations. It outlines common incidents involving these emergency responders, such as drug labs, explosives incidents, and SWAT operations. Special consideration is provided for the law enforcement officer, and sections on evidence collection as well as personal protective equipment allow for an effective and safe response in these situations.

Features of This Book

■ *Street Stories* drawn from HAZMAT responders from around the country opens each chapter. A recurrent theme runs through each of them. Each of the stories has a lot of lessons learned, all very personal to that writer. I am very thankful that these professionals, all good friends, have chosen to share with you some personal stories that influenced their lives and their careers. The authors of these Street Stories are the best in the HAZMAT business, but years ago they sat where you are sitting now, starting their careers. Whether you are starting out in a career organization or a volunteer force, I advise you to read each of these stories about experiences in which the authors and their crews could have been seriously injured or killed. When reading the chapter, review in your mind the clues that were present that would have alerted you to the presence of hazardous materials. When you are done with that chapter, go back and reread the Street Story. These true stories offer tremendous potential for learning, and you should take advantage of it.

■ *Case Studies* are integrated into each chapter and bring real-life experiences to the student. These case studies come from years of experience, and nothing is held back. If mistakes were made, they are discussed; learning from others' mistakes is a good way to learn.

■ *Safety & Caution Notes* are highlighted throughout the book for easy identification and reinforce important safety information for emergency responders.

■ *Key Terms* bolded in the text are also listed at the end of each chapter along with definitions. This provides an excellent resource for review and refresher training. A comprehensive list of these terms and definitions is listed at the back of the book in the *Glossary*.

■ *Review Questions* at the end of each chapter evaluates student comprehension of the content and provides a reference for further study and mastery of important concepts.

New to This Edition

The learning process involved with hazardous materials is an ever-expanding process. Each day the number of chemicals used by industries and transported across the country is on the increase, and the complexity of our job is increasing as well. Terrorism adds a unique and terrible twist to the emergency response field. The task of responding to terrorism events in most cases has landed on the HAZMAT teams in conjunction with the local police department.

The intent behind the release may differ, but handling hazardous materials incidents can be easily accomplished by a well-trained and equipped HAZMAT team. For this reason, I have worked to incorporate many of the new challenges and considerations in hazardous materials response today, to help train and prepare responders to achieve a successful outcome to HAZMAT incidents:

■ Updated to NFPA Standard 472, 2008 Edition
■ Reflects new International Placarding System
■ Highlights new Safety Information
■ Introduces and defines new D.O.T. terminology
■ New sections on the Threat of Radiation:
  ■ Radiation Detection
  ■ Radioactive Material Containers
  ■ Radiation Hazards and Proper Protection
■ Current information on Chemicals and Associated Health Hazards
■ Current information on Air Monitoring
■ New step-by-step guidelines for Assessing Terrorist Threats
■ All new photos with Current Technology & Identification Systems
■ Includes Metric with U.S. Customary Units

Supplement to This Book

In addition to this book, an *e.resource* CD-ROM is provided for instructors teaching this hazardous materials course. The CD provides many tools for presenting information in the classroom, as well as evaluating student comprehension of the content. This CD includes the following:

■ *Instructor's Manual* containing *Lesson Plans* correlated to the accompanying *PowerPoint presentations* and *Answers to Review Questions* provides instructors with the necessary information for classroom instruction.

■ *PowerPoint Presentations* outline important concepts in each chapter and correlate to the accompanying *Lesson Plans*. Each presentation integrates photos and graphics from the book to enhance classroom presentation and reinforce critical information.

■ *Quizzes* in Word format provide instructors with an option for evaluating student comprehension of content. Organized by chapter, each quiz contains approximately 20 questions covering the content of the chapter. The Word format enables the instructors to add, delete, or edit questions based on the needs of students.

■ *An Image Library* is also available to instructors as an option for visually enhancing classroom presentations. Images are organized by chapter for accessibility, and can be used to incorporate into PPT or to emphasize critical points.

■ *A Correlation to NFPA 472, 2008 Edition* is also provided on this CD for instructor reference. Each job performance requirement is linked to a specific page number in the book, outlining Standard-related content.

ACKNOWLEDGMENTS

It is a unique experience to write a text such as this, and one of the best parts of the job is acknowledging some of the people who have made an impact in my life or have assisted in some fashion with this text. I can never acknowledge everyone who I would like to, but I will do the best I can. These acknowledgments are in no particular order.

I would like to thank the authors of the Street Stories that begin each chapter, and am eternally grateful that John Eversole wrote the Foreword. All of them have been of great assistance through the years, and I have learned a lot from them. I am very thankful for their contributions to this project. To all of my friends with whom I teach or travel and to those who have shared their knowledge, I am most appreciative. This text is not my effort alone. I have a great phone book, and I haven't hesitated to use it. When I teach, I always learn, and this project also taught me a lot. I have had a unique career and have made some great friends. We have had a lot of fun (my rule #1) and tried to avoid getting killed (my rule #2).

Several individuals have made an impact on my career, and I always wanted to write a text to thank them for their assistance. The first is Dave Barker from Johnstown, Pennsylvania, my first chief and instructor. His work ethic, drive, and desire pushed me to try to become not only an instructor, but a great instructor just as he was. The other officers at Richland Township provided a lot of leadership through my years there as well. The officers and firefighters at Monroeville #4 (Pennsylvania) expanded my learning to a higher level, and it was there that I decided to become a career firefighter. The live-in crew and my EMS partners during that time were great to work with and a lot of fun, too. In the city of Durham, North Carolina, where I first got interested in hazardous materials, Luther Smith, now a retired battalion chief, was the best officer I have ever worked for—and a good friend. It was with the crew of Engine 3 that I had the experiences of a lifetime. I am thankful for the opportunity given to me in Baltimore County by Fire Chief John F. O'Neill and Deputy Chief Gary Warren. Chief Warren allowed me to grow and try a lot of these ideas out on our team. I was able to accomplish a lot in my career due to the efforts of Chief Warren, another good friend. All of the shifts at HAZMAT 114 have been of great assistance to me. Thanks to stations 13 and 14 B shift for being in the photographs. One of the benefits our HAZMAT team has is the assistance of chemist Gene Reynolds, who has been a motivator to learn more about this business, a great relief at many incidents, and a good friend. Buzzy Melton, who is also a chemist and a friend, added his talents to many of our combined training programs. It was during the development of a training program that Buzzy and I developed the risk-based response theory. Buzzy provided the "show me" theory of HAZMAT response. In this realm there are two groups that shine when it comes to HAZMAT response; both provide strategies and methodologies light years ahead of others. The knowledgeable and experienced responders at the Maryland Department of the Environment are some of the best HAZMAT responders in the business. As you will see throughout this text they have supplied a large number of photographs, which demonstrates some of their unique responses. They are always willing to share and educate others in addition to responding to chemical situations throughout the state. I am forever indebted to them for their assistance and guidance. The FBI's Hazardous Materials Response Unit (HMRU) also has some of the best and the brightest in the HAZMAT world. Steve Patrick has provided me invaluable assistance, advice, and friendship over the years, and is responsible for a lot of the concepts in this text. The team members in HMRU keep our country safe by handling dangerous evidence, materials, and hazardous crime scenes worldwide. I am thankful to all of those in HMRU who answer my questions and assist in this and other projects.

My wife and children deserve a lot of thanks for putting up with this and many other projects. My mother planted the idea of pursuing my dreams, and I am lucky enough to have found a wife who shares those dreams with me. Some other folks who are behind the scenes include Jeanne Mesick, my first editor and friend, who helped me out considerably. She was the ring leader in my inner saga circle, who always helped me out. The saga circle includes Trish, who always models for me, as well as Lisa and Stephanie. They are the ones who kept me focused on rule #1. HAZMAT has given me the opportunity to travel the world, and to be with the brightest and the best. I will leave you with some simple rules: Always follow the science and you will survive. Stay flexible, and when told of something, always respond "show me." Be safe, apply rules #1 and #2, and always continue to learn.

REVIEWERS

The author and Delmar gratefully acknowledge the time and effort of the many individuals who reviewed this material and offered their suggestions and expertise in an objective manner. A special thanks to those reviewers:

Andy Byrnes, Hazardous Material Programs Coordinator, Institute of Emergency Services and Homeland Security, Utah Fire & Rescue Academy

Bob Royall, Jr., Sr. Captain/Assistant Coordinator, Hazardous Materials Response Team, Houston Fire Department

Donald E. Ricketts, Project Manager

Eugene Reynolds, Chemical Engineer, FMC Agricultural Chemicals

Glen Rudner, Hazardous Materials Officer, Virginia Department of Emergency Management

Matt Woody, EHS Resource Group LLC

Richard Price, Fire Chief, Fairview Texas Fire & Rescue

Toby Bevelacqua, Division or Batallion Chief, Orlando Fire & Rescue

Tom Creamer, Special Operations Coordinator, Worcester, MA Fire Department

ABOUT THE AUTHOR

 Chris Hawley is a Deputy Program Manager for Computer Sciences Corporation (CSC) and is responsible for several WMD courses within the DOD/FBI/DHS International Counterproliferation Program. This program provides threat assessment, HAZMAT and Anti-Terrorism training, and full-scale exercises worldwide, and is focused on Eastern Europe and Central Asia. Previous to the international work, Chris retired as a Fire Specialist with the Baltimore County Fire Department. Prior to this he was assigned as the Special Operations Coordinator, reporting to the Division Chief of Special Operations, and was responsible for the coordination of the Hazardous Materials Response Team and the Advanced Technical Rescue Team along with two team leaders. Before this assignment he was one of four shift instructors at the Fire Rescue Academy. As a shift instructor he was responsible for all countywide training on his shift. He has been a HAZMAT responder for over 19 years. Chris has 25 years of experience in the fire service, and prior to service in Baltimore County he was a Hazardous Response Specialist with the City of Durham, NC Fire Department.

Chris has designed innovative programs in hazardous materials and anti-terrorism and has assisted in the development of many other training programs, including those provided by the National Fire Academy, Federal Bureau of Investigation, U.S. Secret Service, and many others. Chris has presented at numerous local, national, and international conferences, and writes articles on a regular basis. He serves on a variety of pivotal committees and groups at the local, state, and federal levels. He also works with local, state, and federal committees and task forces related to hazardous materials, safety, and terrorism.

Chris has published four texts on hazardous materials and terrorism response for Delmar Cengage Learning. He has co-authored a text titled *Special Operations for Terrorism and HAZMAT Crimes,* with Mike Hildebrand and Greg Noll through Red Hat Publishing. He also assists a variety of publishers with the review and development of emergency services texts and publications.

INTRODUCTION TO HAZARDOUS MATERIALS

Hazardous materials response is one of the specialty fields within the emergency services. When the responders are called to a hazardous materials release, the actual handling of chemical releases is usually handled by a well-trained team whose function is to handle such emergencies. In almost all cases firefighters and other emergency responders are required to assist in the effort as well.

NOTE Many times at hazardous materials incidents the actions taken in the first 5 minutes determine how the next 5 hours/days/months will go.

Emergency responders are bombarded with exposures to hazardous materials each and every day in their regular duties. A large number of emergency responders are exposed to blood-borne pathogens each day, sometimes suffering fatal effects from these exposures. By the nature of our jobs we are exposed to numerous toxic and cancer-causing agents, many times in nonemergency situations. When confronted with emergency situations, the exposure and risk increase as well. This text covers some additional hazardous materials risks and offers some fundamental response profiles that will help keep you safe. As the emergency services look to further their efforts at improving health and safety, the response to chemical accidents is one area where extra precautions are needed.

SAFETY In a fire situation the injuries can be acute, as when a building collapses during firefighting operations; by comparison, a chemical exposure can also kill responders immediately with a deadly dose or cause serious illness or injury that prolongs death for 10 to 20 years.

As the community became more environmentally aware, so did the emergency services. Hazardous materials response is still a very young field, and most communities have only had fire department hazardous materials response teams for 16 years or less. The Jacksonville, Florida, team is the oldest in the country at 25 years old. This field is constantly changing, and the response to hazardous materials emergencies has undergone three distinct changes in response methodology. Tactics employed to handle incidents 10 years ago certainly were appropriate then, but might bring criticism if used today. Technology is rapidly changing as well; air monitoring devices that were used strictly by hazardous materials response teams now are common on fire engines, medic units, bomb squads, and tactical police units. On the hazardous materials response team side, technology has also increased to a level that requires a high degree of training and expertise. The response to chemical accidents is a rapidly changing field and brings new experiences every day. At building fires, tank truck fires, clandestine drug labs, shipboard incidents, pressurized cylinder leaks, overturned tank trucks, and attacks of terrorism, hazardous materials response teams face new challenges on every response. This text is to provide you with the practical framework to keep you and your fellow responders alive in situations that involve hazardous materials.

The next few sections will describe in detail the many aspects of hazardous materials and your actions at those incidents.

In this text we cover the objectives as required by the Hazardous Materials Section, NFPA 1001. This section requires that the student receive hazardous materials training at operations level, which is outlined in NFPA 472, *Competence of Responders to Hazardous Materials/Weapons of Mass Destruction Incidents Standard*. This text will identify which objectives are awareness (A) or operations (O) level. In some cases the objective exceeds the operations level, and in most cases would be at the technician level; however, some of the objectives provided here exceed any NFPA levels. This information is crucial for your survival but may not be covered by an NFPA objective, or it may be an objective that exceeds the operations level.

NFPA 472 CORRELATION GUIDE

Competencies at the Awareness Level

NFPA 472 Objective		Description	HMI Chapter Reference	HMI Page Reference
	4.1	General	Not Applicable	
4.1.1		Introduction	Not Applicable	
4.1.1.1		Introduction	1–7	1–260
4.1.1.2		Introduction	1–7	1–260
4.1.1.3		Introduction	1	3–14
4.1.2		Goal	1	3–141
4.1.2.1		Goal	1–7	1–260
4.1.2.2		Goal	1–7	1–260
	4.2	Analyzing the Incident	1, 2, 3,	5–8, 19–79, 85–109
4.2.1		Detecting the Presence of Hazardous Materials	2	19–79
4.2.2		Surveying Hazardous Materials/WMD Incidents	1, 2, 3, 7	5–8, 19–79, 85–109, 213–239
4.2.3		Collecting Hazard Information	1, 2, 3, 6, 7	5–8, 19–79, 85–109, 198–209, 213–239
	4.3	Planning the Response	Reserved	
	4.4	Implementing the Planned Response	1, 5	3–14, 139–153
4.4.1		Initiating Protective Actions	1, 5	3–14, 139–153
4.4.2		Initiating the NotificationProcess	1, 5	3–14, 139–153
	4.5	Evaluating Progress	Reserved	
	4.6	Terminating the Incident	Reserved	

Competencies at the Operational Level

NFPA 472 Objective		NFPA 472 Description	HMI Chapter	HMI Pages
	5.1	General	Not Applicable	
5.1.1		Introduction	Not Applicable	
5.1.1.1		Introduction	1–7	1–260
5.1.1.2		Introduction	1–7	1–260
5.1.1.3		Introduction	1	3–14
5.1.2		Goal	1–7	1–260
5.1.2.1		Goal	1–7	1–260
5.1.2.2		Goal	1, 2, 3, 7	3–14, 19–79, 85–109, 213–239
	5.2	Analyzing the Incident	1, 2, 3	3–14, 19–79, 85–109
5.2.1		Surveying the Hazardous Materials Incident	1, 2	3–14, 19–79
5.2.1.1		Example Materials	1, 2, 7	3–14, 19–79, 213–239
5.2.1.1.1		Example Tank Cars	2	58–62
5.2.1.1.2		Example Intermodals	2	56–58
5.2.1.1.3		Example Cargo Tanks	2	43–55
5.2.1.1.4		Example Storage Tanks	2	62–67
5.2.1.1.5		Example Non-Bulk	2	37–42
5.2.1.1.6		Example Radioactive Packages	2	42–43

Competencies at the Operational Level *(continued)*

NFPA 472 Objective		NFPA 472 Description	HMI Chapter	HMI Pages
5.2.1.2		Container Markings	2	20–37
5.2.1.2.1		Shipping Papers	3	103–106
5.2.1.2.2		Facility Container Markings	2	32–36
5.2.1.3		Identify Pipeline and Radiation Materials	2, 5	34–37, 76–78, 157–158
5.2.1.3.1		Pipeline Markings	2	34–36
5.2.1.3.2		Pesticide Label	2	36–37
5.2.1.3.3		Radiation Label	2	37
5.2.1.4		Survey Conditions	2, 5	19–79, 139–149
5.2.1.5		Verify Information	2	19–79
5.2.1.6		WMD Hazards	7	213–239
5.2.2		Collecting Hazard and Response Information	1, 2, 3, 6, 7	3–14, 19–37, 85–109, 191–209, 213–239
5.2.3		Predicting the Behavior of a Material and Its Container	2, 5	19–79, 139–173
5.2.4		Estimating the Potential Harm	2, 4, 5, 7	19–79, 113–124, 139–185, 213–234
	5.3	Planning the Response	1, 2, 4, 5, 6, 7	5–11, 19–79, 113–134, 139–185, 191–209, 213–239
5.3.1		Describing Response Objectives for Hazardous Materials/WMDIncidents	5, 6, 7	139–185, 191–197, 213–234
5.3.2		Identify Action	5, 6, 7	139–185, 191–197, 213–234
5.3.3		Determine Suitability of PPE	4, 5	124–134, 143–153
5.3.4		Identify Decontamination Issues	5	173–185
	5.4	Implementing the Planned Response	5, 6, 7	139–185, 191–197, 213–234, 238–239
5.4.1		Establishing and Enforcing Scene Control Procedures	5, 7	139–173, 212
5.4.2		Preserving Evidence	7, 8	221, 233, 234–239, 257–260
5.4.3		Initiating the Incident Command System	5, 7	139–153, 238–239
	5.4.	Using Personal Protective Equipment	4	113–134
	5.5	Evaluating Progress	5	139–153
5.5.1		Evaluating the Status of Defensive Actions	5	139–153
5.5.2		Communicating the Status of the Planned Response	1, 5	6, 139–153
	5.6	Terminating the Incident	Reserved	

Competencies for Operations Level Responders Assigned Mission Specific Responsibilities

NFPA 472 Objective		NFPA 472 Description	HMI Chapter	HMI Pages
	6.1	General	Not Applicable	
6.1.1		Introduction	Not Applicable	
6.1.1.1		Introduction	Not Applicable	
6.1.1.2		Introduction	Not Applicable	

NFPA 472 Objective	NFPA 472 Description	HMI Chapter	HMI Pages
6.1.1.3	Introduction	Not Applicable	
6.1.1.4	Introduction	Not Applicable	
6.1.1.5	Introduction	Not Applicable	
6.1.2	Goal	Not Applicable	
6.1.3	Mandating of Competencies	Not Applicable	
6.1.3.1	Ability to Perform	Not Applicable	
6.1.3.2	Minimum Requirements	Not Applicable	

Mission Specific Competencies: Personal Protective Equipment

NFPA 472 Objective	NFPA 472 Description	HMI Chapter	HMI Pages
6.2	Mission Specific Competencies: Personal Protective Equipment	4	113–134
6.2.1	General	Not Applicable	
6.2.1.1	Introduction	Not Applicable	
6.2.1.1.1	Introduction	Not Applicable	
6.2.1.1.2	Introduction	Not Applicable	
6.2.1.1.3	Introduction	Not Applicable	
6.2.1.1.4	Introduction	Not Applicable	
6.2.1.2	Goal	Not Applicable	
6.2.1.2.1	Goal	Not Applicable	
6.2.2	Analyzing the Incident	Reserved	
6.2.3	Planning the Response	Not Applicable	
6.2.3.1	Selecting of PPE	4, 5, 8	113–134, 143–148, 253–260
6.2.4	Implementing the Planned Response	Not Applicable	
6.2.4.1	Using PPE	4, 5, 8	113–134, 143–148, 253–260
6.2.5	Terminating the Incident	1	11
6.2.5.1	Reporting and Documenting	1	10–11

Mission Specific Competencies: Mass Decontamination

NFPA 472 Objective	NFPA 472 Description	HMI Chapter	HMI Pages
6.3	Mission Specific Competencies: Mass Decontamination	5	175–185
6.3.1	General	Not Applicable	
6.3.1.	Introduction	5	175–185
6.3.1.1.1	Introduction	Not Applicable	
6.3.1.1.	Introduction	Not Applicable	
6.3.1.1.	Introduction	Not Applicable	
6.3.1.1.4	Introduction	Not Applicable	
6.3.1.2	Goal	5	175–185
6.3.1.2.1	Goal	5	175–185
6.3.1.2.2	Mass Decon Tasks	5	175–185
6.3.2	Analyzing the Incident	Reserved	
6.3.3	Planning the Response	Not Applicable	
6.3.3.1	Selecting PPE	4, 5	113–134, 173–185
6.3.3.2	Selecting Decontamination Procedures	5	113–134

Mission Specific Competencies: Mass Decontamination *(continued)*

NFPA 472 Objective	NFPA 472 Description	HMI Chapter	HMI Pages
6.3.4	Implementing the Planned Response	Not Applicable	
6.3.4.1	Performing Incident Management Duties	5	139–141, 146–150
6.3.4.2	Performing Decontamination Operations	5	113–134
6.3.5.1	Evaluating Mass Decon Process	5	113–134
6.3.6	Terminating the Incident	Not Applicable	
6.3.6.1	Reporting and Documenting	1	10–11

Mission Specific Competencies: Technical Decontamination

NFPA 472 Objective		NFPA 472 Description	HMI Chapter	HMI Pages
	6.4	Mission Specific Competencies: Technical Decontamination	5	173–175, 181–185
6.4.1		General	Not Applicable	
6.4.1.1		Introduction	5	173–175, 181–185
6.4.1.1.1		Introduction	5	173–175, 181–185
6.4.1.1.2		Introduction	5	173–175, 181–185
6.4.1.1.3		Introduction	5	173–175, 181–185
6.4.1.1.4		Introduction	5	173–175, 181–185
6.4.1.2		Goal	Not Applicable	
6.4.1.2.1		Goal	5	173–175, 181–185
6.4.1.2.2		Tasks	5	173–175, 181–185
6.4.2		Analyzing the Incident	Reserved	
6.4.3		Planning the Response	Not Applicable	
6.4.3.1		Selecting PPE	4, 5	113–134, 173–175
6.4.3.2		Selecting Decon Procedures	5	173–175, 181–185
6.4.4		Implementing the Planned Response	Not Applicable	
6.4.4.1		Performing Incident Management		139–141, 146–150
6.4.4.2		Performing Decon Operations	5	154–165
6.4.5		Evaluating Progress	5	154–165
6.4.5.1		Evaluating Effectiveness of Decon	5	154–165
6.4.6		Terminating the Incident	1	10–11
6.4.6.1		Reporting and Documenting	1	10–11

Mission Specific Competencies: Evidence Preservation and Sampling

NFPA 472 Objective		NFPA 472 Description	HMI Chapter	HMI Pages
	6.5	Mission Specific Competencies: Evidence Preservation and Sampling	7, 8	221, 233, 234–239, 257–260
6.5.1		General	Not Applicable	
6.5.1.1.1		Introduction	7, 8	221, 233, 234–239, 257–260
6.5.1.1.2		Introduction	7, 8	221, 233, 234–239, 257–260
6.5.1.1.3		Introduction	7, 8	221, 233, 234–239, 257–260
6.5.1.1.4		Introduction	Not Applicable	

NFPA 472 Objective	NFPA 472 Description	HMI Chapter	HMI Pages
6.5.1.	Introduction	Not Applicable	
6.5.1.2.1	Introduction	7, 8	221, 233, 234–239, 257–260
6.5.1.2.2	Tasks	7, 8	221, 233, 234–239, 257–260
6.5.2	Analyzing the Incident	Not Applicable	
6.5.2.1	Notifications	7, 8	221, 233, 234–239, 257–260
6.5.3	Planning the Response	Not Applicable	
6.5.3.1	Identify Unique Aspects	7, 8	221, 233, 234–239, 257–260
6.5.3.2	Selecting PPE	4	113–134
6.5.4	Implementing the Planned Response	7, 8	221, 233, 234–239, 257–260
6.5.4.1	Implementing the Planned Response	7, 8	221, 233, 234–239, 257–260
6.5.4.2	Decontamination	5	175–185
6.5.5	Implementing the Planned Response	Reserved	
6.5.6	Terminating the Incident	Reserved	

Mission Specific Competencies: Product Control

NFPA 472 Objective	NFPA 472 Description	HMI Chapter	HMI Pages
6.6	Mission Specific Competencies: Product Control	5, 6	153–173, 191–197
6.6.1	General	5, 6	153–173, 191–197
6.6.1.1	Introduction	5, 6	153–173, 191–197
6.6.1.1.1	Introduction	5, 6	153–173, 191–197
6.6.1.1.2	Introduction	5, 6	153–173, 191–197
6.6.1.1.3	Introduction	5, 6	153–173, 191–197
6.6.1.1.4	Introduction	5, 6	153–173, 191–197
6.6.1.2	Goal	5, 6	153–173, 191–197
6.6.1.2.1	Introduction	5, 6	153–173, 191–197
6.6.1.2.2	Tasks	5, 6	153–173, 191–197
6.6.2	Analyzing the Incident	Reserved	
6.6.3	Planning the Response	Not Applicable	
6.6.3.1	Identifying Control Option	6	191–197
6.6.3.2	Selecting PPE	4	113–123
6.6.4	Implementing the Planned Response	Not Applicable	
6.6.4.1	Performing Control Options	5, 6	153–173, 191–197
6.6.4.2	Describe Technical Decon	5	173–175
6.6.5	Evaluating Progress	Reserved	
6.6.6	Terminating the Incident	Reserved	

Mission Specific Competencies: Air Monitoring and Sampling

NFPA 472 Objective	NFPA 472 Description	HMI Chapter	HMI Pages
6.7	Mission Specific Competencies: Air Monitoring and Sampling	6, 7	197–209, 234–237
6.7.1.1	Introduction	Not Applicable	

Mission Specific Competencies: Air Monitoring and Sampling (continued)

NFPA 472 Objective	NFPA 472 Description	HMI Chapter	HMI Pages
6.7.1.1.1	Introduction	Not Applicable	
6.7.1.1.2	Introduction	Not Applicable	
6.7.1.1.3	Introduction	Not Applicable	
6.7.1.1.4	Introduction	Not Applicable	
6.7.1.2	Goal	6, 7	197–209, 234–237
6.7.1.2.2	Tasks	6, 7	197–209, 234–237
6.7.2	Analyzing the Incident	Reserved	
6.7.3	Planning the Response	Not Applicable	
6.7.3.3	Selecting PPE	4	113–123
6.7.3.4	Selecting PPE	4	113–123
6.7.4	Implementing the Planned Response	Local AHJ	
6.7.4.1	Field Testing	Local AHJ	
6.6.4.2	Decontamination	Local AHJ	
6.6.5	Evaluating Progress	Reserved	
6.6.6	Terminating the Incident	Reserved	

Mission Specific Competencies: Victim Rescue/Recovery

NFPA 472 Objective	NFPA 472 Description	HMI Chapter	HMI Pages
6.8	Mission Specific Competencies: Victim Rescue/Recovery	5, 7	142–146, 225–227
6.8.1	General	Not Applicable	
6.8.1.1	Introduction	Not Applicable	
6.8.1.1.1	Introduction	Not Applicable	
6.8.1.1.2	Introduction	Not Applicable	
6.8.1.1.3	Introduction	Not Applicable	
6.8.1.1.4	Introduction	Not Applicable	
6.8.1.2	Goal	5, 7	142–146, 225–227
6.8.1.2.1	Introduction	5, 7	142–146, 225–227
6.8.1.2.2	Tasks	5, 7	142–146, 225–227
6.8.2	Analyzing the Incident	Reserved	
6.8.3	Planning the Response	5, 7	142–146, 225–227
6.8.3.1	Victim Recovery Scenarios	5, 7	142–146, 225–227
6.8.3.2	Selecting PPE	4	113–123
6.8.4.1	Tasks	5, 7	142–146, 225–227
6.8.5	Evaluating Progress	Reserved	
6.8.6	Terminating the Incident	Reserved	

Mission Specific Competencies: Respond to Illicit Laboratories Incidents

NFPA 472 Objective	NFPA 472 Description	HMI Chapter	HMI Pages
6.9	Mission Specific Competencies: Respond to Illicit Laboratories Incidents	8	247–250
6.9.1	General	Not Applicable	
6.9.1.1	Introduction	8	247–250
6.9.1.1.1	Introduction	8	247–250
6.9.1.1.2	Introduction	Not Applicable	

NFPA 472 Objective	NFPA 472 Description	HMI Chapter	HMI Pages
6.9.1.1.3	Introduction	Not Applicable	
6.9.1.1.4	Introduction	Not Applicable	
6.9.1.2	Goal	Not Applicable	
6.9.1.2.1	Introduction	Not Applicable	
6.9.1.2.2	Tasks	8	245–260
6.9.2	Analyzing the Incident	7, 8	213–239, 245–260
6.9.2.1	Potential Lab Operation	8	245–260
6.9.3	Planning the Response	Not Applicable	
6.9.3.1	Determining Response Options	5, 6, 7, 8	139–185, 191–209, 213–239, 245–260
6.9.3.2	Identify Unique Aspects	7, 8	213–239, 245–260
6.9.3.3	Identify Law Enforcement Jurisdiction	Local AHJ	
6.9.3.4	Identify Unique Tasks at Labs	7, 8	213–239, 245–260
6.9.3.4.1	Unique Tasks	7, 8	213–239, 245–260
6.9.3.4.2	Lab Scenarios	7, 8	213–239, 245–260
6.9.3.5	Selecting PPE	4	113–123
6.9.4	Implementing the Planned Response	Local AHJ	
6.9.4.1	Implementing the Planned Response	Local AHJ	
6.9.4.1.1	Tasks	5, 7	139–185, 213–239
6.9.4.1.2	Potential Situations	5, 6, 7, 8	139–185, 191–209, 213–239, 245–260
6.9.4.1.4	Processing and Remediation	Local AHJ	
6.9.4.1.5	Decontamination	Local AHJ	
6.9.5	Evaluating Progress	Reserved	
6.9.6	Terminating the Incident	Reserved	

Competencies at the Technician Level

NFPA 472 Objective		NFPA Description	HMI Chapter Reference	HMI Page Reference
	7.1	General	Not Applicable	
7.1.1		Introduction	Not Applicable	
7.1.1.1		Risk Based Response	5	143–148
7.1.1.2		Introduction	Not Applicable	
7.1.1.3		Training	Not Applicable	
7.1.1.4		Local Competencies	Local AHJ	
7.1.2		Goal	Not Applicable	
7.1.2.1		Introduction	Not Applicable	
7.1.2.2		Technician Tasks	1–8	3–260
	7.2	Analyzing the Incident	1, 2, 3	3–14, 19–79, 85–109
7.2.1		Surveying Incidents	1, 2	10–11, 19–79
7.2.1.1		Containers	2	37–67
7.2.1.1.1		Rail Cars	2	58–62
7.2.1.1.2		Intermodal Tanks	2	56–58
7.2.1.1.3		Cargo Tanks	2	43–55
7.2.1.1.4		Facility Containers	2	62–67
7.2.1.1.5		Non Bulk Packages	2	37–42
7.2.1.1.6		Radiation Packages	2	42–43
7.2.1.2		Facility Container Capacities	2	62–67
7.2.1.2.1		Transport Container Capacities	2	37–62
7.2.1.2.2		Specialty Container Capacities	2	56–58

Competencies at the Technician Level *(continued)*

NFPA 472 Objective	NFPA Description	HMI Chapter Reference	HMI Page Reference
7.2.1.3.3	Monitoring Technology	6, 7	197–209, 234–237
7.2.1.3.4	Monitoring Capabilities/Limitations	6, 7 (parts)	197–209, 234–237
7.2.1.3.5	Testing Potential Hazards	Local AHJ	
7.2.1.3.6	Field Maintenance	Local AHJ	
7.2.1.4	Radiation Label	2	37
7.2.1.5	Sample Collection	8	257–260
7.2.2.2	Terms	2 (part)	68–79
7.2.2.3	Cryogenic Spill	2, 5	51–54, 158–161
7.2.2.5	Pressure of Containers	2, 5	40, 49–54, 157–161
7.2.3.1	Design of Containers	2	37–67
7.2.3.1.1	Design of Bulk Containers	2	62–67
7.2.3.1.2	Non Bulk Containers	2	37–42
7.2.3.1.3	Radiation Containers	2	42–43
7.2.3.2	Petroleum Pipelines	2	34–36
7.2.3.3	Petroleum Pipelines	2	34–36
7.2.5.2	Determining Hazards	4	113–124
7.2.5.2.1	Exposure Values	4	212–124
7.2.5.2.2	Predicting Area of Harm	3	86–100
7.2.5.3	Estimating Outcomes	5	142–147, 150–155
7.2.5.4	Public Response Options	5	142–146, 150–153
7.3	Planning the Response	Not applicable	
7.3.3	Selecting PPE	4	124–134
7.3.3.1	Levels of PPE	4	124–134
7.3.3.2	PPE Options		
7.3.3.3	Respiratory Protection	4	124–126
7.3.3.4	PPE Selection Factors	4	124–134
7.3.3.4.1	PPE Terms	4	124–134
7.3.3.4.5	PPE Selection	4	124–134
7.3.4	Selecting Decontamination Procedures	5	173–185
7.3.5.5	Evidence Collection	7, 8	221, 233, 234–239, 257–260
7.4.2	Using PPE	4	124–134
7.4.5	Performing Decon	5	173–185

HAZARDOUS MATERIALS: LAWS, REGULATIONS, AND STANDARDS

OUTLINE

▎ **Objectives**

▎ **Introduction**

▎ **Laws, Regulations, and Standards**

▎ **Emergency Planning**

▎ **OSHA HAZWOPER Regulation**

▎ **Standards**

▎ **Standard of Care**

▎ **Additional Laws, Regulations, and Standards**

▎ **Summary**

▎ **Key Terms**

▎ **Review Questions**

(Photo courtesy Mike Austin)

STREET STORY

We arrived on the scene to find a patient lying beside the hot dusty road, baking in the summer sun. He was a male in his mid-thirties, wearing torn blue jeans, old tennis shoes, and no shirt. A local farmer was on the scene. He was driving home when he noticed the man beside the road. It was obvious he was in distress, so the farmer dialed 911 from a cellular phone.

When we got off the fire engine, it was apparent the man was struggling to breathe. He was sunburned, sweaty, and unable to respond to any of our questions. He was shaking slightly, but it did not look like a typical epileptic seizure. We put the man on high-flow oxygen and began to assess his vital signs. It was the size of his pupils that tipped us off. In addition to the pinpoint pupils, the patient was incontinent and producing volumes of frothy sputum. We stripped off his clothes, decontaminated him with water, and alerted the incoming ambulance by radio that we might have organophosphate poisoning. Fortunately, we recently had some hazardous materials training on pesticide exposures. Had we not been armed with that information, we might not have picked up on the clues so quickly.

The transport ambulance arrived what seemed like hours later, and by that time we were assisting his respirations with a bag valve mask and suctioning his airway. An IV was started, and atropine was administered in large doses to counteract the effects of a suspected pesticide. Unfortunately, we were never able to identify the substance he was exposed to or even know the duration of the exposure.

Sometimes, despite the greatest efforts and best technology available, people do not survive. He fought hard to breathe and even harder to live, but in the end, neither we nor the hospital staff could save his life. He died an hour or so after we first saw him, probably never knowing why he became so sick.

I think about him from time to time, wondering if the call could have gone differently, but I always arrive at the same conclusion: You do your best, learn from any mistakes or situations you find yourself in, and hope you have a better outcome the next time around.

*We could have suffered a secondary exposure because we were not properly protected, but we did not. We could have had an inhalation exposure from his contaminated clothing, but we did not. We could all have ended up as patients, but we did not. It was only luck that brought us through without harm. Would I do it differently next time? Absolutely. Our first responsibility as firefighters is to not make the problem worse when we show up! You are no good to anyone if you become part of the problem or end up being a victim yourself. Use common sense, wear the right **personal protective equipment (PPE)** for the right situation, and, above all, think before you act. Do not count on luck to bring you back to the station.*

—Street Story by Rob Schnepp, Firefighter/Paramedic, Alameda County Fire Department, Alameda County, California

OBJECTIVES

After completing this chapter, the reader should be able to explain:

◼ The local emergency response plan and standard operating guidelines (A)

◼ The student's role within these documents at the awareness level (A)

◼ The notification process to request assistance (A)

◼ The role of the Local Emergency Planning Committee (O+)

◼ The role of the SARA Title III regulation and emergency response (O+)

◼ Other regulations related to fire department activities (O+)

INTRODUCTION

NOTE There are two levels of attainment for those receiving hazardous materials emergency response training: the **Occupational Safety & Health Administration (OSHA)** level and the **National Fire Protection Association (NFPA)** level. The requirements vary from state to state, but the minimum level is the OSHA training. The term *certification* is commonly used, and typically only the employer can certify the training. Neither OSHA nor the NFPA certifies responders; their fire departments certify that they have met the training objectives outlined by OSHA or the NFPA. The NFPA requirements are much more stringent, but better reflect the current emergency responders' actual duties. If trained to meet the NFPA training objectives a responder would meet the OSHA requirements. When firefighters receive their basic training it is typically based on the NFPA training objectives for firefighters. The NFPA Standard 1001, also known as the Standard for Firefighter Professional Qualifications, has been revised and the 2008 edition was published July 2007. As part of the revision to be certified as a Firefighter I, the student will also have to be certified at the operations level according to NFPA 472, which is the Competence of Responders to Hazardous Materials/Weapons of Mass Destruction Incidents Standard. In addition they will have to be certified with the Mission Specific Competency: Product Control of NFPA 472. All of this information is important for firefighters' survival regardless of in which state the training is provided; firefighters are required to have some form of hazardous materials training. The training can be based on the OSHA regulations or the NFPA standards. The information in this text exceeds the requirements for OSHA training and covers both awareness and operations levels and, in some cases, exceeds the operations level of the NFPA requirements. See the discussion on NFPA standards and OSHA's HAZWOPER for more information on Awareness and Operations level training.

To establish the groundwork for these sections, it is necessary to provide a definition for **hazardous materials (HAZMAT).** The text that follows provides these standard definitions. Out of these definitions, the following may be considered the best definition: "A hazardous material is any substance that jumps out of its container when something goes wrong, and hurts and harms the things it touches." Ludwig Benner, Jr., who worked for the National Transportation Safety Board, provided this early definition. It is a perfect definition for this field.

As you will learn, almost everything solid, liquid, or gas is hazardous to human health if it is used improperly or escapes its container, as is demonstrated in **Figure 1-1.** The most toxic material is not hazardous as long as it is used correctly and stays in the container in which it was intended to be transported.

Figure 1-1 An example of hazardous material that ignites when it comes in contact with the air after escaping its container. A material that is air reactive is known as pyrophoric.

Figure 1-2 Members of a HAZMAT team investigate the source of a chemical release in the back of a tractor trailer carrying a mixed load of hazardous materials and ordinary consumer goods. *(Courtesy of the Maryland Department of the Environment ERD)*

Another important principle is that even if the material escapes its container, it must have the potential to harm humans, property, or the environment to be considered hazardous. Hazardous materials are packaged to be transported every day and are intended to stay in that package, as shown in **Figure 1-2.** When responders become involved in an incident that involves such a package, their increased level of concern is understandable. But as long as the material remains in the package and the responders do not open it, the risk to them and the environment is minimal. The material may be toxic, flammable, corrosive, or reactive, but it is only harmful if it escapes or is forced from the container.

DEFINITIONS OF HAZARDOUS MATERIAL

Department of Transportation (DOT) hazardous material: Any substance or material in any form or quantity that poses an unreasonable risk to the safety, health, and property when transported in commerce. The DOT also provides more exacting definitions with each hazard class.

Environmental Protection Agency (EPA) hazardous substances (40 CFR300.5): A chemical released into the environment that could be potentially harmful to the public's health or welfare.

OSHA hazardous chemicals (29 CFR1910.1200): Those chemicals that would be a risk to employees if exposed in the workplace.

LAWS, REGULATIONS, AND STANDARDS

First responders should have a basic understanding of the legislative history of hazardous materials. The first regulations applying to emergency responders specified how they would respond and how firefighters would be trained for HAZMAT incidents. Many environmental and safety regulations affect how we respond to emergencies. This section will provide an overview of the major federal laws, regulations, and standards. Readers should consult their local environmental and OSHA offices for more information about state and local laws and regulations.

Development Process

To enhance understanding of this section it is important to know the difference between laws, regulations, and standards. **Laws** are legislation passed by Congress; once signed by the president, legislation becomes law. **Regulations,** on the other hand, are developed by government agencies, like OSHA or the EPA. Regulations have the weight of law but are not passed by Congress or signed by the president. In some cases laws require the development of regulations and provide a framework for the government agency to follow. **Standards** are developed by a nongovernmental consensus committee, such as the NFPA; after a public review, recommendations from the organization become standards. These do not have the weight of law, but could be applied by a regulating agency or in court. Some standards are being applied with the weight of law, typically by OSHA. As citizens and an industry, it is important for emergency responders to participate in the development and review of laws, regulations, and standards. Many states now have adopted NFPA standards into health and safety regulations, which cover not only industry but now can be applied as a regulation to a fire department. The catchall for occupational safety regulators is known as the **general duty clause,** which means that all employers have a general duty or an obligation to provide a workplace free from hazards. To apply the general duty clause the enforcement agency demonstrates that the standard is a consensus standard for the fire service and is for the most part universally adopted as a national standard. Failure to follow the standard, which provides objectives that may have prevented the injuries or deaths, allows the enforcement agency to apply the general duty clause. The regulatory agency does not need a standard to apply the general duty clause, but it certainly strengthens its case against the alleged violator.

EMERGENCY PLANNING

The first law that regulated how emergency services organizations respond to emergencies was the **Superfund Amendments and Reauthorization Act,** commonly referred to as **SARA.** This law was passed in 1986 to protect emergency responders and the community with the intent to inform emergency responders and area residents of chemical hazards within their community. The main component of the law is the **Emergency Planning and Community Right to Know Act (EPCRA).** It is divided into two sections: planning for emergencies and a mechanism to provide chemical storage information to emergency responders.

State and Local Emergency Response Committees

The planning portion established a requirement to provide **State Emergency Response Committees (SERCs)** in each state. The SERC is responsible for ensuring the state has the resources necessary to respond safely and effectively to chemical releases. The SERC also provides the framework to implement **Local Emergency Planning Committees (LEPCs).** The LEPC is a group comprised of the representatives of the community, emergency responders, industry, hospitals, media, and other government agencies. Most LEPCs are set up on a county basis, although larger municipalities may establish their own committees. In some states each county has an LEPC, while

Figure 1-3 The LEPC should be a cross section of the community: emergency responders, citizens, and industry. The size of the community should dictate the frequency of the meetings, but they should be held several times a year.

some LEPCs represent entire states. If the fire department is not involved with the LEPC, then the local emergency management office should be the point of contact for this committee. The LEPC, as shown in **Figure 1-3,** is responsible for the development of a HAZMAT emergency plan and its annual revision.

Local Emergency Response Plans

The emergency plan is an important component of a successful response. The emergency plan should outline emergency contacts and procedures. Target

LOCAL EMERGENCY PLANNING COMMITTEES

The importance of emergency services or the HAZMAT team's involvement in an LEPC cannot be understated. The LEPC meeting is a low-key event promoting interaction with safety and environmental professionals in the community. The LEPC has a legal mandate to fulfill regardless of the HAZMAT team's involvement. The emergency plan could dictate how the HAZMAT team performs its job on a chemical release. The LEPC is required to exercise the emergency plan every year, and this is a great time to showcase the HAZMAT team's talents to the community. Many LEPCs are involved with community environmental and outreach events, further improving community relations. For example, the Baltimore County Department of Environmental Protection conducts a Household Hazardous Waste Day on an occasional basis.

The LEPC, bomb squad, and HAZMAT teams are involved in assisting with the collection of hazardous

waste. In addition to providing a public showcase, it is a great learning opportunity for the emergency responders as a variety of hazardous waste usually shows up. The Baltimore County LEPC has taken a proactive approach with the business community to make sure all businesses are compliant with state and federal regulations. The fire department is not the enforcement authority in these matters but assists in the completion of forms and conducts inspections to reveal any problem areas. The fire department works with businesses to achieve compliance and avoid their being fined by other agencies. This work is done under the guidance of the LEPC and is well received by the business community. The LEPC also conducts annual seminars prior to filing deadlines for chemical inventory reporting and hosts an annual business seminar on environmental and safety issues.

facilities such as those that store extremely hazardous substances are to be included with specific information and response tactics. A decision tree for evacuation and in-place sheltering is usually included. If the decision is made to evacuate citizens, the plan provides the mechanism to shelter and take care of evacuees. It is crucial that personnel understand this plan and know how to access its resources during an emergency.

The LEPC is also responsible to make sure local resources are adequate to respond to a chemical release in the community and to serve as the focal point for community awareness. Ensuring that local responders receive the proper amount of training and that the emergency plan is evaluated annually by conducting an exercise (drill) are also part of the committee's responsibility. It is important for the emergency services to be an integral player in the LEPC. Decisions regarding your response to chemical incidents may be planned at the LEPC level. Many of the facilities that you respond to will have representatives at these meetings, and it is better to meet these representatives before, rather than during,

an emergency. Although each state has different policies, in many locations funding for training and planning is available through an LEPC, as most federal HAZMAT grants are generally provided through the LEPC system.

One of the most widely known sections of the local emergency plan is incident levels. Some jurisdictions have outlined incident levels as described in **Table 1-1** to indicate the severity of incidents. This is usually outlined in the local emergency plan and duplicated in the HAZMAT team policy.

Chemical Inventory Reporting

The other section of EPCRA, usually referred to as SARA Title III, covers the chemical reporting portion of the act. It requires certain facilities to report chemical information to the state, LEPC, and the local fire department. Failure to report this information on an annual basis can result in fines of $25,000 a day. To qualify as a reporting facility the entity has two methods of reaching a reporting threshold. One criterion to meet the reporting threshold is storing more than

TABLE 1-1

Incident Levels	
Level	**Incident scale**
1	Small-scale incident that can usually be handled by the first responders but may also require minimal additional resources. Notifications are usually local and may only be internal to the fire department. The highest level of PPE required is firefighters' turnout gear and self-contained breathing apparatus (SCBA), and that may not be necessary. The material is not toxic by skin absorption, although it may have an inhalation hazard. The size of the spill is usually small and will have minimal environmental impact. Incidents of this type include natural gas or propane leaks and small fuel spills.
2	Incidents at this level usually require additional assistance, such as by the HAZMAT team. The incident may require additional notifications to environmental or emergency management agencies on a local or state basis. The amount of material may be larger or more hazardous. The type of PPE will probably be chemical protective clothing or other clothing not carried by the first responders. The first responders may have switched from an active role to a support role. A level 2 incident may require a small evacuation and a larger isolation area. The incident will also probably hamper the ability of the jurisdiction to respond to other emergencies, depending on available resources. An example incident would be an overturned gasoline tanker or a leaking propane tanker. A leaking drum in the back of a tractor trailer would be classified as a level 2 incident.
3	A level 3 incident requires substantial local resources and the assistance of other local and state agencies. Resources on the federal level may also be required or, at a minimum, notified. The incident will require the evacuation of the affected area and a substantial isolation area. The release is large or the material is extremely toxic. Examples of a level 3 incident include a train derailment with chlorine railcars leaking or a substantial leak from an ammonia tank truck.

10,000 pounds of a chemical. The 10,000-pound limit is for a single chemical and is not the total of all chemicals on site. A facility that stores drums of ammonia and water is only required to report the ammonia, not the total weight of the drum. It would need to account for all of the ammonia at the facility, no matter the location or container. Retail gasoline stations are exempt from reporting gasoline or diesel fuel as long as the amount stored is less than 75,000 and 100,000 gallons, respectively. If the gas station stores more than 10,000 pounds of propane or any other chemical, it is required to report. Other facilities that store or use gasoline or diesel are required to report these materials if the quantity exceeds 10,000 pounds (1,300 gallons).

Another method of meeting the threshold of reporting is to store one of 387 chemicals that the EPA considers **extremely hazardous substances (EHSs)**. These substances have separate reporting requirements and lower reporting thresholds. Some require reporting at 100 pounds. If these chemicals are released, the facility has a whole host of responsibilities to comply with, including immediate notification to the LEPC, usually by dialing 911 or other listed emergency contact numbers as determined by the LEPC. Common EHSs are chlorine, sulfur dioxide, anhydrous ammonia, and sulfuric acid. As EHS materials may present an extreme threat to the community, it is important for emergency responders to become familiar with the facilities that use them and to use caution when responding to potential releases involving these materials. The EPA maintains the listing of EHSs and other reporting requirements in their publication, "The List of Lists," also available on their Web site at *www.EPA.gov.*

Facilities that are required to report as SARA facilities must submit either a Tier 1 or Tier 2 Chemical Inventory Report. The Tier 2 report, as shown in **Figure 1-4,** is the most common and lists the chemical name, storage amount in a range, storage location

Figure 1-4 This form is an example of what facilities are required to submit to the fire department and LEPC on an annual basis.

and information, and emergency contact information, including 24-hour phone numbers. The facility is also required to submit a list of chemicals or **Material Safety Data Sheets (MSDSs).** A site plan outlining the storage areas may also be required to be submitted. EHS facilities are also required to submit a copy of their emergency plan. Depending on the state or locality, reporting requirements may be more stringent, although all must meet the minimum listed for the Tier 2 report. It is important to note, however, that the information contained on a Tier 2 form regarding the chemicals located on site is for the prior year. Facilities report from January 1 to March 1 for the year prior.

The basis for this reporting is to inform emergency responders of the hazard of responding to a SARA facility. It also allows emergency responders to make informed decisions as to site entry, tactical decisions, and potential evacuation decisions. As we will learn, MSDSs are a valuable tool in obtaining chemical information. EPCRA requires their availability to emergency responders. The Hazard Communication Standard (29 CFR 1910.1200) also requires businesses to have MSDSs available for emergency situations.

OSHA HAZWOPER REGULATION

Another section of SARA required OSHA to develop a regulation covering activities that involve hazardous materials. The regulation known as the **Hazardous Waste Operations and Emergency Response** became final on March 6, 1989. It is referred to as **HAZWOPER,** or by its identification number, 29 CFR 1910.120.

NOTE Federal regulations are published in the *Federal Register*, issued daily; they are then annually published in a text called the Code of Federal Regulations (CFR). Each federal agency is identified by a two-digit number (29 for OSHA, 40 for EPA, 49 for DOT), and each regulation is given a number to identify it.

Considerable discussion within the fire service when this regulation was established focused on whether it applied to all of the fire service. The primary unknown was whether this regulation applied to volunteer firefighters. In some states OSHA regulates only the private sector, and not career firefighters, while in other states only the public sector was subject to this regulation. In many cases volunteer firefighters were determined to be local government employees, because they received worker's compensation benefits or safety equipment from this employer. To end the confusion, the EPA adopted the same regulation and issued it on the EPA's behalf.

ETHICS Because employment is not a concern of the EPA, all persons are covered by HAZWOPER regardless of their employment status.

The EPA's version of HAZWOPER is referenced by the identification number 40 CFR 311. It is interesting to note that OSHA now puts a provision in some of its regulations that specifically mandates the coverage of volunteer employees.

This regulation has had far-reaching effects for the emergency services. It requires certain training and the development of standard operating procedures. It spells out specific requirements in handling chemical

CASE STUDY

The HAZMAT company was dispatched to a building fire involving chemicals. The first alarm assignment was already on the scene actively extinguishing the fire in the building. The building housed a computer chipboard manufacturing company. The bulk of the fire was in the production area. When the firefighters entered the building to extinguish the fire, they found several open vats of chemicals. Most of the vats had melted down, releasing a substantial amount of product, most of which was lying on the floor of the building. Fire crews had contaminated themselves and would require some type of decontamination. As decontamination is chemical specific, the HAZMAT crews attempted to locate the owner, who would be needed to identify what types of materials were present. The HAZMAT crews were unable to find any chemical storage information specific to that facility. The owner was located but did not have the required information. The materials that were on the floor and in the vats could not be identified through the use of facility paperwork. The crews were held for several hours, wondering what health threat they faced. Had the proper paperwork been provided, there would have been no delay in getting information.

INSTRUCTIONS FOR TIER 2 FORM

The Tier 2 form contains a lot of valuable information (**Figure 1-4**). Probably the most valuable is the 24-hour contact information. Some jurisdictions have entered the Tier 2 information into a computer database for easy retrieval. When companies are alerted for a response to a SARA facility, the dispatcher informs them that they are responding to a SARA Title III facility. This alerts the responders to the storage of chemicals. Unfortunately a lot of the information is coded on these forms. The parts of the Tier 2 form include:

1. Full name and address for the reporting facility. A street address is required. The Standard Industrial Classification Code (SIC) identifies the type of facility. A Dun and Bradstreet number is a unique identification number for the business, similar to Social Security numbers for individuals.

2. Emergency contact section. Two contact persons who should be knowledgeable about the chemicals listed or at least have information related to the facility are specified. The phone number is the day phone, and the 24-hour number is for nights and weekends.

3. The chemical description box. The Chemical Abstract Registry Service Number (CAS #) identifies a specific chemical. The chemical name must be clearly provided on the lines below the CAS number. The form indicates whether the material is pure or a mixture and its state (solid, liquid, or gas) and also notes whether it is an EHS.

4. Physical and health hazards. All hazard categories that may apply must be checked. The EPA provides the following categories, and the corresponding OSHA guidelines are also provided.

EPA hazard category	OSHA hazard category
Fire Hazard	Flammable
	Combustible
	Pyrophoric
	Oxidizer
Sudden Release of Pressure	Explosive
Reactive	Compressed gases
	Unstable reactive
	Organic peroxide
	Water reactive
Immediate (acute) Health Hazard	Highly toxic
	Toxic
	Irritant
	Sensitizer
	Corrosive
	Other hazardous chemicals having an adverse effect with short-term exposure
Delayed (chronic) Health Hazard	Carcinogens
	Other hazardous chemicals having an adverse effect with long-term exposure

5. Inventory.
   The reporting ranges that must be determined are as follows:

	Reporting ranges	
Range code	From	To (in pounds)
01	0	99
02	100	999
03	1,000	9,999
04	10,000	99,999
05	100,000	999,999
06	1,000,000	9,999,999
07	10,000,000	49,999,999
08	50,000,000	99,999,999
09	100,000,000	499,999,999
10	500,000,000	999,999,999
11	1 billion	1 billion

6. Storage codes and locations. Boxes for codes provide information related to container type, pressure, and temperature. The codes are listed below:

	Storage Types		
Code	Description	Code	Description
A	Aboveground tank	K	Box
B	Belowground tank	L	Cylinder
C	Tank inside building	M	Glass bottles or jugs
D	Steel drum		
E	Plastic or nonmetallic drum	N	Plastic bottles or jugs
F	Can	O	Tote bin
G	Carboy	P	Tank wagon
H	Silo	Q	Railcar
I	Fiber drum	R	Other
J	Bag		

Pressure conditions

1 = Ambient pressure
2 = Greater than ambient pressure
3 = Less than ambient pressure

Temperature conditions

4 = Ambient temperature
5 = Greater than ambient temperature
6 = Less than ambient temperature
7 = Cryogenic conditions

releases. It allows only persons trained in the handling of chemical releases to respond to chemical incidents. Only certain training levels allow operation on working in and around chemical releases. Just to be present at a chemical accident requires training, and as responders' activity levels increase, so do their needs for further training.

Paragraph q

The majority of this regulation covers the employers' responsibilities at hazardous waste sites. The last section, **Paragraph q,** covers emergency response and applies to the emergency services. It establishes five levels of training and a requirement for annual refresher training. It requires that an incident command system be used and that an incident commander be present at chemical releases. At larger incidents a safety officer must be appointed. The higher the training levels, the more detailed the requirements. The OSHA two-in-two-out rule originated from the HAZWOPER regulations and was further defined in the Respiratory Protection Regulation (1910.134). The two-in-two-out rule requires that when employees enter a hazardous environment, known to be **immediately dangerous to life or health (IDLH),** they must use the buddy system, which requires at least two persons go into the environment together with a rescue crew standing by. The minimum backup or rescue crew is two, but the rule does not require that, above two, the rescue crew be one-for-one for those entering the environment. If four persons enter the IDLH atmosphere, a backup crew of two is sufficient; the rule requires that an "adequate" crew be standing by. The thought process is that typically only one person will get trapped and that two will suffice to make a rescue, with the other crew members assisting. The incident commander determines what comprises an adequate rescue crew.

Medical Monitoring

Other sections of this regulation include a requirement for medical monitoring of certain employees. The level of responsibilities dictates whether the

---

OSHA AND NFPA TRAINING LEVELS

The training levels established by OSHA and the NFPA and their basic responsibilities are as follows:

Awareness—Responders that have a potential to come across a possible chemical release, identify the potential for a chemical release, call for assistance, and stand by to isolate the area and deny entry to other persons. Persons trained to the awareness level cannot take any action beyond this. This level is intended for police officers, public works employees, and other government employees.

Operations—Persons trained at the operations level can act in defensive fashion to chemical spills. Acting defensively does not extend to entering a hazardous area, but responders can set up dikes, dams, and other containment measures. Training at the operations level allows persons to assist technicians in setting up the various activities required at a chemical incident. Operations training can be expanded to include specialized activities such as decontamination, so that persons trained at this level can assist with this activity. This level is intended for the fire and EMS service.

Incident Commander (IC)—Designed for a person who has received operations level training and training in incident command procedures. The incident commander will be in charge of the incident. To be the incident commander does not require the highest level of chemical response training, but this individual is the senior response official. The IC must rely on the expertise of other responders such as the HAZMAT team, facility officials, or other technical specialists to make strategic and tactical decisions.

Technician—This is the level at which offensive activities can be completed in the hazard area. Other than some specific restrictions outlined in HAZWOPER, there are no general restrictions as to the activities technicians can perform, as long as the activities fit within the scope of their training. Technicians can stop leaks and complete mitigation of the incident. HAZMAT technicians are expected to mitigate or stop the incident from progressing.

Specialist—This level is identified only in HAZWOPER and is intended to denote a person who has received more training than a technician, or someone who specializes in a specific chemical or area of expertise. The training concentrates on chemistry and the identification of unknown material. In some instances specialists operate at an incident and supervise the technicians. The NFPA removed this level in 1992. The NFPA in the 2002 edition of 472 has several specialty levels, such as tank car specialist. In some cases persons such as heavy equipment operators, chemists, or other persons with specific knowledge may be deemed specialists by the local jurisdiction.

employer is mandated to provide employees with an annual physical. The requirements are as follows:

- Exposure to a chemical above the permissible exposure limit
- Requirement to wear a respirator or coverage by the OSHA respiratory regulation (29 CFR 1910.134)
- Due to a chemical exposure
- HAZMAT team membership

The physician determines the extent of the exam and can establish that exams be given every 2 years. The only emergency service employees or volunteers required to be given a physical are the members of a HAZMAT team, although OSHA 1910.134 (Respiratory Protection) requires that a medical survey questionnaire be answered by every responder. Answers to the questionnaire determine if a physical is required or not. If a responder is exposed to a hazardous material above the OSHA specified level, the department is required to provide a physical exam. The physician determines the extent of the exam and any other future visits or tests. One common misconception of this section is that HAZWOPER requires that a pre-entry medical exam be provided to those persons who must wear chemical protective clothing. The regulation requires an annual (or biannual, at the physician's discretion) physical to determine fitness for duty; it requires only that EMS be standing by for any entry work. Although it would be a recommended practice to provide pre-entry physicals, there are certain instances in which this is not practical. Situations involving victim rescue or rapidly deteriorating situations that require immediate mitigation are two immediate examples. It is always recommended that persons who wear chemical protective clothing (or any protective clothing) be provided a post-entry physical, a concept that should be applied for firefighting operations. To provide some type of recent pre-entry exam many HAZMAT teams have their personnel complete a medical exam as they report to work, which is certainly an option. OSHA requires that the medical records be kept by the employer for a period of 30 years past the last date of employment. This applies to the HAZWOPER regulation, as well as other OSHA regulations. In simple terms it says any medical record generated by the employer or for the employer must be kept and must be available for employees if they request it.

STANDARDS

The group that establishes most of the standards for the fire service is the National Fire Protection Association. The NFPA establishes a variety of committees that develop standards, which are then made available for review and comment by the public. After the review process, each standard is voted on by the NFPA committee and, if passed, then goes to the next public meeting where it is voted on by the group in attendance. Once passed it is sent to the Standards Council, where it is voted on again. If it passes there, it becomes a standard. As with emergency medical services for which an acceptable standard of care applies, typically based on a national or regional standard, the NFPA establishes a standard of care. A person cannot be held criminally liable for violating an NFPA standard, but may be held civilly liable. One of the areas in which the NFPA standards have been applied as having the weight of a regulation is in the hazardous materials arena. OSHA has used this clause to cite employers for violating an NFPA standard.

NOTE OSHA has used what is known as the **general duty clause,** which means that all employers have a general duty or an obligation to provide a workplace free from hazards.

NFPA 472

Prior to the 2008 edition the NFPA had three hazardous materials response standards; NFPA 471, 472, and 473. In the 2008 edition the NFPA elimi-

NFPA 472

Professional Competence of Responders to Hazardous Materials/ Weapons of Mass Destruction Incidents

nated the 471 Standard which covered the response to hazardous materials emergencies and moved much of its information into 472. The NFPA Hazardous Materials committee felt that training standards adequately covered the response objectives, and that only two standard documents were required. NFPA 472, *Professional Competence of Responders to Hazardous Materials/Weapons of Mass Destruction Incidents* is the listing of objectives required to meet the training levels established by the NFPA. OSHA established only basic criteria to meet in order for the employer to certify its employees as to their ability to respond to hazardous materials incidents. This NFPA document expands the requirements in order for the employer to certify its employees. The levels mirror those of the HAZWOPER regulation, but in 1992 the NFPA removed the specialist category. In the 1997 edition they added the following competencies: private sector specialist employee, hazardous materials branch officer, hazardous materials branch safety officer, tank car specialist, cargo tank specialist, and intermodal specialist.

In the 2002 edition the NFPA added objectives related to terrorism response. These changes were part of the 1997 edition but were listed as tentative interim amendments (TIAs). The 2002 version adopts them as part of the standard, and they are covered by this text. The 2008 edition adds several law enforcement competencies for the collection of evidence in hazardous situations. There are several criminal scenarios that involve the potential use of hazardous materials or weapons of mass destruction and present significant risk to investigators. The revision to this standard provides training objectives for response personnel who may be collecting evidence when hazardous situations exist. Also in the 2008 revision is the addition of **mission-specific competencies,** which used to be reserved for hazardous materials technicians. These competencies reflect the realities of real world incidents. Many firefighters have access to and the ability to use air monitoring devices, which used to be a technician-level skill. The revision to the standard provides that, if properly trained and equipped, an operations-level responder can use some air monitoring devices.

NFPA 473

The second (previously third) standard is NFPA 473, *Competencies for EMS Personnel Responding to Hazardous Materials/Weapons of Mass Destruction Incidents*. This standard adds some additional competencies beyond NFPA 472 with regard to EMS issues. It provides EMS Level I and Level II training levels. The competencies in this edition have been updated and use common terminology known to the EMS response community. Instead of Level I and II it now relies on BLS and ALS providers. The standard also requires that to meet the competencies of the standard all EMS personnel will meet the core competencies of Hazardous Materials Operations and some of the Mission Specific Competencies based on their on-scene responsibilities. The Delmar book *Emergency Medical Response to Hazardous Materials Incidents* offers more information on this topic.

NFPA 473
Competencies for EMS Personnel Responding to Hazardous Materials/ Weapons of Mass Destruction Incidents

STANDARD OF CARE

As emergency responders, we must abide by a **standard of care.** In years past it was deemed acceptable to wash spilled gasoline down a storm drain. Anyone who does that today could face local, state, and federal charges for violating the Clean Water Act. This standard of care comprises the laws, regulations, standards, local protocols, and experience

Figure 1-5 The firefighters responding to this tanker fire have a responsibility to follow a standard of care. The response should follow local, state, and national accepted practices, standards and regulations.
(Courtesy Mark Brady PGFD)

that are detailed throughout this text. Violations of this standard of care are evaluated based on three theories: liability, negligence, and gross negligence.

When operating in and around hazardous materials, liability becomes a very real issue. Other than EMS responses like those shown in **Figure 1-5,** any actions at a chemical release are most likely to put responders at risk as personally liable for any wrongdoing. **Liability** is responsibility for one's actions. Environmental and safety regulations are designed to place blame on the person who made the decision, not on the organization itself, unless the organization has policies that violate these regulations. In the earlier example, the person who made the decision to wash the gasoline down a storm drain could end up in jail or pay substantial fines. **Negligence** results from not following the standard of care or an accepted practice. **Gross negligence** is willful disregard for the standard of care. In other words, a conscious decision is made to not follow the standard of care. To avoid becoming liable, you should receive the training required for your position and act only to your level of training. When you are not sure of your next step, it is best to consult with your local HAZMAT team or other local or state environmental protection agency.

ADDITIONAL LAWS, REGULATIONS, AND STANDARDS

Responders should be aware of the items listed here, as they are commonly encountered or applied in chemical releases.

Hazard Communication

The hazard communication regulation issued by OSHA (29 CFR 1910.1200) requires that employers make available Material Safety Data Sheets (MSDSs) for all chemicals located at the facility at quantities above "household quantities." The term *household quantities* provides that small amounts such as those that would be purchased for home use would not require MSDSs. An example would be a gallon of household ammonia for use in an office. If a case of household ammonia is brought in, it would require an MSDS as this is not a normal household quantity. The employer also must provide a list of these MSDSs and specify when the chemicals arrived on site. The employer must provide training on these MSDSs and the hazard communication program. One of the important components of this program is the requirement for the facility to make these sheets available for emergency responders. Emergency responders are also responsible for following this regulation.

RESPONDER FACT Emergency response organizations in many states are required to abide by the federal or state environmental and safety regulations. Emergency response organizations are no different than local industry. Even if an emergency response organization is exempt from following environmental or safety regulations, department administrators should consider the public perception of not following the regulations. In many cases the emergency response organization is a regulating entity, which may issue and enforce regulations that local businesses are required to follow. The department's public image is improved if it voluntarily follows these regulations. An added benefit is that responders are working in a safer workplace and protecting the environment.

Superfund Act

The Comprehensive Environmental Response, Compensation, and Liability Act (CERCLA) is commonly referred to as the Superfund law. This law was established in 1980 for the cleanup of toxic waste sites around the country, and was the forerunner to SARA. It laid the groundwork for regulating the emergency services with regard to response to chemical emergencies. When responding to a Superfund site, some additional concerns and requirements must be noted. The emergency responders should have a copy of the site's existing emergency response plan. The site should have its access limited and hazard zones established; depending on the nature of the work involved, it may have a decontamination corridor. Prior to work beginning on a Superfund site, the local responders should receive notification and should meet with the site supervisor to learn the hazards of the site and any protective measures that may be in place. As Superfund sites vary from old landfills to mercury dumping sites, it would be impossible here to provide a breakdown of the potential site hazards and precautions that the site may hold. Prior to becoming a Superfund site, the area may be listed as a National Priority List (NPL) site. Contact your state environmental agency or federal regional EPA office to learn of any Superfund or NPL sites in your area.

Clean Air Act

The 1990 Clean Air Act Amendments, known as CAAA, require facilities processing threshold quantities of certain chemicals to develop a Risk Management Program (RMP). Since June 1999 facilities have been required to submit plans to the EPA, the state, and other local emergency planning agencies. The plans must evaluate the worst case and most probable scenarios for a chemical accident at the facility. As part of the planning process, the facility must coordinate planning efforts with the local fire department and conduct exercises to test the plan.

Respiratory Protection

OSHA's respiratory protection regulation (29 CFR 1910.134) also affects the emergency services by adding requirements that had previously been exempt. The biggest changes are the inclusion of the two-in-two-out rule, which was in place for chemical incidents but now also includes structural fire situations. OSHA has now classified a fire as an IDLH situation and requires the incident commander to take certain precautions to ensure firefighters' safety. The emergency response organization is now required to fit test all responders and provide a Licensed Health Care Provider's (LHCP's) approval to wear respiratory protection by the use of a medical survey or a physical exam. Record keeping has also changed; specific records must be kept by the response organization regarding the employees and the respiratory protection program. Records include the listing of personnel training, daily equipment checks, periodic maintenance, routine service, review of the respiratory protection program, and other items.

Firefighter Safety

NFPA 1500, *Standard on Fire Department Occupational Safety and Health Program*, is sometimes referred to when discussing

NFPA 1500

Standard on Fire Department Occupational Safety and Health Program

HAZMAT issues. This standard is a broad-based program for providing a safe workplace for firefighters. It has been applied by OSHA using the general duty clause, specifically regarding two-in-two-out.

NFPA Chemical Protective Clothing

NFPA 1991 and 1992 are standards for chemical protective clothing ensembles, including encapsulated suits, splash protective suits, and support garments. NFPA 1994, Standard on Protective Ensembles for Chemical/Biological Terrorism Incident, has three levels of protective equipment that can be used in the event of a chemical or biological attack.

SUMMARY

The maze of laws, regulations, and standards can be confusing and sometimes overwhelming. Most laws and regulations are not easy to read, and it can take years to comprehend every detail of their intent. They are subject to interpretation and change frequently, and emergency responders must keep abreast of those that affect their everyday jobs. In larger departments staff may be assigned to monitor health, safety, and environmental laws and regulations. In smaller departments this may not be possible; thought of as an industry, the responders are the same as the chemical plant or gas station in a community. The responders must follow the same health, safety, and environmental laws and regulations. For your health and safety, it will benefit you to keep abreast of these issues, and it will help make your community a safer place to live and work.

KEY TERMS

Air monitoring devices   Used to determine oxygen, explosive, or toxic levels of gases in air.

Awareness Level   The basic level of training for personnel who during the performance of their job may come across a hazardous materials/WMD release, the basis of which is the ability to recognize a hazardous situation, isolate and deny entry, and call for assistance.

Clandestine drug labs   Illegal labs set up to manufacture street drugs.

Emergency Planning and Community Right to Know Act (EPCRA)   The portion of SARA that specifically outlines how industries report their chemical inventory to the community.

Extremely hazardous substances (EHS)   A list of 366 substances that the EPA has determined present an extreme risk to the community if released.

Firefighters protective clothing (FFPC)   The protective clothing that firefighters use for firefighting, rescue, and other first response activities; also known as turnout or bunker gear.

Gross negligence   Occurs when an individual disregards training and continues to act in a manner without regard for others.

Hazardous Waste Operations and Emergency Response (HAZWOPER)   The OSHA regulation that covers safety and health issues at hazardous waste sites, as well as response to chemical incidents.

Incident Commander Level   A training level that encompasses the operations level with the addition of incident command training. Intended to be the person who may command a chemical incident.

Laws   Legislation that is passed by the House and Senate, and signed by the president.

Liability   The possibility of being held responsible for individual actions.

Local Emergency Planning Committee (LEPC)   A group composed of members of the community, industry, and emergency responders to plan for a chemical incident, and to ensure that local resources are adequate to handle an incident.

Material Safety Data Sheet (MSDS)   Information sheet for employees that provides specific information about a chemical, with attention to health effects, handling, and emergency procedures.

Mission specific competencies   The knowledge, skills, and abilities to perform the mission of the organization. Examples are provided in NFPA 472.

Negligence   Acting in an irresponsible manner, or different from the way in which someone was trained; that is, differing from the standard of care.

Operations Level   The next level of training above awareness that provides the foundation that allows for the responder to perform defensive activities at a chemical incident.

Paragraph q   The paragraph within HAZWOPER that outlines the regulations that govern emergency response to chemical incidents.

Regulations   Developed and issued by a governmental agency and have the weight of law.

Specialist Level   A level of training that provides for a specific type of training, such as railcar specialist; someone who has a higher level of training than a technician.

Standards   Usually developed by consensus groups establishing a recommended practice to follow.

State Emergency Response Committee (SERC)   A group that ensures that the state has adequate training and resources to respond to a chemical incident.

Superfund Amendments and Reauthorization Act (SARA) A law that regulates a number of environmental issues, but is primarily for chemical inventory reporting by industry to the local community.

Technician Level A high level of training that allows specific offensive activities to take place, to stop or handle a chemical incident.

REVIEW QUESTIONS

1. How does the level of response to hazardous materials differ from a structural fire response?

2. How can hazardous materials be easily defined?

3. Who is to receive the chemical reporting required under SARA Title III?

4. Who is responsible to ensure the proper response to a hazardous materials emergency?

5. Which regulation covers all emergency responders who respond to chemical accidents?

6. What is the difference between negligence and gross negligence?

7. Which regulation requires employers to maintain Material Safety Data Sheets?

8. What was the largest change to the 2008 edition of NFPA 472?

9. Which training regulation are most firefighters required to follow?

10. To perform offensive activities at a hazardous materials release what level of training is required?

Suggested Readings

Bevelacqua, Armando S., *Hazardous Materials Chemistry*, 2nd ed., Delmar Cengage Learning, Clifton Park, NY, 2006.

Bevelacqua, Armando S., and Richard H. Stilp, *Hazardous Materials Field Guide*, 2nd ed., Delmar Cengage Learning, Clifton Park, NY, 2007.

Bevelacqua, Armando S., and Richard H. Stilp, *Terrorism Handbook for Operational Responders*, 2nd ed., Delmar Cengage Learning, Clifton Park, NY, 2004.

Hawley, Chris, *HazMat Air Monitoring & Detection Devices*, 2nd ed., Delmar Cengage Learning, Clifton Park, NY, 2006.

Hawley, Chris, *Hazardous Materials Incidents*, 2nd ed., Delmar Cengage Learning, Clifton Park, NY, 2004.

Henry, Timothy V., *Decontamination for Hazardous Materials Emergencies*, Delmar Cengage Learning, Clifton Park, NY, 1999.

Laughlin, Jerry, and David Trebisacci (editors), *Hazardous Materials Response Handbook*, 4th ed., National Fire Protection Association, Quincy, MA, 2002.

Lesak, David, *Hazardous Materials Strategies and Tactics*, Prentice Hall, Upper Saddle River, NJ, 1998.

Noll, Gregory, Michael Hildebrand, and James Yvorra, *Hazardous Materials: Managing the Incident*, 3rd ed., Fire Protection Publications, Oklahoma University, 2005.

Schnepp, Rob, and Paul Gantt, *Hazardous Materials: Regulations, Response & Site Operations,* Delmar Cengage Learning, Clifton Park, NY, 1999.

Stilp, Richard, and Armando Bevelacqua, *Emergency Medical Response to Hazardous Materials Incidents*, Delmar Cengage Learning, Clifton Park, NY, 1997.

RECOGNITION AND IDENTIFICATION

OUTLINE

- Objectives
- Introduction
- Location and Occupancy
- Placards, Labels, and Markings

- Other Identification Systems
- Containers
- Senses
- Chemical and Physical Properties

- Radiation
- Summary
- Key Terms
- Review Questions

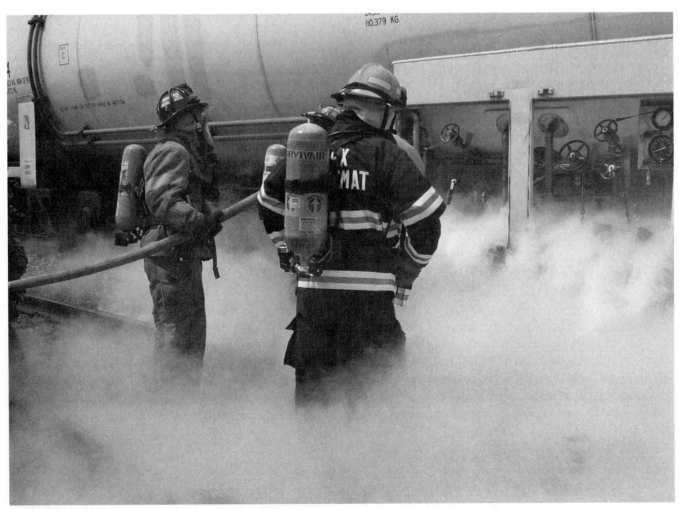

(Courtesy of Mike Austin)

STREET STORY

While the news media tends to focus on the large and spectacular, the reality is that every day the fire service responds to literally hundreds of incidents involving hazardous materials. These incidents usually do not get much attention because they are small in scope and are handled with a minimal amount of fire department resources. Common examples include natural gas leaks and flammable and combustible liquid spills.

We have an old saying in the HAZMAT community that the initial ten minutes of an incident will dictate the tone for the first hour. Clearly, the actions—or inactions—of the first responders will set the tone for an incident. I think back to two incidents in my career that reinforce this point.

The first incident involved an engine company sent to investigate a call regarding an unknown liquid spilled along a highway. Upon arrival, the engine company officer found a very viscous red liquid spilled along a long distance of road, apparently from the back of a tractor trailer. Unsure of the identity and the potential hazards of the liquid, the officer requested that a HAZMAT unit respond to the scene and provide assistance. Using their monitoring and detection equipment, responders were eventually able to determine the liquid posed no threat to the community. In fact, the unknown liquid was eventually identified as strawberry syrup! Although the officer took some ribbing from his peers, he clearly made some proper decisions to ensure the safety of both his personnel and the community.

In the second incident, a police officer was called to provide assistance to a public works road crew that had discovered what appeared to be a 5-pound portable extinguisher wrapped in duct tape with a fuse on top. Although he believed it to be a hoax, the officer still requested the response of the fire department bomb squad. After conducting a thorough risk assessment and requesting the additional assistance of a HAZMAT response team, the bomb squad determined that the perceived hoax was in fact an actual explosive device. When the device was disrupted (i.e., blown up) by the bomb squad, it made a lasting impression on the police officer!

In both of these incidents, recognition and identification by the first responder set the tone for the incident. If you do not know what to do, isolate the area, deny entry, and call for help.

—Street Story by Greg Noll, Emergency Planning and Response Consultant, Hildebrand, Noll Associates, Lancaster, Pennsylvania

OBJECTIVES

After reading this section the student should be able to identify and explain:

■ The hazards associated with the nine hazard classes as defined by DOT (A)

■ The standard occupancies where hazardous materials may be used or stored (A)

■ The standard container shapes and sizes of common products (A)

■ Both facility- and transportation-related markings and warning signs (A)

■ The standard transportation types for highway and rail (A)

■ The use of transportation containers in identifying possible contents (A)

■ The location of emergency shutoffs on highway containers (O)

■ The importance of understanding chemical and physical properties of hazardous materials (O)

Figure 2-1 Basic clues to recognition and identification are location and occupancy, placards, labels, markings, container types, and senses. Here, the location is an industrial complex; the containers are visible upright storage tanks, which are labeled with their contents.

INTRODUCTION

This chapter is of primary importance to the emergency responder.

SAFETY By honing your skills in recognition and identification (R & I), you can improve your ability to stay alive.

The inability to recognize the potential for chemicals and identify a chemical hazard can place you and others in severe danger. Public safety is inherently dangerous, but the response to hazardous materials release places an additional risk. Not only can there be immediate effects from some materials, but multiple exposures can have long-reaching effects. In rare cases, fires have killed hundreds of people. Equally rare but even more deadly, releases of extremely hazardous substances have resulted in the deaths of thousands of people in one incident. In 1984 the release of methyl isocyanate in Bhopal, India, reportedly killed several thousand persons. The exact numbers are in dispute and range from 4,000 to 20,000 people killed. It is also known that many thousands more were injured, and many still suffer today from the chemical exposure. This tragedy was the basis for the Emergency Planning and Community Right to Know Act (EPCRA). Several facilities in the United States use this material.

SAFETY There are four basic clues to recognition and identification: location and occupancy; placards, labels, and markings; container types; and the senses.

A mere suspicion in any of these areas should be enough to place a first responder on guard for the possibilities of a chemical release and associated hazards, as shown in **Figure 2-1.**

LOCATION AND OCCUPANCY

CAUTION The size of the community does not determine the potential for hazardous materials. Every community has hazardous materials.

Most communities have a gas station or a hardware store. The average home has a large amount of hazardous materials, which can cause enormous problems during a response. In rural communities, farms present unique risks due to the storage of pesticides and fertilizers, as shown in **Figure 2-2.** All of these locations and occupancies provide the potential for the storage of hazardous materials. In

Figure 2-2 Agricultural supply stores have a large quantity of hazardous materials, including pesticides, herbicides, and fertilizers. In many cases they also stock propane and other fuel.

general the more industrialized a community, the more hazardous materials the community will contain. Communities adjacent to industrialized areas or along a major transportation corridor (interstate highway, rail, water, air, and pipeline) will also have the same hazards as these materials will be traveling in the community, as shown in **Figure 2-3.** Buildings that typically store hazardous materials include hardware stores, hospitals, auto part supply stores, dry cleaners, manufacturing facilities, print shops, medical offices, photo labs, agricultural supply stores, semiconductor manufacturing facilities, electronics manufacturing, light to heavy industrial plants, marine terminals, rail yards, airport terminals and fueling areas, pool chemical stores, paint stores, hotels, swimming pools, food manufacturing facilities, and many others.

PLACARDS, LABELS, AND MARKINGS

In this section we will examine the first concrete evidence of the presence of hazardous materials. A number of systems are used to mark hazardous materials containers, buildings, and transportation vehicles. The systems are part or combinations of laws, regulations, and standards. As an example, International Codes Council (ICC), a new code development organization, requires the use of the NFPA 704 marking system for certain occupancies, as shown in **Figure 2-4.** The ICC is a combination of the Building Officials and Code Administrators (BOCA), International Conference of Building Officials (ICBO), and the Southern Building Code Congress International (SBCCI), which used to be separate code development organizations.

Placards

This is the most commonly seen system for identifying the location of hazardous materials in transportation. The **Department of Transportation (DOT)** regulates the movement of hazardous materials (dangerous goods in Canada) by air, rail, water, roadway, and pipeline, per 49 CFR 170-180. After meeting certain guidelines a shipper must placard a vehicle to warn of the storage of chemicals on the mode of transportation (see **Figure 2-5**).

NOTE The quantity of hazardous materials that must be carried to require placarding is 1,001 pounds (454 kg), unless it is a DOT Table 1 material that requires placarding at any amount.

Figure 2-3 If a road goes through your community, the potential for a hazardous materials incident exists. One of the most common chemical releases is a gasoline spill.

Figure 2-4 The NFPA has developed its 704 system of identifying potential chemical hazards in a building.

Figure 2-5 The DOT requires some shippers of hazardous materials to provide warning placards to warn responders of chemicals that may be on the truck. This truck contains several different materials. The corrosive placard with the number "3096" indicates that part of the shipment is in bulk containers.

Table 2-1 and Table 2-2 provide further explanation of the placarding system.

> **NOTE** The DOT has established a system of nine hazard classes and has more than twenty-seven placards to identify a shipment.

Figure 2-6 Corrosive placard.

The idea behind these hazard classes is to identify a shipment per a general grouping and provide some basic information regarding the potential hazards. The placards are signs that are 10 3/4 inches by 10 3/4 inches (27.3 × 27.3 cm); they are to be placed on four sides of the vehicle, like the one shown in **Figure 2-6.** Labels are 3.9 inches by 3.9 inches (10 × 10 cm) and are affixed near the shipping name on the container. The labels are smaller versions of the placard and are designed to provide warnings on the package's contents.

The system is designed so that materials designated by the DOT that present a capacity to harm the environment, humans, and animals are easily identified. Materials must present an unreasonable risk to the safety, health, or property upon contact. The DOT has established two tables, with different guidelines for both tables. Materials listed in Table 1 are those that are most hazardous and require the use of placards regardless of the quantity of material being carried. Table 2 materials are those that require placarding at 1,001 pounds (454 kg). The criteria establishing this 1,001-pound (454-kg) rule are not clearly defined, but responders should be aware that a spill of 999 pounds (453 kg) of a material can be as hazardous as 1,001 pounds (454 kg). The shipper uses the hazardous materials table (49 CFR 107.101) to determine which labels and placards are required.

TABLE 2-1

Materials That Require Placarding at Any Amount (DOT Table 1)	
Hazard Class or Division	**Placard Type**
1.1	Explosives 1.1
1.2	Explosives 1.2
1.3	Explosives 1.3
2.3	Poison gas
4.3	Dangerous when wet
5.2 Organic Peroxide, Type B liquid or solid, temperature controlled	Organic peroxide
6.1 (assigned to Packing Group (PG) I, inhalation hazard only) Zone A or B	Poison inhalation hazard
7 (radioactive label III only)	Radioactive

TABLE 2-2

Materials That Require Placarding at 1,001 Pounds (454 kg) (DOT Table 2)	
Class or Division	**Placard Type**
1.4	Explosives 1.4
1.5	Explosives 1.5
1.6	Explosives 1.6
2.1	Flammable gas
2.2	Nonflammable gas
3	Flammable
Combustible Liquid	**Combustible**
4.1	Flammable solid
4.2	Spontaneously combustible
5.1	Oxidizer
5.2 (other than Type B liquid or solid, temperature controlled)	Organic peroxide
6.1 (Packing Group (PG) I or II, other than hazard Zone A or B)	Poison
6.1 (PG III)	Keep away from food
6.2	None
8	Corrosive
9	Class 9
ORM-D	None

The table may also list a packing group for the material, which indicates the danger associated with the material being transported.

■ Packaging Group I—Greatest danger
■ Packaging Group II—Medium danger
■ Packaging Group III—Minor danger

Packaging groups are assigned only to classes 1, 3 through 6, 8, and 9 as shown in **Table 2-3, Table 2-4, Table 2-5,** and **Table 2-6.** These are determined based on flash points, boiling points, and toxicity. See **Table 2-3** for more information. At this point in the text these items may be confusing; these terms are

further explained toward the end of this chapter, but to assist with this section, they can be defined as:

■ Flash point—the temperature of a liquid that, if an ignition source is present, provides sufficient vapors to start a fire.
■ Boiling point—the temperature of a liquid that is changing from the liquid state to the gaseous state.
■ Toxicity—degree of harmfulness, as provided through exposure limits. These exposures are provided in lethal concentrations (LC) and lethal doses (LD), which measure the amount of material needed to produce toxic effects.

TABLE 2-3

Packing Groups: Flammability Class 3		
Packing Group (PG)	**Flash Point**	**Boiling Point**
I		95°F (35°C)
II	73°F (23°C)	95°F (35°C)
III	73°F (23°C) and 141°F (60.5°C)	95°F (35°C)

TABLE 2-4

Packing Groups: Poisonous Gas Division 2.3	
Hazard Zone	**Inhalation Toxicity—Lethal Concentration (LC_{50})***
Hazard Zone A	LC_{50} less than or equal to 200 ppm
Hazard Zone B	LC_{50} greater than 200 ppm and less than or equal to 1,000 ppm
Hazard Zone C	LC_{50} greater than 1,000 ppm and less than or equal to 3,000 ppm
Hazard Zone D	LC_{50} greater than 3,000 ppm and less than or equal to 5,000 ppm

*LC_{50} is the lethal concentration to 50% of an exposed population (gases).

TABLE 2-5

Packing Group Division 6.1			
For packing materials that are toxic by a route other than inhalation, the following are used:			
Packing group	**Oral toxicity—Lethal Dose (LD_{50})* (mg/kg)**	**Dermal toxicity—Lethal Dose (LD_{50}) (mg/kg)**	**Inhalation by dusts and mists (LC_{50}) (mg/L)**
I	≤ 5	≤ 40	≤ 0.5
II	$>5, \leq 50$	$>40, \leq 200$	$>0.5, \leq 2$
III	Solids: $>50, \leq 200$; liquids: $>50, \leq 500$	$>200, \leq 1000$	$>2, \leq 10$

*LD_{50}, lethal dose to 50% of the exposed population (solids and liquids).

TABLE 2-6

Packing Group Division 6.1	
Poisonous by inhalation	
Packing group	**Vapor concentration and toxicity**
I (Hazard Zone A)	$V \geq 500\ LC_{50}$ and LC_{50} 200 mL/M^3

MEASURING TOXICITY

The DOT uses several terms to provide a measure of toxicity. The DOT establishes ranges for the shipment of toxic materials. The determination of how toxic a material is can be based upon exposure limits established for many chemicals. LC_{50} and LD_{50} are terms used to describe a concentration or amount of a material that causes lethal effects to 50 percent of an exposed population. LC is lethal concentration, used to measure toxicity of gases, and LD is a lethal dose of solids or liquids. To determine a LC_{50} or LD_{50}, researchers expose a population of test animals to a given amount of a substance. The level at which 50 percent of the test population dies establishes the LC_{50} or LD_{50}. When discussing toxicity, two common units of measure for the amount of material required to cause toxic effects are parts per million (ppm) and milligrams per cubic meter (mg/M^3). The use of ppm usually is reference to a gas and represents a part of a million pieces, as one marble is one part per million of a million marbles. Usually mg/M^3 refers to the amount of a solid or liquid material. It is a comparison of the amount of chemical in milligrams (mg) as compared to a cubic meter of air, as that amount of material could cause toxic effects within that cubic meter. The use of mg/L is the reference to the amount (mg) that is within the space of a liter. More information related to toxicity is provided in Chapter 4.

Placards provide four useful methods of communicating the hazard class they represent. They have distinct colors and include a picture at the top of the square depicting a representation of the hazard.

Placards state the hazard class across their middle and display the hazard class and division number at the bottom. As an example, the placard as shown in **Figure 2-6** is black and white, which represents the corrosive class. The top of the placard has a picture of a hand and a steel bar being eaten away by the corrosive material being poured on it. The middle of the placard reads, "corrosive," and at the bottom is the class number 8. The DOT also requires the addition of a four-digit number known as the United Nations/North America (UN/NA) identification number either on a placard or on an adjacent orange rectangle. This identifies a bulk shipment of over 119 gallons (450 liters) and specifies what the material is. A tank truck carrying gasoline, which would be considered a bulk shipment, would display a flammable liquid placard with the number 1203 either in the middle of the placard or on an orange rectangle adjacent to the placard, as shown in **Figure 2-7.** A tank truck placarded for gasoline could also haul diesel fuel or fuel oil under the same placard without having to change placards. The DOT assumes that if the truck is placarded for the worse hazard, responders will be adequately warned. This panel with the DOT ID number, as shown in **Figure 2-7,** provides an additional bit of information to the responder, as before the only information would be from the placard indicating that a flammable liquid was on board.

Figure 2-7 There are three ways to signify the four-digit ID number for bulk shipments. The most common is the four-digit DOT identification number in the middle of the placard. The four-digit number in the middle of the placard or on an adjacent panel indicates that the shipment is in bulk quantity.

Figure 2-8 "Explosive" placards and labels.

The nine hazard classes and subdivisions[1] are as follows:

■ **Class 1 Explosives** (placards and labels shown in **Figure 2-8**)

  ■ Division 1.1—mass explosion hazard, such as black powder, dynamite, ammonium perchlorate, detonators for blasting, and RDX explosives.

  ■ Division 1.2—projectile hazard, such as aerial flares, detonating cord, detonators for ammunition, and power device cartridges.

  ■ Division 1.3—fire hazard or minor blast hazard; examples include liquid fuel rocket motors and propellant explosives.

  ■ Division 1.4—minor explosion hazard, which includes line throwing rockets, practice ammunition, detonation cord, and signal cartridges.

  ■ Division 1.5—very insensitive explosives that present mass explosion potential but during normal shipping would not present a risk. Ammonium nitrate and fuel oil (ANFO) mixtures are an example of this division.

  ■ Division 1.6—very insensitive explosives with no mass explosion potential. Materials that present an unlikely chance of ignition are part of this grouping.

The placards for divisions 1.1, 1.2, and 1.3 will have the division number and compatibility group located just above the class number for any quantity, as shown in **Figure 2-8.** For divisions 1.4, 1.5, and 1.6 the placard will have a compatibility group number when a vehicle is transporting more than 1,001 pounds (454 kg). The DOT provides compatibility charts that dictate that certain high hazard shipments cannot be mixed with other hazardous materials. The compatibility charts are designed to prevent the mixing of certain materials in the event of an accident. Some materials when mixed can create substantial issues such as explosions, fires, or release of toxic gases.

Incidents involving explosives can be very dangerous, especially when involved in fire. Making a tactical decision to attack a fire involving explosives can endanger the responders, especially if the fire has reached the cargo area of the vehicle. The recommendations in the DOT **Emergency Response Guidebook (ERG – Guide 112)** should be followed, paying particular attention to the isolation and evacuation distances. When explosives are involved in traffic accidents not involving fire, the actual threat is minimized depending on the circumstances. As long as the explosives were transported legally and as they were intended to be transported, the responders should face little danger. Some cities require an escort and transportation of explosives at nonpeak hours. Spilled explosive materials may present a health hazard if inhaled or absorbed through the skin.

When explosives are involved in a fire situation, the best course of action would be the immediate withdrawal from the area as well as isolating and denying entry. The area should be evacuated following recommendations from the current version of the DOT ERG. Many chemicals used in explosives are powerful oxidizers which if ignited can be very aggressive, will burn with considerable energy, and in some cases are explosive in nature.

■ **Class 2 Gases** (placard and label shown in **Figure 2-9**)
  ■ Division 2.1—flammable gases that are ignitable at 14.7 psi (1 bar) in a mixture of 13 percent or less in air, or a **flammable range** with air of at least 12 percent regardless of the **lower explosive limit.** Propane and isobutylene are examples of this division.

CASE STUDY

On November 29, 1988, an engine company from the Kansas City, Missouri Fire Department was dispatched to a reported pickup truck fire. While en route to the incident, the engine company was told to use caution as there were reports of explosives involved. When firefighters arrived, they found two separate fires, one in the pickup truck and the other in a trailer. The first engine began to extinguish the fire in the pickup truck and requested a second engine for assistance and the district battalion chief. They attempted to contact the second engine to warn them of the explosives on fire on top of the hill. The second engine arrived and began to attack the fire on top of the hill. That crew requested the assistance of the first engine and also requested a squad for water. From the radio communications with the battalion chief, the crews thought that the explosives had already detonated.

The battalion chief was a quarter of a mile away where he had stopped to talk with the security guards when the explosives detonated. The explosion moved the chief's car 50 feet (15 m) and blew in the windshield. The blast was heard for 60 miles (87 km) and damaged homes within 15 miles (24 km). The chief requested additional assistance and staged the responding companies. There was a report of more explosives that had not yet detonated on the hill. Luckily, the chief did not let any other responders onto the scene; shortly after pulling back, a second explosion went off, reportedly larger than the first one. The next morning a team of investigators went to the site to begin the investigation. They discovered that all six firefighters were killed in the blast, and the explosion was very devastating. Only one engine was recognizable; the other was reduced to the frame rail. The first blast was from 17,000 pounds (7711 kg) of ammonium nitrate, fuel oil, and aluminum mixture. It also had 3,500 pounds (1588 kg) of ammonium nitrate and fuel oil, commonly referred to as ANFO. The second explosion involved 30,000 pounds (13,608 kg) of the mixture. The use of ANFO is common throughout the United States for blasting purposes. Ammonium nitrate is commonly used as a fertilizer in the agricultural business and is used on residential lawns. The Oklahoma City Murrah Building explosive was devised of ammonium nitrate and nitromethane, very similar to ANFO. The World Trade Center bombing in 1993 used an explosive made up primarily of urea nitrate and three cylinders of hydrogen. Both explosives are comparable to each other, and when used improperly or for criminal purposes, the results can be devastating.

Figure 2-9 "Gas" placards and labels.

■ Division 2.2—nonflammable, nonpoisonous, compressed gas, including liquefied gas, pressurized **cryogenic** gas, and compressed gas in solution. Anhydrous ammonia, carbon dioxide, liquid argon, and nitrogen are examples.

SAFETY Ammonia is listed by the DOT as being nonflammable, but in reality it is flammable. DOT has criteria to meet the flammability standard, and ammonia does not fit into that criteria. Responders have been killed in fires involving ammonia releases. Ammonia should be considered by responders as being flammable, especially when confined in a building.

■ Division 2.3—poisonous gases that are known to be toxic to humans and would pose a threat during transportation. Chlorine and liquid cyanogen are common examples of this division. Gases in this division are also assigned a letter code identifying their toxicity levels. You will learn more about these levels in the section on toxicology. The hazard zones associated with this division are:

■ Hazard zone A—LC_{50} less than or equal to 200 ppm

■ Hazard zone B—LC_{50} greater than 200 ppm and less than or equal to 1,000 ppm

■ Hazard zone C—LC_{50} greater than 1,000 ppm and less than or equal to 3,000 ppm

■ Hazard zone D—LC_{50} greater than 3,000 ppm and less than or equal to 5,000 ppm

The hazard zones are a quick way to determine how toxic a material is, as hazard zone A is more toxic than B, and so forth. The shipping papers will identify these by the addition of Poison Inhalation Hazard (PIH) zone A, B, C, or D.

CAUTION The difference between a flammable liquid and a combustible liquid is based on the material's flashpoint. The flashpoint is the temperature of the liquid at which point it is producing enough vapors mixed with air to start a flash fire if an ignition source is present. The fire service usually refers to a flammable liquid as one that has a flash point of 100°F (37.7°C) or lower. The DOT classifies any liquid with a flashpoint of 141°F (60.5°C) or lower as being a flammable liquid. Any liquid with a flashpoint greater than 141°F (60.5°C) is considered combustible by the DOT. It can be confusing when using the terms "flammable" and "combustible" to describe a liquid material. Most references use 100°F (37.7°C) as the criteria for flammable and combustible, with the exception of the DOT.

NOTE Remember that many of these gases have multiple hazards. Based on information that can be obtained from a Material Safety Data Sheet (MSDS) and the DOT Emergency Response Guide (ERG), some of these products such as those in Division 2.2 can also be flammable and cause unforeseen problems.

■ **Class 3 Flammable liquids** (placard and label shown in **Figure 2-10**) are those liquids that have a flash point of less than 141°F (60.5°C) . Gasoline, acetone, and methyl alcohol are examples.

■ Combustible liquids are those with flash points above 141°F (60.5°C) and below 200°F (93°C). The DOT allows that liquids with a flash point of 100°F (37.7°C) can be shipped as a combustible liquid. Examples include diesel fuel, kerosene, and various oils.

Figure 2-10 "Flammable" and "combustible" placards and labels.

Figure 2-11 Class 4 placards and labels.

Figure 2-12 Class 5 placards and labels.

■ **Class 4 Flammable solids** (placard and labels shown in **Figure 2-11**) also include spontaneously combustible materials and dangerous-when-wet materials.

■ Division 4.1 includes wetted explosives, self-reactive materials, and readily combustible solids. Examples include magnesium ribbons, picric acid, and explosives wetted with water, alcohol, or plasticized.

■ Division 4.2 comprises spontaneously combustible materials, including pyrophoric materials or self-heating materials. An example is zirconium powder.

■ Division 4.3 dangerous-when-wet materials are those that when in contact with water can ignite or give off flammable or toxic gas. Calcium carbide when mixed with water makes acetylene gas, which is very flammable and unstable in this form. Sodium and lithium are other examples that can, when wet, ignite explosively. Aluminum and magnesium are not as explosive as sodium but will react with water. If on fire they will react violently if water is used in an attempt to extinguish the fire.

■ **Class 5 Oxidizers and organic peroxides** (placard and labels shown in **Figure 2-12**)

■ Division 5.1 is the class assigned to materials that can produce oxygen, which in turn increases the chance of fire and during fires makes the fire burn more intensely. Ammonium nitrate and calcium hypochlorite are examples.

■ Division 5.2 includes the organic peroxides that can explode or **polymerize,** which if contained is an explosive reaction. These are further subdivided into seven types:

■ Type A—can explode upon packaging; these are forbidden by the DOT, which means they cannot be transported and must be produced on site.

■ Type B—can thermally explode; considered a very slow explosion.

■ Type C—neither detonates nor **deflagrates** rapidly, and will not thermally explode.

■ Type D—only detonates partially or deflagrates slowly; has a medium or no effect when heated and confined.

■ Type E—shows low or no effect when heated and confined.

■ Type F—shows low or no effect when heated and confined, and has low or no explosive power.

■ Type G—is thermally stable and is desensitized.

■ **Class 6 Poisonous materials** (placards and labels shown in **Figure 2-13**) are divided into two groups: poisonous materials other than gases and infectious substances.

■ Division 6.1—materials that are so toxic to humans they would present a risk during transportation. Examples include arsenic, aniline, and some pesticides.

POLYMERIZATION

A chemical that polymerizes is one that presents an extreme danger to firefighters. Polymerization is a runaway chain reaction that once started cannot be stopped. If the material is contained, it will rupture the container in a violent manner, sending pieces of the container for a considerable distance. Polymerization can be best described as a chain. It starts off as one link, and once the reaction is started, many other links develop and attach themselves to the other links. Each link develops a great deal of heat, and for each 41°F (5°C) of heat that develops, the reaction rate doubles. Thus they quickly duplicate themselves, eventually rupturing the container. The reaction can result in a fire or explosion, as many of the chemicals that can polymerize are also flammable liquids. A container of spray foam insulation that is used for sealing spaces in walls around doors and windows is a good example. When the foam hits the air, it reacts and rapidly expands to a much greater volume. This reaction is uncontrollable, and will continue to expand for quite some time until the polymerization is completed. A small container that holds a few ounces does not present much risk, but a railcar full of a chemical that may polymerize presents an extreme risk to the community.

■ Division 6.2—microorganisms or their toxins that can cause disease to humans or animals. Anthrax, rabies, tetanus, and botulism are examples.

Hazard zones associated with Class 6 materials are:

■ Hazard zone A—LC_{50} less than or equal to 200 ppm.

■ Hazard zone B—LC_{50} greater than 200 ppm and less than or equal to 1,000 ppm.

■ **Class 7 Radioactive materials** are those determined to have radioactive activity at certain levels. Although there is only one placard, the labels shown in **Figure 2-14** are further subdivided into Radioactive I, II, and III, with level III being the highest hazard. The designations I, II, and III are dependent on two criteria: the transport index and the radiation level coming from the package. The transport index is the degree of control the shipper is to use and is based on a calculation of the radiation threat the package(s) presents. Responders should understand that packages labeled radioactive may have a level of radiation emitting from the package. These emissions are legal within certain guidelines. The levels are as follows:

■ Radioactive I label—less than 0.005 mSv/hr (0.5 mR/hr).

■ Radioactive II label—more than Radioactive I and less than 0.5 mSv/hr (50 mR/hr).

■ Radioactive III label—more than Radioactive II and less than 2 mSv/hr (200 mR/hr).

■ Fissile material label—for radioactive material that can be used as a fuel. Common examples include uranium-233, uranium-235, uranium-239, and uranium-241.

■ Packages emitting more than 2 mSv/hr require special handling and transportation and are subject to additional regulations.

Figure 2-13 Class 6 placards and labels.

Figure 2-14 Class 7 placards and labels.

Figure 2-15 Class 8 placards and labels.

NOTE mSv/hr is milliseivents per hour; mR/hr is millirem per hour. Both are used to show the absorbed dose. Seivents are the metric system and REM is used in the United States.

■ **Class 8 Corrosives** (placard and label shown in **Figure 2-15**) include both acids and bases. The DOT describes corrosives as materials capable of causing visible destruction in skin or corrosion of steel or aluminum. Examples include sulfuric acid, sodium hydroxide, and ammonia.

■ **Class 9 Miscellaneous hazardous materials** are a general grouping comprising mostly hazardous waste. Dry ice, molten sulfur, liquid asphalt, and polymeric beads are examples that would use a Class 9 placard, shown in **Figure 2-16.** This is known as a catchall category: If a substance does not fit into any other category and presents a risk during transportation, it becomes Class 9.

■ **Other Placards and Labels**

NOTE A "Dangerous" placard is used when the shipper is sending a mixed load of hazardous materials if more than 1,001 pounds (454 kg) of more than one of the materials listed in Table 2.

■ If the shipper sends 2,000 pounds (907 kg) of corrosives and 1,000 pounds (453.5 kg) of a flammable liquid, then instead of displaying two placards the shipper can display a Dangerous placard, as shown in **Figure 2-17.** If any of the items exceeds 2,205 pounds (1000 kg) picked up at one location, the shipper must

Figure 2-16 Class 9 placard and labels.

Figure 2-17 "Dangerous" placard.

Figure 2-18 "Chlorine" placard.

display the placard for the material that exceeds 2,205 pounds (1000 kg) in addition to the Dangerous placard.

■ Specific Name placards or placards with "Inhalation Hazard" shown in **Figure 2-18** are sometimes used in place of the Poison Gas placard.

■ The PG III placard and label that has replaced the "Stow Away from Foodstuffs" is a placard that indicates that a poisonous material is being transported, but is not so poisonous as to meet the rules to be placarded as a poison. Chloroform is an example that would use the placard like the one shown in **Figure 2-19.**

■ ORM-D (Other Regulated Material)–Class D is left over from a previous DOT regulation. It is a subdivision that includes ammunition and consumer commodities, such as cases of hair spray. The package will have the printing "ORM-D" on the outside. The previous regulation listed ORM-A through ORM-E, which are now grouped together in Class 9 Miscellaneous hazardous materials.

■ "Marine Pollutant" is displayed on shipments that would harm marine life if the material were released into a waterway,. The placard is shown in **Figure 2-20.**

■ Elevated temperature material will have a "HOT" label either to the side of or on the placard

Figure 2-19 "Harmful" placard.

Figure 2-20 "Marine pollutant" placard.

Figure 2-21 "Hot" placard.

NOTE The DOT has changed some terminology for Other Regulated Materials, commonly referred to as ORMs. Typically these were shipments of consumer goods, which are now known as Materials of Trade (MOT). Some common examples are shown in **Table 2-7.** The DOT has changed some of the rules for transporting smaller quantities of hazardous materials when used in a business. With the exception of tanks containing diluted mixtures of Class 9 materials a person cannot transport more than 440 pounds (200 kg) of a material. If the material is a high hazard material (packing group I) the maximum amount of material allowed in one package is 1 pound (0.5 kg) for solids and 1 pint (0.5 liters).

as shown in **Figure 2-21** if it meets one of the following criteria:

Liquid above 212°F (100°C)
Liquid that is intentionally heated and has a flash point above 100°F (37.7°C)
Solid at 464°F (240°C) or above

▮ "Infectious Substances" is a label like the one shown in **Figure 2-22** that is sometimes used on the outside of trucks. It is not required by the DOT but may be required by other agencies, such as the Department of Health and Human Services or a state agency. The DOT now calls medical waste "regulated medical waste" or RMW. An infectious substance is a material known or reasonably expected to contain a pathogen. A pathogen is a microorganism (including bacteria, viruses, ricksettia, parasites, fungi) or other agent such as proteinaceous infectious particle (prion), that can cause disease in humans or animals. They have also split this category into two subdivisions. Category A is an infectious substance in a form capable of causing permanent disability or life threatening, or fatal disease in otherwise healthy humans or animals when an exposure occurs. A category B material

TABLE 2-7

Examples of Materials of Trade (MOT)	
Class or Division	**Example**
Flammable gases (Division 2.1)	Acetylene, propane
Non-flammable gases (Division 2,2)	Oxygen, nitrogen
Flammable or combustible liquids (Class 3)	Paint, paint thinners, gasoline
Flammable solids (Division 4.1)	Charcoal
Dangerous-when-wet materials (Division 4.2)	Some fumigants
Oxidizers (Division 5.1)	Bleaching compounds
Organic peroxides (Division 5.2)	Benzoyl peroxides
Poisons (Division 6.1)	Pesticides
Some infectious substances (Division 6.2)	Diagnostic specimens
Corrosive materials (Class 8)	Muratic acid, drain cleaners, battery acid
Miscellaneous hazardous materials (Class 9)	Asbestos, self-inflating life boats
Consumer commodities (ORM-D)	Hair spray, spray paints

Figure 2-22 "Biohazard" labels.

Figure 2-24 White square placard.

presents a lesser hazard but the infectious substance is not in a form generally capable of causing permanent disability or life threatening or fatal disease in otherwise healthy humans or animals when exposed.

■ The "Fumigated" marker, like the one shown in **Figure 2-23,** is used when a trailer or railcar has been fumigated or is being fumigated with a poisonous material. This placard is most commonly seen near a port where containers are frequently fumigated after arriving from a foreign port.

■ A white square background, as shown in **Figure 2-24,** is used in the following situations:
  ■ On the highway for controlled radioactive III shipments
  ■ On rail for:
    ■ Explosives 1.1 or 1.2
    ■ Division 2.3 Hazard Zone A (poison gas) materials

■ Division 6.1 Packing Group I Hazard Zone A (poison)
■ Division 2.1 (flammable gas) in a DOT 113 tank car
■ Division 1.1 or 1.2, which is chemical ammunition that also meets the definition of a material that is poisonous by inhalation

Problems with the Placarding System

The placarding system relies on a human to determine the extent of the load, determine the appropriate hazard classes, and interpret difficult regulations to determine if a placard is required. The placard then must be affixed to all four sides of the vehicle before shipment. Placards are required only for shipments that exceed 1,001 pounds (454 kg), except for DOT Table 1 materials, which require placarding at any amount. Authorities estimate that about 10 percent of the trucks traveling the highways are not placarded or not placarded correctly.

CAUTION Given the restrictions of many cities on bridges and tunnels where hazardous materials are not allowed, it can be anticipated that many trucks are not carrying the proper designations.

A placard can come off during transport and legally need not be immediately replaced. The fact that an incident involves a truck or train should alert the first responder to the potential for hazardous materials; when a placard is involved, extra precautions should be taken.

DANGER

THIS UNIT IS UNDER FUMIGATION WITH _____ APPLIED ON
FUMIGANT NAME
DATE _____
TIME _____
DO NOT ENTER

Figure 2-23 "Fumigation" label.

Figure 2-25 The DOT adds the "Poison Inhalation Hazard" label to those materials that present severe toxic hazards.

Labeling and Marking Specifics

Package markings must include a shipping name of the material, the UN/NA identification number, and the shipper and receiving companies' names and addresses. Packages that contain more than a ~~reportable quantity (RQ)~~ of a material must also be marked with RQ near the shipping name. Packages that are listed as ORM-D materials should be marked as such. Some packages containing liquids must use orientation arrows. For materials that pose inhalation hazards, an Inhalation Hazard label must be affixed next to the shipping name as shown in **Figure 2-25.** Hazardous wastes will be marked waste or will use the EPA labeling system to identify these packages. Labels are identical to placards, other than their size.

NOTE Materials that have more than one hazard may be required to display a primary hazard label and a subsidiary label.

Figure 2-26 The primary hazard placard will be on top and the subsidiary or secondary hazard placard will be slightly below.

The primary label will have the class and division number in the bottom triangle and the subsidiary label will be slightly below the primary label as shown in **Figure 2-26.** As an example, the material acrylonitrile, inhibited, is required to be labeled flammable with a subsidiary label of poison.

OTHER IDENTIFICATION SYSTEMS

NFPA 704 System

This system is designed for buildings, not transportation, and alerts the first responders to the potential hazards in and around a facility. The system is much like the placarding system and relies on a triangular sign that is divided into four areas, as shown in **Figure 2-27.** The four areas are divided by color

NFPA 704
One of the other more common systems used to identify the presence of hazardous materials is the NFPA 704 system.

Figure 2-27 NFPA 704 marking.

as well, and use a ranking system to identify severity. The four areas and colors are:

- Health hazard—Blue
- Fire hazard—Red
- Instability hazard—Yellow
- Special hazards—White

The system uses a ranking of 0 to 4 with 0 presenting minimal risk, while a ranking of 4 indicates severe risk. The specific listing is as follows:

- Health is based on a limited exposure to the materials using standard firefighting protective clothing as the protective clothing for the exposure.
  4—Severe health hazard
  3—Serious health hazard
  2—Moderate hazard
  1—Slight hazard
  0—Minimal hazard
- Flammability is the ability of the material to burn or be ignited.
  4—Flammable gases, volatile liquids, pyrophoric materials
  3—Ignites at room temperature
  2—Ignites when slightly heated
  1—Needs to be preheated to burn
  0—Will not burn
- Reactivity is based on the material's ability to react, especially when shocked or placed under pressure.
  4—Can detonate or explode at normal conditions
  3—Can detonate or explode if strong initiating source is used
  2—Violent chemical change if temperature and pressure are elevated
  1—Unstable if heated
  0—Normally stable
- Special hazards are used to indicate water reactivity and oxidizer, which are included in the NFPA 704 system. In some cases other symbols may be used such as the tri-foil for radiation hazards, ALK for alkalis, and CORR for corrosives. The presence of a slashed W indicates an accompanying ranking structure for water reactivity

in addition to the hazards listed in the other triangles:
  4—Not used with the slashed W and a reactivity ranking of 4
  3—Can react explosively with water
  2—May react with water or form explosive mixtures with water
  1—May react vigorously with water
  0—Slashed W is not used with a reactivity ranking of 0

One potential problem with the NFPA 704 system is that it groups all of the chemical hazards listed in a building into one sign. If the sign is placed on a tank that contains one material, the system does a good job of warning about the contents of the tank, but it does not provide the name of the product. For a facility with hundreds if not thousands of materials, the system will warn only of the worst-case scenario. As an example, dramatically different tactics are used to handle a flammable gas incident versus a flammable liquids incident, but both can be classified as fire hazard 4. The system is best used to alert the first responder to the presence of hazardous materials and warn of the worst-case scenario.

Hazardous Materials Information System

Commonly referred to as HMIS, the Hazardous Materials Information System was designed to provide a mechanism to comply with the Hazard Communication regulation. This system may also be known as Hazardous Materials Information Guide (HMIG). The Hazard Communication regulation requires that all containers be marked with the appropriate hazard warnings and the ingredients be provided on the label, like those shown in **Figure 2-28.** Many products that come into the workplace are missing adequate warning labels. The HMIS is not a uniform system, but can be developed by the facility or by the manufacturer of the labels. Thus, one system may vary from another. Most systems are similar to the NFPA 704 system and use blue, red, and yellow colors with a numbering system that provides an indication of hazard. The colors may be used in a triangle format or, in most cases, in stacked bars. The numbers are usually 0–4, the same as the NFPA system, but in rare cases may differ from the NFPA system. The facility manager or other representative should have the key to the symbols, or there should be a chart indicating the symbols and warning levels somewhere in the facility, usually with the Material Safety Data Sheet (MSDS). In most cases the MSDS and other hazard communication information should be in a central location.

Figure 2-28 HMIS label.

Some systems include a picture of the level of the personal protective equipment (PPE) that is required for the substance.

Each HMIS system is different, and responders should not assume what the actual hazard level is until warning levels can be determined. When conducting inspections or preincident walk-throughs, responders should determine the location of the MSDS information and make sure it is accessible during emergencies. In some cases the MSDSs may be stored in or near the chemical storage area, which during an emergency may be inaccessible. Responders should suggest that another set be provided at a guardhouse or in a locked box on the outside of the building. One large industrial complex with a large number of individual labs had a fire in one of these labs. The responding firefighters correctly asked for the MSDSs for the materials in the lab, only to be told that they were in the lab. The MSDSs were placed inside the lab for the employees' use, and their location was convenient for that purpose. After the fire emergency responders requested that the plant store a master copy of all MSDSs in the security office and move the MSDSs for the individual labs to the outside of the lab. MSDS boxes were placed outside each lab area, so that employees would have easy access and, in the event of an emergency, firefighters could also reach the MSDS for that lab without having to enter the lab.

Military Warning System

Whenever possible, the military uses the DOT placarding system, but in some cases it may employ its own system.

SAFETY Within the DOT's Emergency Response Guidebook is an emergency contact number for responding to an incident involving a military shipment.

In most cases, for extremely hazardous materials, arms, explosives, or secret shipments it would be safe for emergency responders to assume prior to their arrival that the military is aware of the incident and probably is responding. The higher the hazard, the more likely there will be an escort for the shipment. There may be shipments for which the driver of the truck is not allowed to leave the cab of the truck, and that may provide warnings to stay away from the truck.

SAFETY On high security shipments the driver and the personnel in the escort vehicle are armed. If you have an incident involving one if these vehicles, obey the commands of the escorting personnel and determine if they have made the appropriate notifications.

If the driver and escort crew are killed or seriously injured in the accident, it would be advisable to notify the military about the incident, although with satellite tracking, help is probably on the way. The phone number to contact the military is in the Department of Transportation (DOT) *Emergency Response Guidebook* (ERG) along with **Chemtrec's** and other emergency contact numbers. Other incidents involving fuels, food, or military equipment may require you notifying the military of the incident. The military typically uses its own marking system at its facilities to mark the buildings. Military facilities use a series of symbols and a numerical ranking system shown in **Figure 2-29.**

Pipeline Markings

At any place an underground pipeline crosses a mode of transportation, the pipeline owner is required to place a sign like the one shown in **Figure 2-30** that indicates the wording "warning," "caution," or "danger," the words "petroleum pipeline," or the hazardous contents of the pipe, and the owners name and telephone number.

The pipeline contents may be general as in "petroleum product," as the pipeline may ship fuel oil, gasoline, motor oil, or other products in the same pipe. Dedicated pipelines, which only carry one product, will be marked with that specific product.

Figure 2-29 Military placarding system.

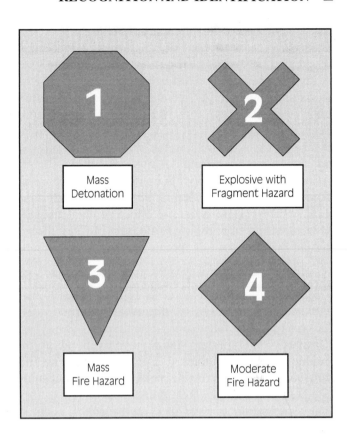

The pipeline should be buried a minimum of 3 feet (1 m) and should be adequately marked. Many larger pipelines, such as the Colonial pipeline, originate in Texas and end in New York; the Colonial is 32 to 36 inches (81–91 cm) in diameter along the main

Figure 2-30 The owner of a pipeline is required to provide the wording "warning, caution or danger," the words "Petroleum Pipeline" or the hazardous contents of the pipe, and the owner's name and telephone number.

pipeline. In the event of a release involving the pipeline the line will be immediately shut down. Even with this immediate shutdown, there is the potential to lose several hundred thousand gallons of hazardous materials if the distance to the shutoff is substantial. An incident involving a pipeline can be a serious event; do not underestimate the need for considerable local, state, and federal resources. Just within the responders' own agency considerable resources may be required, such as command staff, logistic support, communications, and tactical units. A fuel oil pipeline rupture in Reston, Virginia, resulted in the loss of more than 400,000 gallons (1,514,164 liters) of fuel, requiring considerable resources from several states to control the spill. The resources included emergency response organizations, local, state, and federal assistance, and a considerable number of private cleanup companies.

NOTE Some pipelines move one type of product, while others move several different types each day.

The product in the pipelines varies from liquefied gases and petroleum products to slurried material. Pipeline companies are required to conduct in-service training and tours for emergency responders in the communities their pipelines transverse. When you have an incident on or near any pipeline, it is advisable to notify the pipeline owner of the incident, even if you are fairly certain the pipeline was not damaged. A train derailment in California caused a pipeline to shift, and the pipeline did not release any product until several days after the original derailment occurred. Most pipeline operators would like to have the opportunity to check the line to head off the potential for a catastrophic release several days later. After several catastrophic pipeline releases the pipeline industry has been required to establish contact with local responders. This contact usually involves a tour of pumping stations, storage facilities, and an overview of the operator's system.

Container Markings

Most containers such as drums are marked with the contents of the drum, like the one shown in **Figure 2-31.** Cylinders should have the name of the product stenciled on their side. In bulk shipments the bulk container will have the name of the product stenciled on the side. In bulk shipments the DOT requires that some products have their own placard, such as the chlorine placard. Trucks that are dedicated

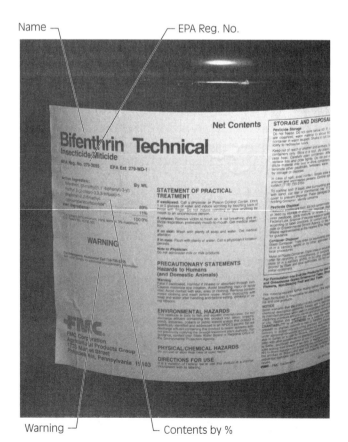

Figure 2-31 The label describes the contents of the drum and provides precautions to be used when the product is handled. The EPA signal word is a significant clue to the hazards of the product.

Figure 2-32 Pesticide label.

haulers will also stencil or mark the product's name on four sides of the vehicle.

Pesticide Container Markings

Due to their toxicity, required markings for pesticides are regulated by the EPA. The label on a pesticide container such as the one in **Figure 2-32** will have the trade name of the pesticide, which is chosen by the manufacturer and is not usually the chemical name for the product. The product name will usually not offer any clue as to the chemical makeup of the product, and in some cases may be difficult to research. Chapter 26 provides a discussion on chlordane and its multiple names.

NOTE The pesticide label will also contain a signal word: Danger, Warning, or Caution. The assignment of the signal word is the clue to how dangerous the material actually is. If the label indicates "danger" then extreme care should be used as the material is highly toxic. The designations of "warning" and "caution" present lesser dangers, but still should be considered very hazardous.

In the United States the EPA issues a registration number, and in Canada the label will have a pest control number. The label will also include a precautionary statement and a hazard statement; examples include "Keep from waterways," and "Keep away from children." The active ingredients will be listed by name and percentage, which in most cases for active ingredients are a small percentage of the product. Inert

Figure 2-33 The IAEA is using a new label that accurately reflects the dangers of radiation. The old label does not indicate any potential hazards except to alert a knowledgeable person to the presence of a radiation source. *(Figure provided by IAEA)*

ingredients are also listed but not specifically named. For liquid pesticides the inert ingredient is usually water or petroleum solvents, such as xylene, plus emulsifying agents.

Radiation Source Labeling

In 2007 the International Atomic Energy Association (IAEA), a group that sets nuclear and radiation standards worldwide, issued a new radiation warning label. The old label using the trefoil, commonly referred to as a "propeller," isn't really associated with radiation or nuclear material. The new label shows energy waves coming from a trefoil with skull and crossbones, with a human running away. The label shown in **Figure 2-33** more accurately represents the hazards associated with radiation. It is not mandated for use by DOT, and the DOT will still require the DOT placards and labels, but the new label will be more common in the future. Devices or containers that store radiation sources will be using this type of label. For devices that use a radiation source and are relatively large, the label is to be placed near the access panel to the radiation source. Obviously responders should not try to access the source, and should request radiation specialists to assist them.

CONTAINERS

General Containers

Hazardous materials come in a variety of containers of many shapes and sizes, from 1-ounce bottles to tanks to ships carrying hundreds of thousands of gallons. Considering the materials you have in your home, you see a wide variety of storage containers. You have compressed gas cylinders holding propane,

steel containers holding flammable and combustible liquids, and bottles, jars, and small drums holding various products. You can also locate cardboard boxes, plastic-lined cardboard boxes, and bags of various types. The type of material and the end use for the product determine the packaging. Packaging used to store household or consumer commodities is usually different from the industrial version. In some cases the industrial version may be full-strength undiluted material, while the household version is only a small percentage of that strength mixed with a less hazardous substance such as water. The type of container is usually a good clue as to the contents of the package.

The more substantial, durable, and fortified a container is, like the container shown in **Figure 2-34,** the more likely the material inside is dangerous. On the other hand, materials that are transported in fiberboard drums are usually not of significance in regard to human health, although they may pose a risk to the environment.

During the recognition and identification process, first responders should be alert for anything unusual

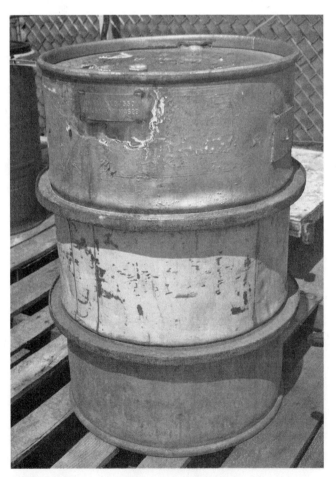

Figure 2-34 The type of container can offer some clues as to its contents. As this drum is reinforced, it has a high likelihood of containing an extremely hazardous material.

when arriving at an incident. When on an EMS call to a residential location, it would be unusual to find a 55-gallon (208-liter) drum in a bedroom along with glassware associated with a lab environment. These types of recognition and identification clues should alert the first responders that additional assistance may be required. Arriving at an auto repair garage and finding 55-gallon (208-liter) drums and compressed gas cylinders should not be unexpected, however.

CARDBOARD BOXES

With the popularity of shopping clubs and discount warehouses, more and more homeowners are purchasing cases of materials, when in the past they only bought a gallon or less. Cardboard boxes are one method of shipping and containing hazardous materials. They can hold glass, metal, or plastic bottles. In some cases they may have a plastic lining such as the box shown in **Figure 2-35,** which holds sulfuric acid. Many household pesticides, insecticides, and fertilizers are contained in cardboard boxes. With the exception of these products and the sulfuric acid shown in **Figure 2-35,** most products contained in boxes are usually not extremely toxic to humans, but they may present an environmental threat. Materials in transport to suppliers may be transported in larger cardboard boxes and then broken down at the retail level. Responders should note any labels on these packages, but the absence of any labels does not indicate that hazardous materials are not present.

BOTTLES

From 1-ounce bottles to 1-gallon (3.8-liter) bottles, the variety is endless, and the types of products that are contained are too numerous to mention. In recent times manufacturers have taken care in packaging their materials for safe transport and use. In the years past, most chemicals were shipped in glass bottles. When shipped, these bottles are usually packed in cardboard boxes and are insulated from potential damage. One-gallon glass containers are usually shipped in what is known as carboys, like those shown in **Figure 2-36.** If the container were dropped, these protective covers should ensure that the bottle will survive the fall and prevent any potential damage during transportation. Carboys are usually seen in laboratories and in smaller chemical production facilities.

Ensuring compatibility with the container is important, but the one area that usually results in a release is the use of an improper cap. The chemical must be compatible with glass and the material used to construct the lid. There are a variety of materials that are used in the manufacture of lids. Like the example in **Figure 2-37,** many new glass containers are coated with plastic to keep the bottle from breaking when dropped. Even if the bottle is cracked, the contents should remain sealed within the plastic coating.

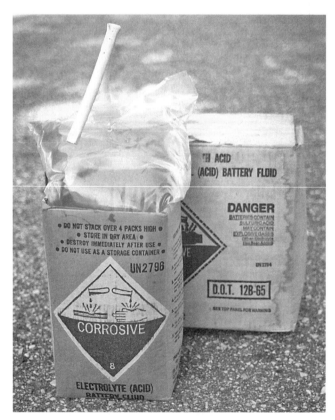

Figure 2-35 Typically chemicals that can cause harm are not packaged in cardboard. This sulfuric acid is one example of a product that can cause harm and is in cardboard.

Figure 2-36 To protect a glass bottle that holds corrosive material, a carboy is used to protect the bottle.

Figure 2-37 This glass jar is coated with plastic that will not allow the liquid to spill out if the glass is broken or dropped.

BAGS

From paper bags to super sacks that weigh several tons, bags also contain a variety of materials. Bags can be simple paper bags, plastic-lined paper bags, fiber bags, plastic bags, or reinforced super sacks or tote bags. Even liquids are being transported in containers

Figure 2-38 It can be quite surprising to open the back of a tractor-trailer and find these super sacks. They contain solid materials, some of which may be toxic.

called bladders, which mimic the super sacks. It can be surprising to open the back of a trailer and find four super sacks like those shown in **Figure 2-38** carrying a material that is classified as a poison. Bags carry anything from food items to poisonous pesticides, and the method of transportation varies widely.

DRUMS

When discussing hazardous materials, these are the containers most responders call to mind. They vary in size from 1 gallon (3.8 liters) up to a 110-gallon (416-liter) overpack drum. The construction varies from fiberboard to stainless steel. The typical drum contains 55 gallons (208 liters) and weighs 400 to 1,000 pounds (181–453.5 kg). It is possible to get an idea of what a drum may contain by the construction of the drum. **Table 2-8** provides an indication of

TABLE 2-8

Drum Contents	
Type of drum	**Possible contents (in order of likelihood)**
Fiberboard (cardboard) unlined	Dry, granular material such as floor sweep, sawdust, fertilizers, plastic pellets, grain, etc.
Fiberboard—plastic lined	Wetted material, slurries, foodstuffs, material that may affect cardboard or could permeate the cardboard
Plastic (poly)	Corrosives such as hydrochloric acid and sodium hydroxide, some combustibles, foodstuffs such as pig intestines
Steel	Flammable materials such as methyl alcohol, combustible materials such as fuel oil, motor oil, mild corrosives, and liquid materials used in food production
Stainless steel	More hazardous corrosives such as oleum (concentrated sulfuric acid)
Aluminum	Pesticides or materials that react with steel and cannot be shipped in a poly drum such as concentrated hydrogen peroxide, or beer

Figure 2-39 Cylinders present additional risks to responders. Not only can the contents be hazardous, but if the cylinder is involved in a fire, it may explode.

potential drum contents, but this is not an absolute listing. Contents can and do vary drum to drum.

CYLINDERS

Cylinders like those shown in **Figure 2-39** come in 1 pound (0.45 kg) sizes up to several thousand pounds and carry a variety of chemical products. The product and its chemical and physical properties will determine in what type of container the product is stored.

> **CAUTION** Other than the hazard of the chemical, a major concern that all cylinders pose is the fact that they are pressurized.

The pressures range from a low of 25 psi to a high of 5,000 psi (1.7–345 bar). One of the most common cylinders firefighters run across is the propane tank, found in sizes from 1 pound up to millions of gallons. When used in residential homes, you will find the 17-pound (7.71-kg) cylinder for barbecues and cylinders of 100 to 250 pounds (45–113 kg) used as a fuel source for the home. In some areas it is not uncommon to find 1,000-gallon (453.5-kg) cylinders, and many times now they are being buried underground.

Specialized cylinders that hold cryogenic gases (extremely cold) appear to be high pressure, but in reality are low pressure. The bulkiness of the cylinders is a result of the large amount of insulation required to keep the material cold. It is not uncommon to find cryogenic cylinders in convenience stores or fast-food restaurants, which presents an unusual hazard. They typically are storing carbon dioxide (CO_2) for the soda-dispensing machines. If the system were to leak, a large amount of CO_2 would be released presenting an asphyxiation risk. Cylinders usually have **relief valves** or **frangible disks** in the

event they are overpressurized or are involved in a fire. The United States is one of the few countries that mandates relief valves, and most cylinders used in other countries do not have this feature. Incidents aboard ships may involve these types of cylinders as the cylinders are being trans-shipped and are destined for delivery in another country. Communities of all sizes have cylinders of chlorine and sulfur dioxide used in water treatment. These cylinders come in 100- and 150-pound (45- and 68-kg) and 1-ton (907-kg) cylinders, which could create a major incident if they were ruptured or suffered a release.

There was an attempt to allow the use of composite propane cylinders, both inside and outside of buildings. The draft version of NFPA 58, which is the Liquefied Petroleum Gas code, had language that would have allowed composite propane cylinders to be used in small portable heaters. The final version only allows the use of composite cylinders outside of buildings. When this standard comes up for revision in 2010 this issue will come up again. These cylinders, which are constructed much like composite breathing apparatus cylinders, can be used in transportation through a special DOT permit. Composite cylinders can carry flammable gases, oxygen, and many other gaseous materials. Composite cylinders are in common use in Europe and are starting to become more popular in the United States. Two major tests have been conducted on these cylinders to determine the outcome if the cylinders were involved in a fire situation. From the initial testing composite cylinders appear to react to fire situations better than steel cylinders. As is discussed later in this section, BLEVEs are a major life threat to responders and are common with steel cylinders. None of the composite cylinders BLEVE'd or had a violent rupture during the testing process.

> **NOTE** The area affected by a 100-pound chlorine cylinder release can be several miles, causing serious injuries if not fatalities. **Figure 2-40** depicts the vapor cloud that could result from such a release.

Totes and Bulk Tanks

Both **totes** and **bulk tanks** are becoming more common as they are sized between drums and tank trucks; they are used for a variety of purposes. These totes and tanks are also called intermediate bulk containers and have capacities between 119 gallons and 793 gallons (450 and 3,001 liters). They hold flammable, combustible, toxic, and corrosive materials. Totes and bulk tanks are used in industrial and food

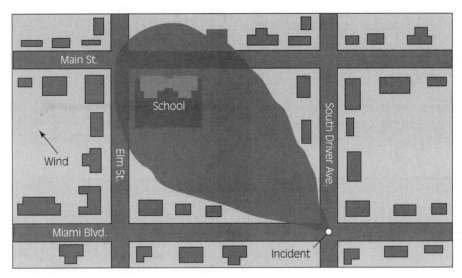

Figure 2-40 The type of vapor cloud commonly referred to as a plume varies with the terrain and buildings in the vicinity. The plume here represents one of the most common types.

applications. They are constructed of steel, aluminum, stainless steel, lined materials, poly tanks, and other products, such as those shown in **Figure 2-41.** The usual capacity is approximately 300 gallons (1136 liters). They are transported on flatbed tractor-trailers or in box-type tractor-trailers. A common HAZMAT incident involving totes occurs when tanks are bottom offloaded through a swinging valve, such as the one shown in **Figure 2-42.** It is a common occurrence for this valve to swing out during transport and get knocked off during movement, as shown in **Figure 2-43.** An unusual tote is the type designed to transport calcium carbide, a material that, when it gets wet, forms acetylene gas, which is reactive and very flammable. A calcium carbide tote is shown in **Figure 2-44.**

Figure 2-42 The most common type of spill occurs when this valve is knocked off, releasing the contents.

Pipelines

Pipelines vary in size and pressure. They are typically from 1/2 inch (1.27 cm) to in excess of 6 feet (1.9 meters) in diameter and are commonly buried underground. The most common products are natural gas, propane, and assorted liquid petroleum products. The larger petroleum pipelines originate in Texas and Louisiana and proceed up throughout the East Coast. The West Coast also has its share of larger pipelines, with a large majority originating in Alaska. Pipelines can originate from any bulk storage facility and can cross many states. Some type of pipeline system is found in every state. Because the amount in the pipelines varies, it is important that first responders know the location of the pipelines and emergency contact names and phone numbers so if there is a suspected problem, they can notify the pipeline owner immediately.

Figure 2-41 The use of 55-gallon drums is decreasing and the use of these portable bulk tanks is increasing. Like super sacks, these can be hidden away in the back of a trailer.

Figure 2-43 The tote in the middle is the nitrocellulose tote. The spill emptied the contents of the tote and covered the floor of the warehouse. On top of the spill is a layer of foam, which did not reduce the flammable vapors. The LEL readings were 90 percent of the LEL at some points during the operation, a very dangerous situation. Complicating the release is the fact that nitrocellulose becomes very shock sensitive as it dries and can detonate. This photo shows numerous other totes containing nitrocellulose, and most of the drums contain flammable waste. It was dangerous to operate in this environment, but the resulting drying was even more dangerous so speed was necessary to reduce the potential for explosion. *(Photo by the Maryland Department of the Environment)*

Figure 2-44 Being carried on this railcar are specialized totes used to carry calcium carbide. When transported, calcium carbide is placarded "Dangerous when wet." The perception is that the addition of water to a material placarded in this fashion would cause an explosion. In the case of calcium carbide the addition of water would result in the production of acetylene gas, which is very flammable and would present a fire hazard. *(Photo by Greg Socks)*

Radioactive Material Containers

The transportation of radioactive or nuclear materials is highly regulated by the DOT and the **Nuclear Regulatory Agency (NRC)**. There are several types of containers that are used to transport these types of materials, all strengthened to keep the material inside the container. Radioactive material that is classified as low-level radioactive material, typically waste material, is required to be transported in what is called a strong, tight container. This container can be made of metal or wood, but should be reinforced to prevent damage to an inner container holding the radioactive material. These materials would present little risk if they escaped the container and emit radiation only slightly above background levels. Another type of package is the **Excepted packaging** shown in **Figure 2-45** which is designed to survive the normal conditions of transport. This packaging is designed to hold materials that have **low specific activity** (emitting low levels of radiation) or **Surface Contaminated Objects** (SCO). These are normally small quantity shipments that contain natural or depleted uranium, or natural thorium. In some cases the package does not have to be labeled as the material does not meet the DOT requirement for labeling and placarding. Industrial packaging is a container that is designed to be more substantial and has to pass some basic testing procedures, such as the drop test. Industrial packaging shown in **Figure 2-46** is usually metal boxes or drums. Industrial packaging is also used to transport low specific activity (LSA) or (SCO) materials. Commonly shipped materials are slightly contaminated tools, medical and research isotopes, and soil waste. For higher levels of radiation a container known as a **Type A** container shown in

Figure 2-45 Excepted Packaging is used to transport low risk radioactive materials. The packaging is exempted from DOT regulations. *(Photo by IAEA)*

Figure 2-46 Industrial packaging is used to transport low risk radioactive materials, typically waste. *(Photo by Kirstie Hansen, IAEA)*

Figure 2-47 is used. The Type A container can be made of cardboard, wood, or most commonly metal drums. The Type A containers are used to transport radioactive material (RAM) and are tested to withstand minor accidents. The level of radiation is still low as only materials that would not present a significant health risk may be placed in a Type A container.

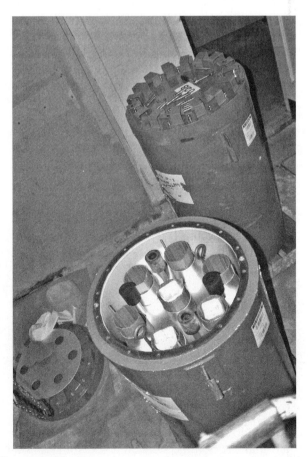

Figure 2-47 Type A packaging is designed to hold low risk radioactive material and offers moderate protection against accidents. *(Photo by John Mairs, IAEA)*

Figure 2-48 Type B containers such as the container being loaded on a ship is a container designed to hold high risk radioactive materials. It is a hardened transport case capable of withstanding severe accidents. *(Photo by Pat Pavlicek, IAEA)*

Materials transported in Type A packages are usually medical isotopes, or industrial, agricultural, or research materials. The highest level of protection is offered by a **Type B** container shown in **Figure 2-48** and is used for the transportation of larger or more dangerous types of radioactive materials. The Type B containers must meet all types of testing procedures and must be able to withstand a severe accident. The Type B container is either a metal drum or a large, heavily shielded transport container. The Type B container may have 10 inches of lead shielding. Spent nuclear fuel is transported in Type B containers.

Highway Transportation Containers

The type of vehicle provides some important clues as to the possible contents of the vehicle. Four basic tank truck types carry hazardous materials, with some additional specialized containers. The most common truck involved in chemical releases are tractor-trailers or box trucks, based on the high numbers of trucks traveling the highways. Tractor-trailers and box trucks carry the whole variety of hazardous materials and portable containers. They can carry loose material that is not contained in any fashion other than the truck itself. They can carry portable tanks that hold 500 gallons or bulk bags that weigh several tons.

CAUTION When dealing with tractor-trailers, the rule is: Expect the unexpected.

Nothing is routine. Until you interview the driver, review the shipping papers, and finally examine the cargo, you cannot confirm or deny any hazardous materials. Sometimes the signage on the trailer is an indication of the possible contents, and a trailer that has several placard holders is a likely candidate for transporting hazardous materials. If a tractor-trailer is refrigerated, like the one shown in **Figure 2-49,** and

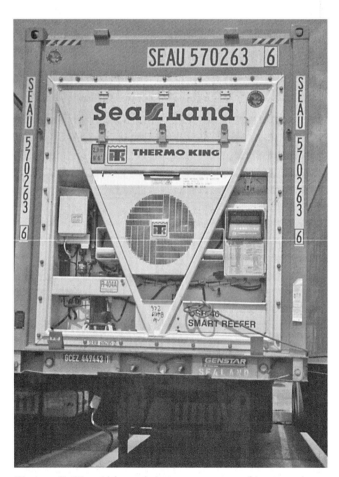

Figure 2-49 Although in most cases refrigerated trailers are carrying food, the possibility exists that they may have chemicals that require refrigeration to remain stable.

is carrying hazardous materials, these require extra precautions as they may require the cold temperature to remain stable.

A common box trailer that is involved with leaking containers is known as an **intermodal container,** more commonly called a **sea container.** Throughout the United States these containers are used for transport of materials. Although common in and around marine terminals, these containers can be found anywhere. They are typically used on ships, then offloaded onto a tractor-trailer or loaded directly onto a flatbed railcar. These containers come from all over the world and can contain any imaginable commodity. The types of containers that are shipped in these trailers may be bags, boxes, drums, bulk tanks, or cylinders. The reasons they are involved in many incidents are the number of containers on the highway and the many stresses subjected on these materials during transport. Although the driver is supposed to have the shipping papers for the contents, on occasion the paperwork is missing or is sealed in the back of the trailer. Finding out the contents of the trailer can be very difficult and frustrating.

Tank trucks carry several hundred gallons up to over 11,000 gallons (41,639 liters). The DOT allows maximum loads by weight, not by gallons, so the actual gallonage varies from state to state. The most common tank truck is the gasoline tank truck, which usually carries 5,000 to 10,000 gallons (18,927–37,854 liters). The DOT changed its regulations in 1995 covering tank trucks, and you may see two systems for identifying tank trucks, like those shown in **Figure 2-50.** In the past the DOT wrote specifications as to how the manufacturer would build a tank truck. Today, the federal agency has established performance-based standards for vehicle

Figure 2-50 Many of the differences between a DOT-306 and a 406 are internal. The biggest difference is that the dome covers on a 406 are less likely to open during a rollover, although the skin of the tank is thinner. *(Courtesy of Maryland Department of the Environment)*

construction. The DOT allows trucks that were manufactured before the new regulation to remain on the road, as long as they meet the applicable inspection requirements. The four basic types of tank trucks are:

■ DOT-406/MC-306 Gasoline tank truck[2]
■ DOT-407/MC-307 Chemical hauler
■ DOT-412/MC-312 Corrosive tanker
■ MC-331 Pressurized tanker

The DOT numbers are the more recent manufactured tanks, and the MC (motor carrier) numbers identify those tanks manufactured prior to September 1995. There are some differences in construction of the two types—some in favor of emergency responders, others in favor of lower cost to the shipping company—but overall the newer tanks hold up better during accidents and rollovers. If you are not sure of the type of tank, all tank trucks have **specification (spec) plates** that list all the pertinent information. The spec plate is located on the driver's side of the tank near the front of the tank. In some cases, you may find shipping papers or material safety data sheets in a paper holder as shown in **Figure 2-51.**

DOT-406/MC-306

This is the most common tank truck on the road today, as shown in **Figure 2-52** and **Figure 2-53.** These trucks are involved in the most accidents mainly due to the sheer number of shipments and trucks on the road. Although the most common products carried on these trucks are gasoline and diesel fuel, almost any flammable or combustible liquid can be found on these types of trucks. This truck is known for its elliptical shape and is usually made of aluminum. The maximum pressure that this truck can

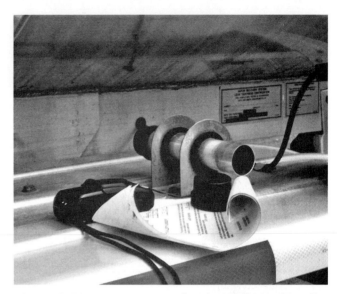

Figure 2-51 In most cases the shipping papers will be with the driver in the cab of the truck, but they may be in a special tube located on the trailer.

hold is 4.25 psi (0.29 bar). The trucks generally have three or four separate tanks called compartments, but two to five tanks are not uncommon. The valving and piping are contained on the bottom of the tank, as shown in **Figure 2-54,** and the number of compartments is indicated by the number of outlets and the number of manhole assemblies located on top of the tank. The compartments are separated by bulkheads, which tend to fail during a rollover situation. In the event that the shipper wishes to carry mixed loads they may add a double bulkhead to protect against the mixing of chemical products. During a violent rollover even this double bulkhead is likely to fail. Although in most states the shipper is not supposed

Black Dot

Hand Railing and Vapor Recovery

Figure 2-52 The MC-306/DOT-406 is the most common truck on the highway today and is referred to as a gasoline tanker. It can and frequently does carry other types of flammable and combustible liquids.

Figure 2-53 The silhouette of an MC-306/DOT-406 is known as being elliptical from the ends and circular from the sides. There is no way of knowing the difference between a 306 and a 406 from a distance, and the best way to confirm the construction type is to check the specification plate found near the forward running gear.

Figure 2-54 Most 306/406 trucks have more than one compartment, and the number of compartments is indicated by the number of valves or the number of dome covers on top of the truck. *(Courtesy of Maryland Department of the Environment)*

to have a mixed load of flammable and combustible materials, responders may encounter a widely differing load compartment to compartment.

CAUTION During a rollover, the bulkheads on a tank truck may shift, allowing all the products to mix and form one big tank.

By an initial examination it may appear only one compartment is ruptured and leaking, but it is possible to lose the whole contents of the tank through a leak in one compartment, as shown in **Figure 2-55.** In the majority of rollovers, experience has shown that at least one bulkhead has separated in almost every accident, resulting in product mixing. In most cases this is not a serious problem as it may be the mixing of grades of gasoline. Within the individual tanks are baffles that limit the movement of the product within the tank.

The emergency shutoff valves, as shown in **Figure 2-56,** are located on the driver's side near the front of the tank, and one is located near the

Figure 2-55 When a truck rolls over on its side, the bulkheads and baffles usually shift, allowing chemicals to mix throughout the tank. Bulkheads separate the various tanks within the truck, while baffles limit the movement of the product within the individual tank. During a rollover these will separate from the sides of the tank. Shown in the photo where most of the sides of the truck have been cut away, you can see the separation of the baffle (hole in the middle indicates a baffle) from the side wall of the tank.

piping on the passenger side. Newer style tanks have considerable vapor recovery systems and rollover protection; in some cases these are combined. Older style tanks used to be made of steel, which presented an explosive situation known as a **boiling liquid expanding vapor explosion (BLEVE).**

DOT-407/MC-307

These are the workhorses of the chemical industry, as shown in **Figure 2-57.** They carry a variety of materials, including flammable, combustible, corrosive, poisonous, and food products. There are two basic types of chemical tanks, one that is insulated and one that is not. The insulated tank can have a number of additional concerns that do not apply to an uninsulated

Figure 2-56 Most trucks have at least one emergency shutoff, and some have two. The most common location is near the driver's door, and the other is usually located near the valve area.

Figure 2-57 This is an uninsulated MC-307 tank on the right. It has reinforced stiffening rings and is a single tank. It does not have separate compartments.

tank (see **Figure 2-58** and **Figure 2-59**). These tanks usually hold 2,000 to 7,000 gallons (7,571–26,498 liters), lower amounts as compared to the 306/406 trucks because most of the products they carry are heavier than petroleum products. The average amount found in these tanks is 5,000 gallons (18,927 liters).

The uninsulated tank is round and has stiffening rings around the tank, and the offloading piping is located on the bottom or off the rear of the tank, as shown in **Figure 2-60** and **Figure 2-61.** The loading piping and manhole are usually on the top in the middle. These tanks comprise only one compartment and do not have separate bulkheads or pots. These tanks generally do not hold up as well as the insulated version during rollovers and accidents. The shell is made of stainless steel and can hold pressures up to 40 psi (2.8 bar).

The insulated tank is a covered version of the uninsulated tank, although in some cases it has a slightly smaller inner tank. The inner tank is made of stainless steel with about 6 inches (15.24 cm) of insulation, and the outer shell is made of aluminum. The inner tank may also be lined with fiberglass or other liner depending on the chemical that is carried. Due to the aluminum outer shell and insulation, these tanks hold up remarkably well during rollovers.

SAFETY   One of the major problems with this insulated tank is that, in the event of a leak, the point where the material leaks out of the outer shell is usually nowhere near the leak on the inner shell.

Figure 2-58 The insulated version of the tank is identical to the uninsulated version but with an aluminum cover and insulation. Note the difference with these two insulated tanks. The one on the right has safety railings around the manway. Although not an absolute rule, the truck on the right tends to carry a more toxic product and has added safety features. Most are not required but were added by the trucking company.

With Metal Skin

Horseshoe
Shape from Rear

Figure 2-59 The silhouette of an insulated MC-307/DOT-407 on the right is horseshoe shaped from the front and rear and circular from the side. The truck on the left is an MC-307/DOT-407 non-insulated.

Figure 2-60 This is an uninsulated MC-307/DOT-407, commonly referred to as a chemical hauler. *(Photo by Greg Socks)*

Within the insulation there can be heating and cooling lines, depending on the product being carried, such as those shown in **Figure 2-62.** Products such as paint are shipped at 170°F (77°C) and need to be heated to that temperature to offload the paint. Some products need to remain at certain temperatures to remain stable, and first responders need to be aware of any special requirements.

In general, both insulated and uninsulated tanks have rollover protection, similar piping, and relief valves that serve two purposes: overpressurization and vacuum protection. The emergency shutoffs are located near the front of the tank on the driver's side and near the offloading piping.

DOT-412/MC-312

Tankers like those shown in **Figure 2-63** and **Figure 2-64** carry a wide variety of corrosives, both acids and bases. These tankers are round and smaller than the 306s and 307s, due to the weight of the corrosives they carry. Most petroleum products weigh about 8 pounds (3.6 kg) per gallon (3.8 liters), while some corrosives weigh up to 15 pounds (6.8 kg) per gallon (3.8 liters). Because of the weight, the stiffening rings that are used are generally bulkier than the ones used on DOT-407 tanks. These tankers are constructed of a single tank that carries up to 7,000 gallons (26,498 liters), with most tanks holding 5,000 gallons or less (18,927 liters). The tanks are made of stainless steel and are usually lined with a rubber or ceramic coating to protect against corrosion. The piping can be on top of the tank or in the middle but is usually located on the end of the tanker. The piping is usually contained within a housing that has the

Without Metal Skin

Round Shape
from Rear

Figure 2-61 The silhouette of an uninsulated MC-307/DOT-407 is circular from the side and the rear. In some cases the side view may present what is known as a double conical, which can be described as two ice cream cones butted together to form the tank with the thickest part in the middle.

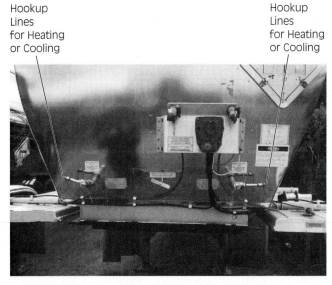

Figure 2-62 Products carried in an insulated 307/407 either need to remain heated or cooled. Some products may need to be heated via heater coils, which run around the tank, before they can be offloaded.

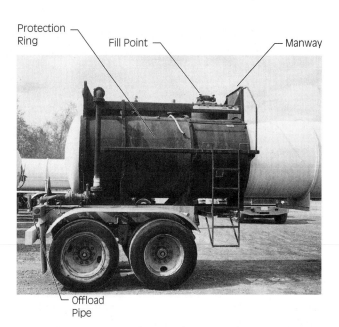

Figure 2-65 One indicator that a truck is carrying a corrosive is the ring that circles a tank around the manhole. It is used to protect the tank from spillage.

Figure 2-63 The MC-312/DOT-412 is designed to carry corrosives. Its design is similar to the uninsulated 307, although smaller. The inner tank may be lined with a variety of materials to prevent the corrosive from attacking the tank.

manhole and offload piping; this housing protects the piping in the event of a rollover, as shown in **Figure 2-65.** The area around the manhole is usually coated with a material, usually a black, tarlike coating, that resists the chemical being carried.

DOT-331

These tanks look like bullets and are noted for their rounded ends and smooth exteriors as shown in **Figure 2-66** and **Figure 2-67.** They carry gases that are liquefied by pressure; one of the most common products carried in these tanks is propane. These tanks may contain a wide variety of liquefied gases, such as ammonia, butane, and other flammable and corrosive chemicals; they can carry up to 11,500 gallons (45,532 liters) and have a general pressure of 200 psi (14 bar), although it can be as high as 500 psi (34.4 bar). The relief valves are located on top of the tank at the rear of the trailer and can be damaged during a rollover. There is a possibility during a

Figure 2-64 The silhouette of an MC-312/DOT-412 is circular from the side and the rear and has a smaller profile than an uninsulated MC-307/DOT-407. The tank hauls corrosive materials, which can weigh upward of 15 pounds per gallon as compared to approximately 8 pounds per gallon for fuels.

Level Indicator

Control Panel

Figure 2-66 MC-331 tanks carry liquefied compressed gases such as propane and ammonia. They are made of steel and carry a variety of products.

rollover that the relief valves will fail to function or, once activated, may malfunction. The steel tanks are uninsulated and heavily fortified with heavy bolts used in the piping and manholes. The tanks are usually painted white or silver to reduce potential heating by sunlight. The tanks contain a liquid along with a certain amount of vapor. The maximum allowable liquid is 80 percent to allow for expansion when heated. The liquid in the tanks is at atmospheric temperature but upon release can go below 0°F (−18°C) and can even frostbite upon contact. The pressure in these tanks is of concern when responding to incidents involving these trucks.

NOTE When a propane tank is emptied and, hence, the pressure reduced, the temperature of the propane drops below 0°F (−18°C). Temperature, pressure, and volume are interrelated. Think of an SCBA bottle. When it gets filled it becomes hot because the pressure and volume are increasing. When the SCBA bottle is used, it becomes cold because it is losing pressure and

volume. Any time one of the parameters is changed, there is a corresponding change in the other properties. When a gas is pressurized to a point that it becomes a liquid, as is the case with propane, it allows for a large quantity of propane to be stored in a small container. When released, however, the temperature will drop because the pressure is decreasing in the container. This is known as autorefrigeration.

Response to Potential BLEVE Situations

When tanks, trucks, tank cars, or other containers are involved in a fire situation there are a number of hazards. One very deadly hazard is known as a boiling liquid expanding vapor explosion or BLEVE. The type of container and the product within the container will dictate how bad a BLEVE may be. A BLEVE occurs when the pressure inside the container increases and overcomes the designed pressure of the

Figure 2-67 The silhouette of an MC-331 is bullet shaped, with rounded ends.

container. The contents are violently released, and if the material is flammable an explosion or large fireball occurs. In the recent past there have been several incidents involving BLEVEs that resulted in emergency responder deaths and injuries, thus emphasizing the need to recognize this event before it occurs. Another phenomenon that transpires with containers is known as a **violent tank rupture (VTR)**, which occurs with nonflammable materials. The concept is the same as a BLEVE, but there is not a characteristic fireball and explosion. With both a BLEVE and a VTR there is some form of heat increase inside the container, typically from a nearby fire. The fire heats the container, which heats the contents. The contents will boil, which creates expanding vapors, which in turn increases the pressure inside the container. In most containers the relief valve will activate, attempting to relieve the pressure. In some cases the pressure inside the tank becomes greater than the relief valve can handle and continues to increase.

One of two possibilities can occur with a BLEVE or a VTR. One possibility is that the relief valve cannot handle the increase in pressure and the tank fails. The other possibility is that the fire or heat source that is creating the problem weakens the container shell, and the resulting increasing pressure vents at this weakened portion of the container. If the heat source is a fire and the container product is flammable, when the relief valve activates, the raw product coming out of the container typically ignites. This may increase the temperature of the tank, increasing the pressure. It is never advisable to extinguish the fire coming from a relief valve, as that is a safety mechanism. You can cool the top of the tank near the relief valve with an unstaffed hose stream.

The difference between a BLEVE and a VTR is manifest when the container fails. A BLEVE occurs with flammable liquids, such as propane, and when the container fails, the vapors of the flammable liquid ignite creating the explosion. How the container ruptures and the amount of product released will determine how bad the explosion will be. With a VTR the container will fail, but because the product is nonflammable, the product will not ignite. The only event is a rupture of the container, spilling its contents. The container can still rocket, and the resulting release of pressure can be violent. A VTR can occur with a container of water or other "nonhazardous" material. As an example, a 55-gallon (208-liter) steel drum of water, when heated, can travel several hundred feet depending on where the release point is on the container. In most cases the bungs (screw top caps) will release, flying a considerable distance, and the container will remain mostly intact.

The failure of the container when impinged from a fire usually occurs as the fire is impacting the tank in the vapor space. When heat is applied to a section of the tank that has no internal mechanism to provide cooling, failure of the steel can occur. When fire impinges on the liquid portion of the tank, the liquid will distribute the heat internally and the steel tank typically won't weaken. The problem is that responding firefighters do not usually know the liquid level, and one cannot easily predict when a tank may fail. Any fire impingement on a tank is a serious problem, and withdrawing from the scene may be the best course of action.

A casual observation of BLEVEs indicates that most BLEVEs and firefighter deaths occur within the first few minutes of arrival. The clock starts ticking on the BLEVE timebomb from the first minute heat is applied to the tank, not with the 911 call or the arrival of firefighters. The critical time for safety may have already passed before responders arrive on the scene.

If you arrive at an incident involving a propane tank on fire or being impinged by substantial fire, you are in extreme danger. When significant heat is on the tank and if the relief valve is not operating, the danger is even more pronounced. Operating in close proximity to a tank in this situation can be a fatal mistake. Firefighters should follow this risk/benefit analysis: We risk a lot to save a lot, and we risk a little to save a little.

SAFETY If the pressure increases at a rate higher than the relief valve can handle, the tank will explode. These explosions have been known to send pieces of the tank up to a mile away. Typically the ends of the tank travel the farthest, although any part is subject to becoming a projectile.

A large number of responders have been killed by propane tank BLEVEs; when a BLEVE such as the one shown in **Figure 2-68** occurs, it usually results in more than one firefighter's death.

Specialized Tank Trucks

MC-338 CRYOGENIC TANKERS

These tankers are a specialized construction, as the example in **Figure 2-69** shows. They are tanks with an outer shell. The inner container is steel or nickel, with a substantial layer of insulation; the exterior is aluminum or mild steel. The space between the shells is placed under a vacuum to serve as an insulation barrier. The ends of the tank are flat, and the piping is contained usually at the end of the tanker in a double door box. Relief valves are located on top of the tank to the rear of the tank. The best way to describe this tanker is to compare it to a Thermos™ bottle on wheels.

Fire impinging on a propane tank.

As heat increases inside the tank, the pressure also increases. The liquid will begin to boil.

As the pressure increases, the relief valve will open, releasing propane. The propane, being heavier than air, will sink.

The vapor will reach the fire and ignite.

The relief valve will ignite, also causing heat to be increased on the tank by that flame. The pressure will increase in the tank.

As the pressure in the tank is increasing, the tank may discolor and the pitch of the relief valve will get higher. Eventually the tank will rupture and the contents will ignite explosively. This is known as a BLEVE.

Figure 2-68   Diagram of a BLEVE.

Figure 2-69   The MC-338 carries refrigerated liquefied gases. The tank resembles a rolling Thermos™ bottle.

CASE STUDY

One recent event involved the death of two firefighters and injuries to seven other emergency responders. The fire was at a farm; the propane tank was on fire. The relief valve on the tank was operating, and the vapors from the relief valve were on fire. About 8 minutes after the firefighters arrived the tank exploded into four parts. The parts went in four different directions. The two firefighters who died were 105 feet away and were struck by one piece of the tank, killing them instantly.

In two other farm incidents, firefighters also lost their lives. In 1993 a fire and BLEVE in Ste. Elisabeth de Warwick, Quebec, Canada, killed three firefighters. In 1997 in Burnside, Illinois, two firefighters lost their lives in a fire and BLEVE. One important note is that in all three of these fatal BLEVE events, the relief valves were operating upon arrival of the fire department.

One other event worth noting is the 1984 PEMEX LPG Terminal fire in Mexico City, Mexico. An 8-inch 20-cm) pipeline broke while filling a tank at the terminal. The resulting vapor cloud, which was 650 feet × 572 feet × 8 feet (198 meters × 174 meters × 2.4 meters), ignited, resulting in an explosion that included a ground shock and major fire. The terminal had two large spheres, four small spheres, and forty-eight horizontal tanks. About 15 minutes into the fire the first BLEVE occurred, and for the next 90 minutes tanks continued to BLEVE. Tanks and liquid propane rained down on the adjacent community. The death toll exceeded 500 people, and thousands were injured.

In 1983 five Buffalo, New York, firefighters lost their lives when a three-story building collapsed. Nine other firefighters were injured in the massive explosion. The force of the explosion caused the first arriving apparatus, including a ladder truck, to be blown across the street. A propane tank inside the building had been struck and was leaking. An unknown ignition source sparked the explosion minutes after fire crews arrived, which caused the building to collapse.

In January 2003 a propane truck (MC-331) went over a guardrail and fell to the ground 35 feet (10.6 meters) below. The tank ruptured, and the leaking propane ignited, resulting in a 600-foot (183-meter) high fireball. The resulting explosion moved the truck several hundred feet away from the first impact area. The driver of the truck was killed.

As can be noted in the stories above, BLEVEs and VTRs can result in injuries and fatalities. Any time containers are under stress, such as in a fire,

they can fail, and many times when they fail, they fail violently and with severe consequences. Some containers have relief valves, while others do not. Materials that are highly poisonous such as chlorine will not have a relief valve except for 90-ton railcars; they will have a fusible metal plug that vents the pressure of the tank. This fusible metal plug is not like a relief valve that shuts off when the pressure inside the tank is decreased. Once a fusible metal plug melts, the contents of the tank are also coming out, no matter the pressure. When a tank is on fire or is being impinged by fire, the tank is being weakened and the contents are being heated, creating increased pressure. Some of the general rules of firefighting and propane tanks (or other containers under pressure) are:

- Withdraw immediately in case of rising sound from venting relief valves or discoloration of the tank.
- Fight the fire from a distance with unstaffed hose holders or monitor nozzles.
- Cool the tank with flooding quantities long after the fire is out; a minimum of 500 gpm (1,893 liters per minute) at the point of flame impingement is recommended by the NFPA.
  - If the water is vaporizing on contact, you are not putting enough water on the tank. Water should be running off the tank if you are cooling the tank.
- Do not direct water at relief valves or safety devices as icing may occur, which would block the venting material and could cause an increase in pressure inside the tank.
- The tank may fail from any direction, and any tank that is being exposed to a fire can fail at any moment.
- For a massive fire use unmanned hose holders or monitor nozzles. If this is impossible, withdraw from the area and let fire burn.
- The dangers associated with a BLEVE are:
  - The fireball can engulf responders and exposed structures.
  - Metal parts of the tank can fly considerable distance.
  - Liquid propane can be released into the surrounding area and be ignited.
  - The shock wave, air blast, or flying metal parts created by a BLEVE can collapse buildings or move responders and equipment.

Cryogenic materials are gases that have been compressed and cooled to a temperature that converts them to liquids. Unlike liquefied gases, which use pressure to reduce them to liquids, cryogenics are pressurized and cooled to the point of becoming liquid. To remain liquid the material must be kept cool. To be a cryogenic material the liquid must be at least –150°F (–101° C) and can be as cold as –456°F (–271° C). The most common products are nitrogen, carbon dioxide, oxygen (liquid oxygen or LOX), argon, and hydrogen. The material inside is kept liquid by the cooling effect and when on the road, the tank has a maximum of 25 psi (1.7 bar). As the truck travels, the sun heats the material and the pressure increases. As the pressure increases, the relief valve will open up, relieving any excess pressure into the atmosphere.

NOTE It is not uncommon to see a white vapor cloud like that shown in **Figure 2-70** coming from the relief valves while the truck is traveling on the highway or sitting alongside the road. This is a normal occurrence and is not reason for alarm.

When the truck makes a delivery, the pressure must be increased to push the liquid out of the tank. The driver opens the piping and allows the material to flow into a heat exchanger, which is located just in front of the rear wheels; once in the exchanger the material will heat up and expand, increasing the pressure in the tank and forcing the liquid into the receiving tank. During transportation and offloading a large amount of ice will be built up on the piping and valves.

TUBE TRAILERS

Tube trailers contain several pressurized vessels, constructed much like the MC-331 tank shown in **Figure 2-71.** The DOT now refers to these as **multiple element gas containers** (MEGC). They are constructed of steel and have pressures ranging from 2,000 to 6,000 psi (138-414 bar). They hold pressurized gases such as air, helium, oxygen, and other gases.

Figure 2-70 When transporting liquids, it is not uncommon to see vapors coming from the relief valves. The truck can only travel with the tank at 25 psi or less. As it heats, the pressure increases, triggering the relief valve. This is a normal situation, and the tank will vent until below the 25 psi. When offloading, the pressure can be increased to assist with the offloading procedure.

Figure 2-71 The tubes on this trailer contain pressurized gases. The pressure can vary from 2,000 to 6,000 psi.

TEMPERATURES AND PRESSURIZED CONTAINERS

When a propane tank is emptied, the temperature of the propane drops below 0°F as the pressure is reduced. Temperature, pressure, and volume are interrelated. Think of your SCBA bottle: When it gets filled, it becomes hot as you are increasing the pressure and volume. When you use the SCBA bottle, the bottle becomes cold, as it is losing pressure and volume. Any time you change one of the parameters, there is a corresponding change in the other properties. When a gas is pressurized to a point that it becomes a liquid, as is the case with propane, it allows for a lot of propane to be stored in a small container. When released, however, the temperature will drop as the pressure is decreasing in the container. This is known as *autorefrigeration.*

Manway

Discharge Points

Figure 2-72 Dry bulk tanks carry a variety of products; some examples include fertilizers, explosives, and concrete.

The piping and controls are usually located on the rear of the trailer, but could be in the front. The typical delivery mode is that the driver will drop a full trailer off at a facility and pick up the empty one for refilling. Although they are not subject to a BLEVE because they only contain a gas, if involved in a fire these trailers can have a VTR and rocket in the same fashion as a BLEVE.

DRY BULK TANKS

These slightly resemble large uninsulated MC-307s in shape with bottom hoppers to unload the product, as shown in **Figure 2-72.** The tanks hold dry products and sometimes a slurry, like concrete. The most common products are fertilizers, flour, grain, lime, and other food products. A potential hazard when dealing with these tankers is predominately environmental, but

these tankers may contain toxic materials. They are usually offloaded using air pressure either from the truck itself or at the facility.

HOT MATERIALS TANKER

These tanks come in a variety and can be modified MC-306s, 307s, 406s, 407s, or dry bulk containers such as the one shown in **Figure 2-73.** They may carry the hot material with a mechanism to keep the material hot or be loaded hot and then heated prior to offloading. Common products are tar, asphalt, molten sulfur, and fuel oil #7 and #8. If the material is allowed to cool, it can cause problems for the responders or the shipping company. Tar trucks may be transported with a propane or fuel oil flame ignited to heat the product en route to the job site, although this may be an illegal act in most states.

Figure 2-73 These trucks carry molten products and can be heating the product while driving. This practice is illegal but is found on occasion. The fuels used to heat the product are either diesel/kerosene or propane.

SAFETY The major problem with these tankers is the heated material itself: Anyone coming into contact can be seriously burned. The molten material could ignite the truck or other combustibles.

Intermodal Tanks

These tanks are increasing in use and carry the same types of products as their highway and rail companions. They are called intermodal as they can be used on ships, rail, or highway, like the example in **Figure 2-74.**

CAUTION Smaller intermodal tanks may be found inside of box-type tractor-trailers.

Figure 2-74 This is an IMO101 tank. Like the totes, these are bulk tanks capable of carrying a large quantity of product. These are normally placed on ships, then delivered locally by a truck, although trains can also be used.

Figure 2-75

Figure 2-77

Figure 2-76

Figure 2-78

Figure 2-75 to Figure 2-78 Shown in these photos are all of the types of intermodal tanks that are carried predominately on ships, then moved by highway or rail to their final destination. Similar to highway tanks, they carry a variety of products and will be placarded and labeled if carrying hazardous materials. *(Photos by Dan Law)*

Intermodals follow three basic types: nonpressure, pressure, and high pressure. They are built in two fashions, either seated inside a steel frame, called a box-type framework, or with the tank as part of the framework, called a beam-type intermodal. Like tank trucks, they are assigned specification numbers: IM (Intermodal) 101, IM 102, Spec 51 (specification 51).

When used internationally, they are called IMOs. They are made to be dropped off at a facility and, when empty, picked up for refilling. Examples of intermodal containers are provided in **Figure 2-75, Figure 2-76, Figure 2-77, Figure 2-78, Figure 2-79, Figure 2-80, Figure 2-81, Figure 2-82,** and **Figure 2-83. Table 2-9** describes IMO containers.

Figure 2-79

Figure 2-82

Figure 2-80

Figure 2-83

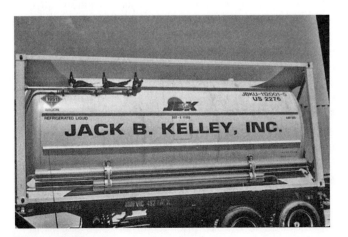

Figure 2-81

Figure 2-79 to Figure 2-83 These photos show all the types of intermodal tanks that are carried predominately on ships, then moved by highway or rail to their final destination. Similar to highway tanks, they carry a variety of products and will be placarded and labeled if carrying hazardous materials. *(Photos by Dan Law)*

TABLE 2-9

IMO Containers			
Container	**Maximum Capacity**	**Pressures**	**Products**
IM 101 or IMO Type 1	6,300	25.4 to 100 psig	Nonflammable liquids, mild corrosives, foods, and other products
IM 102 or IMO 2	6,300 gallons	14.5 to 25.4 psig	Flammable materials, corrosives, and other industrial materials
Spec 51 or IMO type 5	5,500 gallons	100 to 500 psig	Liquefied gases, much like an MC-331
Specialized tanks (cryogenic tanks are also known as Type 7 tanks)	Varies	Varies	Includes cryogenic tanks and tube banks (small tube trailer) that carry the same products as their highway counterparts

The piping for the IM 102 is contained outside of the tank, usually on the ends, with a manhole on top of the tank. The piping for the IM 101 is contained with a housing on top of the tank. For all IMs, the tank and valves are contained within the structure to avoid damage during transport.

Rail Transportation

As with highway transportation there are only a few types of railcars. The piping and shape of railcars may resemble their highway counterparts but that is the extent of the similarities.

RIGHTING OF OVERTURNED TRUCKS

The righting of overturned tank trucks can be controversial at times. The general rule is that anytime a truck is overturned, it must be offloaded prior to the truck being righted. There is substantial risk of tank failure if a loaded truck is righted by a tow truck. Some tow companies use air bags and have been trained to right fully loaded tank trucks, but it is not advisable to allow them to do so. The towing company may be aggressive in wanting to right the truck, but the incident commander is in charge, and if the tank were to fail and the product entered a waterway, the incident commander could be held responsible for violations of the Clean Water Act.

A tank truck is designed to function with its wheels down; when it is wheels up, the strength of the tank is in question. The stresses on a tank during a rollover cannot be seen and when rolled over there are more stresses placed on the tank. Gasoline tankers (MC306/DOT 406) are especially susceptible to failure during a righting operation.

The principle responsible party is required by law to provide response experts, either from within their company or an outside contractor, to mitigate this type of incident. In some cases the local public sector HAZMAT team would remain in command and monitor the scene until the scene is no longer an emergency. In some parts of the country the local HAZMAT team is aggressive in this mitigation and performs the offload. No matter who handles the incident the product needs to be offloaded from the truck. There are several methods of removing the product from the tank, some easier than others. One method is to remove the product through existing valves and piping, which can sometimes be awkward when the truck is upside down. Another method is to attach a valve assembly to the existing valves, and then offload the product. Comparable to this method are flexible bladders that attach to the valves and are connected to piping. Another method is to drill the tank to allow for the offload of the product. Each of the methods has benefits and pitfalls, that may not apply in every case. First responders can speed up this process by calling for another compatible tank truck to transfer product into and calling for a vacuum truck or pump to conduct the transfer. These requests should be made in consultation with the responsible party and the HAZMAT team. In some states box trailers that have any quantity of chemicals, including household commodities, are required to offload prior to righting, as shown in **Figure 2-84.**

(A)

(B)

(C)

Figure 2-84 From this view (A) it would appear easy to put this truck back on its wheels for removal. The view provided by (B) shows that the roof and sides of the truck are severely damaged and will not hold back the cargo. The top and sides of trailers are merely a thin skin and offer no structural integrity. The wreck and resulting shift in cargo caused the roof and side to fail. If there are any hazardous materials in the cargo, their release would move quickly into the storm drain. (C) Any discharge into this drain is a violation of the Clean Water Act. (*Courtesy of Maryland Department of the Environment*)

Railcars come in three basic types: nonpressure (general service), pressure, and specialized cars. Although they are categorized in this fashion, the commodities they carry will determine the ultimate use of the car. As with highway transportation, dedicated railcars are marked with the products they carry.

Railcars carry chemicals, combustible and flammable liquids, corrosives, and slurries.

NOTE The easiest way to determine if a railcar is nonpressure is if the valves, piping, and other appliances are located outside of a protective housing (see **Figure 2-85**).

NOTE In rail transportation the quantities are greatly increased—up to 30,000 gallons (113,562 liters) for hazardous materials and 45,000 gallons (170,344 liters) for nonhazardous materials.

Rail incidents usually involve multiple railcars, while highway incidents usually involve one or two trucks. The incidents may occur in rural areas, away from water supplies and easy access. Rail incidents will involve multiple agencies, and the local community can expect a large contingent of assistance coming from the state and federal government, which in itself can be difficult to manage.

CAUTION The term nonpressure car is actually misapplied as it can have up to 100 psi (7 bar) in the tank.

Figure 2-85 The indication that this is a nonpressure railcar is the fact that all of the valves are on the outside of the tank and not contained within any protective housing.

Figure 2-86 The striping around the tank indicates that the car carries corrosive materials. The coating protects the tank from spillage.

Figure 2-87 The pressure car is indicated by the valves being contained within the protective housing of the rail car and no exposed piping, or valving at the bottom of the rail car.

In some cases, the car will be bottom unloaded, with the ability to load the car through the top. It is possible that the car will be top loaded and unloaded. The cars will usually have a small protective housing, but the relief valves and other piping are located outside of this protective housing. Most of the piping is located on the catwalk on top of the car. Corrosive cars, such as the one shown in **Figure 2-86,** will be smaller than other non-pressure cars and may have a characteristic strip of protective coating around the middle of the car.

Non-pressure cars can be insulated and may have heating and cooling lines around the tank. Prior to offloading the tank may need to be heated to ease the offloading process. There exists a non-pressure railcar that is known for its expansion dome. The piping valves and fittings will be sitting on top of this protective housing, which was constructed to hold any potential expansion. These cars are not in regular service but may be seen in larger industrial facilities that have their own railroad service onsite.

Pressure tank cars, as shown in **Figure 2-87,** also carry a wide variety of products, which include flammable gases like propane and poisonous gases like chlorine and sulfur dioxide. The pressure cars will have pressures in excess of 100 psi up to 600 psi (7–41.3 bar). Most pressure cars that carry flammable materials will be insulated with a spray-on insulation or may be thermally insulated. This insulation is 1 to 2 inches (2.5–5 cm) thick and helps reduce the chance of a BLEVE during a fire situation.

A catwalk around the protective housing provides for relatively easy access. Like the highway tank truck system, railcars are also listed by specification number. These numbers are stenciled on the side of the car and indicate some characteristics of the car. See **Table 2-10.**

TABLE 2-10

Railcar Specifications	
Specification Number	**Type of Car**
DOT105	Pressure
DOT109	Pressure
DOT112	Pressure
DOT114	Pressure
DOT120	Pressure
DOT103	Non-pressure
DOT104	Non-pressure
DOT111	Non-pressure
DOT115	Non-pressure
AAR201	Non-pressure
AAR203	Non-pressure
AAR206	Non-pressure
AAR211	Non-pressure
DOT113	Cryogenic
AAR204	Cryogenic
AAR204XT	Cryogenic
DOT106	Miscellaneous
DOT107	Miscellaneous
DOT110	Miscellaneous
AAR207	Miscellaneous
AAR208	Miscellaneous

NOTE A pressure railcar has all of the valves, piping, and fittings located under the protective housing on top of the railcar.

SPECIALIZED RAILCARS

Specialized railcars have the same characteristics as highway transports; in fact, some highway box trailers are loaded onto flatbed railcars. When transported in this fashion, they are referred to as trailers on flatcars (TOFC). You can find regular box trailers and refrigerated trailers on flatcars. Freight boxcars haul the same products as highway trailers, with the exception of carrying much larger quantities; examples of these cars are shown in **Figure 2-88** and **Figure 2-89**. The boxcars can carry boxes, cylinders, bulk tanks, totes, and super sacks. Refrigerated railcars, called *reefers,* are similar to highway and intermodal boxes and contain their own fuel source. Dry bulk closed railcars are common, and open hopper cars may also be found usually carrying coal or other ones. Tube trailers for rail are rarely used; the more common mode is a highway tube trailer set on a flatcar. Cryogenics are also carried in railcars and will be the same low pressure (25 psi/1.7 bar) as in highway transportation, with the characteristic venting when the pressure increases.

MARKINGS ON RAILCARS

Railroads use the same placarding system as highway, with the exceptions noted in the placard section. The placards are the same size, but additional information on the railcars is better than their counterparts on the highway. The information on the car itself is printed larger as compared to highway transport. By looking at the sides of a railcar, you can tell the specification type, maximum quantities, test pressures, relief valve settings, and other pertinent information. In addition to a placard the name of the hazardous materials will be stenciled on two sides of a **dedicated railcar.** These types of markings are shown in **Figure 2-90.**

Figure 2-88 The specialized railcars are similar to their highway counterparts but transport greater quantities.

High Pressure

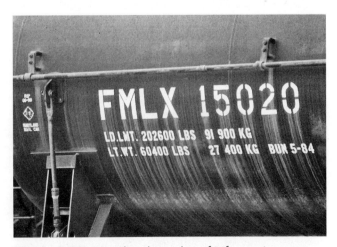

Figure 2-89 This rail car holds liquid carbon dioxide, which if released would be very cold. This rail car holds more than three times the amount than would be carried on the highway.

Figure 2-90 A railcar has a lot of information stenciled on the side of the car.

It was once common to paint a car carrying hydrogen cyanide white with a red stripe running around the middle; it was called a "candy-striper."

The specification information listed on the car is coded to reveal a lot of information about the car. The tank's owner can be determined by the code on the top line, which consists of four letters followed by a series of numbers. If the fourth letter is an *X,* then the car is owned by someone other than the railroad. The numbers indicate the car number and can be cross-referenced to determine the tank's contents when contacting the railroad. An example of the coding on a railcar is as follows:

DOT 111 A 60 AL W 1
The DOT is the authorizing agency.
111 is the class or specification number.
A has information about the rail car (applies to 105's on 17).

The following markings in conjunction with those above will be found only on pressure rail cars. The letters will be found between the authorizing agency and the pressure markings:

- ■ A—has top and bottom shelf couplers
- ■ S—has features of an A car and head puncture resistance
- ■ J—has features of A and S, and jacketed thermal protection
- ■ T—has features of A, S, and J plus spray-on insulation

60 is the pressure rating of the tank.
AL is the material that the tank is made up of:

- ■ No letter—carbon steel
- ■ AL—aluminum alloy
- ■ A—carbon steel
- ■ A-AL—aluminum alloy
- ■ AN—nickel
- ■ B—carbon steel, elastomer lined
- ■ C, D, or E—alloy steel

W is the construction type of the tank.

- ■ W—fusion welded
- ■ F—forged welded
- ■ X—longitudinally welded

1 is a number that describes fittings, materials, and linings, which can be retained in rail-specific text.

Bulk Storage Tanks

Bulk storage tanks come is sizes from 250 gallons (946 liters) up to the millions of gallons; they store a variety of products, the most common being petroleum products. These tanks are seen in residential homes, rural areas, and industrial facilities, such as the example in **Figure 2-91.** In residential homes the most common tank is a 250-gallon (946-liter) home heating oil tank, and some homes or small businesses may have gasoline or diesel fuel tanks that vary from 250 to 500 gallons (946–1,893 liters). There are two basic groupings of tanks: in-ground and aboveground. Since the passage of the EPA underground storage regulations, there has been considerable movement to remove **underground storage tanks (UST)** and

Figure 2-91 The sizes of tanks vary from a few hundred gallons up to several million gallons. A catastrophic failure of the tank may overwhelm the responders' ability to handle the incident.

replace them with **aboveground storage tanks (AST).** If the facility owner elects to keep tanks underground, then there are additional requirements placed upon them for testing, containment, and leak prevention.

> **NOTE** Leaks from a UST can be overwhelming as the leak may not be detected for a period of time and will spread throughout the area, contaminating a large area.

A substantial release from a million-gallon tank of gasoline would require an enormous response from the emergency responders and environmental contractors.

Underground tanks are usually constructed of fiberglass or steel and are coated with an anticorrosion material. New tanks are usually double-walled—a tank within a tank—to prevent any spillage into the environment. Ethanol gasoline has created some problems as the ethanol may not be compatible with the fiberglass tanks and may cause a release. As new formulations for gasoline are developed the true effect on the storage and piping systems may not be known for many years. The piping from the underground system comes up through the ground, and its use will determine the route. A UST at a gasoline station will have the offloading piping come up through the pump, as **Figure 2-92** demonstrates.

The piping is manufactured so that in the event that a car knocks off the pump it will shut off the flow of gasoline and will snap the piping off at or near the ground level. The only spillage of gasoline would be from the amount in the piping aboveground and in the hose, if everything works properly; see the accompanying case study for more information. The loading piping is located separately, and all of the fill pipes are located in the same area for easy transfer from a tank truck. The fill pipes are usually color-coded and marked so that the driver can differentiate between unleaded, unleaded super, and diesel fuel, although the color coding is not standard across the country and varies company to company. At some location on the property will be vent pipes for the tanks, generally away from the pumps and near the property line. Many facilities are being retrofitted with vapor recovery systems so there will not be any vapors released to the air. The vapors travel back into the tank through vapor recovery lines. These systems must be maintained to be effective. Many state regulations require them, and they have the added advantage of reducing fires due to static electricity. There will also be other manhole covers approximately 6 to 8 inches (15.2–20 cm) in diameter with a triangle on the top of the cover. These are inspection wells and typically surround the tank. The holes are drilled at various depths so that leaks can be easily detected by air monitoring of the well or by taking a sample if water is in the well. If a facility has had a leak, there may be a large number of these wells on and around the property. Most gas station tanks hold 10,000 to 25,000 gallons (37,854–94,635 liters); if they are not properly monitored, they can slowly release a substantial amount of product over a short period of time.

> **NOTE** It is not uncommon to find gasoline bubbling up in a basement miles away from a gas station from a release that occurred several years ago.

Figure 2-92 Piping system of an underground gasoline pumping system.

Another common problem occurs when farms are redeveloped to become housing developments; if there were unknown USTs on the property that were not discovered during construction, these tanks may eventually leak and cause serious problems. The tank and environmental industry refers to a **leaking underground storage tank** as a **LUST.**

Aboveground tanks have been used for many years but are becoming more and more common; an example is provided in **Figure 2-93.** They are of two methods of construction: upright and horizontal. ASTs can be nonpressure (atmospheric tanks) or pressure tanks. ASTs hold a wider variety of chemicals than their UST counterparts, but petroleum products are still a leading commodity stored in these tanks. They vary in size from the 250-gallon (946-liter) home heating oil tank to oil tanks of several million gallons. Industrial and commercial applications are required to have a containment area around the tank. The containment area must be able to hold the contents of the

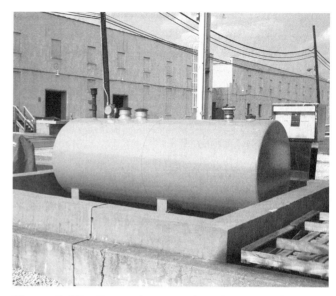

Figure 2-93   Due to increased environmental concerns many tanks are being placed above ground and are called aboveground storage tanks, or AST.

CASE STUDY

A recent incident in Baltimore County, Maryland, demonstrates that no matter how many protection systems are in place, a chemical release off the property is still possible. On an early afternoon a gas station was getting a tank filled with fuel; the tank filling went as expected with no problems. The gas station had installed a state-of-the-art alarm system that would indicate a fault in the system and alert to any potential releases. The system goes through a system check to ensure that the system is intact and there are no releases. This system check is completed after each sale, if no pump is operated for a period of 30 seconds. If a pump is on or the period between sales is less than 30 seconds, then the system check does not begin to work. During a busy afternoon, it could be estimated that the system check would not function for a long period of time, as at least one of the pumps is operating all the time. In addition to the system check, each of the storage tank locations, some piping, and the pumps all have electronic monitors that detect the presence of a liquid. When a liquid is detected, an alarm activates, the same type of alarm as if the system check fails the system. One of the potential problems with many self-service gas stations is that there is only one attendant, who for many reasons cannot leave the work area to check on a problem outside or, depending on the alarm panel location, may not be able to check any alarms. In some gas stations the alarms are in other rooms and cannot be seen or heard by the only attendant. Another problem is that the liquid alarms will activate when it rains, tripped by rainwater.

The pumps that sit on top of the storage tanks can supply upward of 80 psi (5.5 bar) of pressure to all pump station nozzles, although this is governed down to an actual working pressure of 10 psi (0.7 bar) at all pumps. In this particular incident it appears that a flange that comes out of the tank pump failed and separated from the tank pump. From the pump discharge there is a 2-inch (5-cm) diameter opening, which at 10 to 80 psi (0.7–5.5 bar) could pump a considerable amount of fuel in a short time. One theory is that a customer may have set a nozzle down after not being able to pump fuel or the amount of fuel being pumped may have been inadequate. This action would have resulted in the tank pump continuing to pump. A liquid alarm sounded in a number of areas at the gas station. The attendant notified the pump repair company, per protocol, but a response was delayed due to a number of extenuating circumstances. Once the repair company arrived, the crew found the tank empty, and the clerk informed them of the recent delivery. It was later found that an estimated 4,500 gallons (17,034 liters) was lost, causing flammable vapor readings in a several block area, although all the gasoline was later recovered. Within the gas delivery system, this station had the best protection in place and had just retrofitted the station with a new piping system. No matter how many protection systems are in place, it is still possible to have a release.

Figure 2-94 The ordinary tank has a weak roof-to-shell seam, so in the event of an explosion, the roof will come off, but the tank should remain intact.

Figure 2-95 HAZMAT crews were on location where a ship rammed an offshore transfer pipe location. From across the river they saw the roof of this tank fly up into the air, heard the explosion, and saw the tank erupt in flames. Upon arrival it was critical to determine the contents of the tank and adjacent tanks. The tier 2 chemical inventory would have this information. Complicating the extinguishment of this asphalt tank was that it is an insulated tank, with fire in-between the outer skin and the tank walls. *(Photo by the Maryland Department of the Environment)*

largest tank within the containment. Regular inspection of these areas is required to ensure that rainwater, snow, or ice does not cause a buildup of liquid in the containment area, which would reduce the containment's ability to hold its intended amount. Depending on the weather conditions the containment area's gate valve may be left in an open position, which leaves the facility at risk for product to escape the facility through the open drain or gate valve.

Upright storage tanks come in three basic construction types: **ordinary tank, external floating roof tank,** and **internal floating roof tank.** The ordinary tank (see the example in **Figure 2-94**) is constructed of steel and typically has a sloped or cone roof to shed rainwater, snow, and ice. The roof and tank shell interface are purposely made weak so that in the event of an explosion, only the top of the tank is relocated, as is shown in **Figure 2-95.** One of the major problems with this type of tank is when the tank is not full, there is room for vapors to accumulate.

CAUTION Any time vapors are allowed to accumulate, there is potential for a fire or explosion.

In chemical process facilities ordinary tanks are constructed purposely without a roof in place; they are typically used as a safety outlet to dump a chemical product into a holding tank. It is not uncommon to have a storage tank with piping into the tank just to catch overflow or the contents of a system in the event of overpressurization or failure. As the product may be hot, producing considerable vapors, or reacting, it is desirable to leave this tank uncovered. Water treatment areas of chemical plants may contain chemicals that need to evaporate in the wastewater

tank, and in many of these systems the tanks are left uncovered. From a planning perspective it is important to know what type of tanks exist in a facility, as many tanks may be in place for safety reasons. It is important to identify these tanks during preincident surveys so as not to misjudge the severity of an incident. In the event of a fire it would be a waste of resources to try to cool a tank, only to find out that the tank was not holding product and was merely a safety outlet.

External floating roof tanks are used to eliminate the buildup of vapors, as they ride on top of the liquid. Since there is no space, vapors cannot accumulate and create problems. External floaters can be seen from the top of the tanks, and there is a ladder affixed to the roof to allow access, as shown in **Figure 2-96.**

NOTE A common incident with these types of tanks results from a lightning strike that causes a fire in the roof/shell interface area. This type of fire is difficult to extinguish and can result in the roof/tank failure if not controlled quickly.

It is also possible that water used for firefighting could sink the roof, causing further problems.

Internal floaters are constructed in the same manner as the external floaters but have added an additional roof over the top of the tank, as the example

Figure 2-96 A floating roof tank providing a view of the floating roof. As the roof moves up and down the ladder moves along. A common fire problem is a lightning strike; what results is a seal fire where the tank wall meets the roof.

in **Figure 2-97** shows. The type of roof varies but is usually a slightly coned roof with vent holes along the outer edge of the tank shell or a geodesic type of roof that is usually made of fiberglass. The internal floater suffers from the same risk of fire problem as the external, although it is a reduced risk. If a fire were to start in the roof/shell interface, the roof would make it more difficult to extinguish.

SPECIALIZED TANKS

These tanks are a combination of the tanks discussed earlier in this section. They include pressure tanks and cryogenic tanks. The larger pressure vessels are divided into two categories, low pressure and high pressure tanks. The low pressure tanks hold flammable liquids, corrosive liquids, and some gases, up to 15 psig (1 bar). Common high pressure commodities are liquefied propane, anhydrous ammonia, or other gases or liquefied petroleum gases. An acid like anhydrous hydrochloric acid, which has a high vapor pressure, would not be uncommon in this type of tank. These tanks may have an external cover, which appears to be a tank within a tank. The pressure tanks are larger versions of the propane tanks shown earlier and can have a capacity of up to 9 million gallons (34,068,706 liters), although less than 40,000 gallons (151,416 liters) is typical. Liquefied petroleum gases are not only stored in just pressure tanks, such as the one shown in **Figure 2-98,** but these types of gases may be stored in other locations, such as caves carved out of mountains, although this is rare. As these facilities are not required to report under the SARA Title III regulations, responders must contact local gas suppliers to find out how they store their gas products. Upright cryogenic storage tanks are located in almost every community, especially if the community has a hospital or medical center, such as those shown in **Figure 2-99.** Most hospitals are supplied with LOX through the cryogenic tank. Facilities that sell or distribute compressed gases are likely to have cryogenic tanks. Fast-food restaurants and convenience stores are now using cryogenic tanks to supply carbon dioxide to soda-dispensing machines.

Figure 2-97 The internal floater is constructed in the same fashion as the external floating roof. It just has a cover on it to keep out rain and debris.

Figure 2-98 A specialized tank, such as this propane tank, has some of the same properties as transportation equivalents.

is a toxic material, has a unique and offensive rotten egg–like odor. After a few minutes of exposure a person breathing in hydrogen sulfide will no longer have the ability to smell the toxic gas. The gas is still present, and could even be in higher levels, but the exposed person no longer has the ability to smell, as his brain has told the olfactory system to ignore the smell. The odorant in natural gas, known as mercaptan, has the same ability.

CAUTION The lack of an odor cannot be equated to a lack of toxicity, as many severely toxic materials are colorless and odorless.

Responders can use these sensory clues gained from other persons to help identify the spilled material. The smell of materials is an important clue when trying to identify a spilled material. If bystanders or evacuated persons can assist by describing an odor, take advantage of this clue, but do not endanger your life to determine an odor. Many toxic materials can be harmful if touched as the material can be absorbed through the skin. Anytime people handle a chemical, they should wear appropriate chemical protective clothing.

Figure 2-99 This tank holds refrigerated liquefied gases, in particular, oxygen.

SENSES

Although the use of senses is listed as one of the four major recognition and identification categories it is one that presents the most risk. Responders should never use smell, taste, or touch to assist in the hazardous materials identification process. If victims were unfortunately exposed to a material, responders should use the information gleaned from these persons, but only after they have been decontaminated. Responders can use their other senses, hearing, and vision to assist in the recognition and identification process. The ability to recognize potentially hazardous situations is key to survival, and there are typically visual clues available to responders. Each of the other recognition and identification categories relies on the use of the visual sense. The venting of a relief valve may only present an audible indication that there is a problem. As the pitch of the relief valve increases, this is an indication that the pressure is increasing inside of the container. Responders should never place themselves in harm's way to smell, taste, or touch hazardous materials; doing so may be the last act they perform. Many chemicals are in a category known as desensitizers or overwhelm the olfactory (sense of smell) system. Hydrogen sulfide, which

CHEMICAL AND PHYSICAL PROPERTIES

Although not intended to be a full chemistry lesson, this next section covers a key safety component. The chemical and physical properties outlined here are appropriate for your level of response. These identifications and use of these key terms can determine the outcome of the incident—and your well-being. As you progress through the response levels, the need for additional chemistry also increases. When in doubt, consult with your HAZMAT team or other resources such as Chemtrec or a local specialist. One of the important lessons in HAZMAT response is to always "Follow the science and you will survive." In today's society there is a fear of chemicals, and in humans there is an inherent fear of the unknown. Through the use of science, responders can make informed decisions, ones based on fact. Responders need to be strong in the face of adversity, and by following fear and hysteria, responders may not make the best decisions. As you will learn, many of the terms discussed here can be applied through basic firefighting. Fire is a chemical reaction; thus, the better you understand this chemical reaction, the better off you will be and the more you will understand how conditions can rapidly change, which will also benefit the citizens of your community.

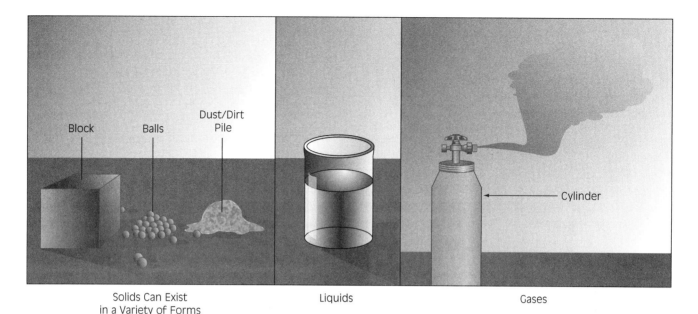

Figure 2-100 States of matter.

States of Matter

The basic chemical and physical properties that are important to understand are the states of matter—solid, liquid, and gases—as shown in **Figure 2-100.** The severity of the incident can be determined by knowing if the material is a solid chunk of material, a pool of liquid, or an invisible gas.

> **NOTE** The level of concern rises with each change of state; relatively speaking a release of a solid material is much easier to handle than a liquid release, and it is nearly impossible to control a gas.

The control methodology for each increases in difficulty from simple controls with a solid to difficult with a gas. Evacuations must be enlarged for releases involving gases, where a minimal evacuation would be required for a solid material.

The route that the chemicals can harm humans also varies with the state of material. Solids usually can only enter the body through contact or ingestion, although inhalation of dusts is possible. Liquids can be ingested, absorbed through the skin, and, if evaporating, inhaled. Gases, on the other hand, can be absorbed through the skin, inhaled, and, to some extent, ingested.

Related to states of matter are melting point, freezing point (shown in **Figure 2-101**), boiling point

Figure 2-101 Melting and freezing points.

(shown in **Figure 2-102**), and condensation point. All of these are points at which a material changes states of matter. The ability of a material to go from a solid to a gas is called **sublimation.** Some common chemicals that sublimate are moth balls and dry ice (carbon dioxide in solid form). The **melting point** is the temperature at which a solid must be heated to transform the solid to the liquid state. Ice has a melting point of 32°F (0°C). The **freezing point** is the temperature of a liquid when it is transformed into a solid. For water the freezing point is 32°F (0°C), which may seem confusing as it is also the **melting point.** The actual temperatures vary by tenths of a degree, but are very close. The **boiling point** is when the liquid is heated to a point that evaporization takes place and the liquid is changed into a gas. Another way of defining boiling point is the temperature of a liquid when the vapor pressure exceeds the atmospheric pressure and a gas is produced. Water boils at 212°F (100°C) and changes into a gaseous state. The important thing to remember about boiling points is the fact that when the liquid approaches this temperature, vapors are being produced, which can cause serious problems.

Vapor Pressure

Out of all a hazardous material's chemical and physical properties, vapor pressure can be critical to a hazardous materials responder. If a product has a high **vapor pressure,** it can be very dangerous to deal with. A material with a low vapor pressure is typically not a major concern. Vapor pressure has to do with the amount of vapors that are released from a liquid or a solid, as shown in **Figure 2-103.** The true definition is the pressure that is exerted on a closed container by the vapors coming from the liquid or solid. You can relate vapor pressure to the ability of a material to evaporate, as the material is not disappearing but moving to another state of matter. Chemicals like gasoline, acetone, and alcohol all have high vapor pressures, while diesel fuel, motor oil, and water all have low vapor pressures. Vapor pressures are measured in three ways: millimeters of mercury (mm Hg), pounds per square inch (psi), and atmospheres (atm). Normal atmospheric conditions are listed here; materials with vapor pressures at or above these levels are usually gases.

■ 760 mm Hg = 14.7 psi = 1 ATM

Although these figures are used to describe normal vapor pressure, they are best described as atmospheric pressure. The standard temperature is 68°F (20°C). If a temperature is not provided, then it can be assumed to be 68°F. Temperature has a direct relationship with vapor pressure. The higher the temperature, the more vapors will be produced. Conversely, the lower the temperature, the lesser amount of vapors that will be produced. In situations such as a chemical spill in a building, the decrease in temperature can aid first responders by reducing the amount

Figure 2-102 Boiling point, the point at which the liquid is turning into a gas.

Figure 2-103 A material with a vapor pressure is pushing against the sides of the container. If the vapor pressure is high and the material is in the wrong type of container, the container could fail.

of vapors. HAZMAT responders can use elevated temperatures to remove a chemical from a building, and this tactic is used occasionally. Chemicals that have a significant vapor hazard are those in excess of 40 mm Hg; they are considered volatile. Chemicals with a vapor pressure of less than 40 mm Hg do not travel throughout the area, but if close enough can harm through inhalation, and they still present extreme toxicity through skin absorption. Chemicals with a vapor pressure above 40 mm Hg can be considered inhalation hazards in addition to any other route of exposure they may possess. When materials become vapors, they present more risk to us and are not easily controlled. The vapor pressure of water is 25 mm Hg, and so we relate vapor pressures to this material. When a spilled material has a vapor pressure less than 25 mm Hg, it will remain as a liquid longer than a given amount of water. If a material has a higher vapor pressure than water, then it will turn to a gas faster than the same amount of water. At 68°F a cup of water will take at least a couple of days to evaporate. The same amount of gasoline will take a few hours to evaporate, which means that vapors (therefore an inhalation risk) are being produced at a faster rate. Ethyl ether will only remain in liquid form for a few minutes as it has a high vapor pressure. Materials such as liquid chlorine, a gas which must be put under pressure to liquefy, will instantly turn to a gas upon release. On the other hand, sulfuric acid has a very low vapor pressure and really only presents a contact hazard. It would take a long time for the acid to turn to the vapor form. Pure sodium hydroxide needs to be heated to 2534°F (1,390°C) for even a small amount of vapor to be released. In reality these two materials present a minor vapor hazard, as they have extremely low vapor pressure. We know as responders that when something is heated to 2534°F (1,390°C), it is probably on fire and there are more issues to contend with other than a small amount of sodium hydroxide. The pesticide ethion has an even lower vapor pressure and will remain liquid for a long period of time, extending to months. The true hazard from ethion is from contact, not from inhalation.

One definition for vapor pressure is the most scientific: it is the force of the vapors as related to overcoming atmospheric pressure that is coming down on the surface of the material, trying to keep the vapors in the material. Vapor pressure is the *force* that is coming off of a material and is related to another term, *volatility*. In many cases these terms are used interchangeably, which is actually a mistake. Volatility is the *amount* of vapors that is coming from the material. Both vapor pressure and volatility are interrelated, but they are not the same thing. In most cases the relationship is consistent, as most high vapor pressure materials also have high volatility, but this does not hold true for all materials. Xylene, for example, has an extremely low vapor pressure (9 mm Hg) but is a flammable liquid that evaporates fairly quickly. There are two reasons for this: fairly high volatility (39,012 mg/M^3) and the makeup of the molecular structure. The chemical structure is a benzene ring which is a major factor in its being a flammable liquid.

In the HAZMAT world there are many gray areas, and we just navigated through one, as this is an example of how vapor pressure offers some information on how to protect your safety. We consider materials to be of concern for inhalation hazards to have a vapor pressure in excess of 40 mm Hg. It is also at this point that we can increase the amount of risk a chemical presents. Chemicals with a vapor pressure in excess of 40 mm Hg have greater potential to move away from the immediate site of the event, can be flammable, or can require increased isolation and evacuation distances. A chemical that has a vapor pressure of less than 40 mm Hg usually can only create harm through touch or ingestion. They may be extremely toxic or corrosive, but the mechanism of harm is through touch or ingestion. **Table 2-11** lists the vapor pressure and boiling point of several chemical compounds.

Vapor pressure along with the use of air monitors is the key to risk-based response, detailed in Chapter 6.

Vapor Density

It is easy to confuse vapor pressure with vapor density, but they have entirely different meanings. **Vapor density** determines whether the vapors will rise or fall, as shown in **Figure 2-104.** When deciding on potential evacuations and other tactical objectives (such as sampling and monitoring), this is an important consideration. One of the primary reasons that natural gas leaks do not ignite or flash back more often is the vapor density of natural gas, which has a tendency to rise up and away. We give air a value of 1. Gases with a vapor density of less than 1 will rise in air. Gases with a vapor density of greater than 1 will sink in comparison to air. This property is important as you need to know where the gases will be found or will be going. Natural gas has a vapor density of 0.5, which means it will rise and dissipate. Propane has a vapor density of 1.56, which means that it will hang low to the ground and seek out the low spots such as gullies and sewers. Propane is likely to flash back as it has greater potential to find an ignition source.

Many reference texts provide vapor densities for gases and for some liquids that move to the gaseous state. The *NIOSH Pocket Guide* lists vapor density as

TABLE 2-11

Common Products and Their Vapor Pressures		
Name	**Vapor Pressure at 68°F**	**Boiling Point**
Water	17–25 mm Hg	212°F
Acetone	180 mm Hg	134°F
Gasoline	300–400 mm Hg	399°F
Ethyl ether	440 mm Hg	94°F
Chlorine	6.8 ATM (5168 mm Hg)	29°F
Diesel fuel	5 mm Hg	304–574°F
Methyl alcohol	100 mm Hg	149°F
Sodium hydroxide	1 mm Hg @ 2534°F	2534°F
Sulfuric acid	0.001 mm Hg	554°F
Ethion (pesticide)	0.0000015 mm Hg	304°F (decomposes)
Sarin nerve agent	2.1 mm Hg	316°F

relative gas density (RgasD) for all of the chemicals occurring as gases. For the other materials and for chemicals in which the vapor density is not provided, we can compare the molecular weight of the gas and compare that to the molecular weight of air. The molecular weight of air is 29. Gases with a molecular weight of less than 29 will rise, and those greater than 29 will sink in comparison to air. For street purposes when dealing with an unknown, most of them are going to be found low in comparison to air. There are only eleven common gases that rise in comparison. The neumonic that is used to remember them is DAMN₂ 3H CAE, which can be described as damn₂ 3H cows ate everything. These gases are provided in

Table 2-12. Chemicals that have vapor densities that are close to the factor of 1 (including 1), such as 1.1, 1.2, 0.9, and 0.8, are going to hang at midlevel and not go anywhere, unless there is wind to move them along. Affecting vapor density is weather, which has a dramatic impact on how a gas distributes. Vapor densities are provided at a standard temperature and pressure on what would be considered an average day. Natural gas should go up, but on days when the humidity is high or there is ground fog, materials with a low vapor density, such as natural gas, have a tendency to stay low. Propane, which has a vapor density of 1.5, is a real risk as it likes to stay low, but heavy materials will rise on hot, low humidity days with lots of sun. These weather

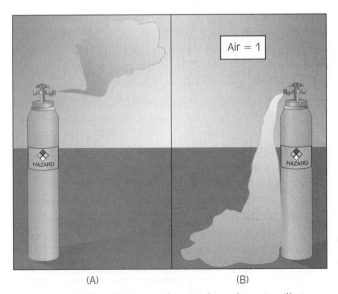

Air = 1

(A)   (B)

Figure 2-104 (A) Vapor density less than 1 will rise in air. (B) Vapor density greater than 1 will sink.

TABLE 2-12

Common Gases That Rise in Air
Diborbane
Acetylene
Methane
Nitrogen
Neon
Hydrogen
Helium
Hydrogen cyanide
Carbon monoxide
Ammonia
Ethylene

Figure 2-105 Specific gravity.

Specific Gravity

conditions are the extreme in most areas of the country, and when dealing outside the norm, do not expect chemicals to react in their normal fashion. This is an example that science is not black and white; there are considerable shades of gray.

Specific Gravity

This chemical property is similar to vapor density, as it determines whether the material sinks or floats in water, as shown in **Figure 2-105. Specific gravity** is of prime concern when efforts are being taken to limit the spread of the spill by the use of booms or absorbent material. Water is given a value of 1, and chemicals that have a specific gravity of less than 1 will float on water. Fuels, oils, and other common hydrocarbons (chemicals comprised of hydrogen and carbon, typically combustible and flammable liquids)

will float on water and have a specific gravity of less than 1. Other materials that have a specific gravity of greater than 1 will sink in water and will make their way to the bottom. It is much easier to recover materials floating on top of the water than those underwater. Carbon disulfide and 1,1,1-trichloroethane are two common materials that sink in water. Any material that sinks in water is especially dangerous when it has the potential of reaching any groundwater, as remediation efforts are difficult and expensive. Also of concern is a material that is water soluble or has the ability to mix with water, not sink or float. Corrosives and many poisons are water soluble and will be difficult to remove from a water source; two examples are provided in the accompanying case study. Alcohol is also water soluble, and as you will learn with firefighting foam, materials that are water soluble are difficult to extinguish.

CASE STUDY

When dealing with water-soluble materials, it is important to protect bodies of water. Cleanup can be difficult, even impossible, if the material enters the water. In one incident more than 10,000 gallons (37,854 liters) of sodium hydroxide, a very corrosive material, entered a small stream. The creek had to be dammed and treated with a neutralizing agent. Any life existing in this mile of stream was killed by the sodium hydroxide, but the environmental damage could have spread for many more miles if it had not been neutralized. Once neutralized, the water was released and eventually led to a larger body of water, where no damage occurred. This took considerable resources and time to accomplish; luckily, the necessary neutralizing agent was located at the site of the spill.

In another incident a very small amount of a pesticide was sprayed near a pond. After a rainfall the pesticide ran into the pond, killing all of the fish in the pond. The only method of removing this pesticide is the addition of other chemicals, which are also hazardous. It will be years before the pond can sustain life.

In Baltimore, Maryland, a railcar was involved in an accident that released 18,000 gallons (68,137 liters) of hydrochloric acid, which eventually ran into a local stream. The addition of the acid actually brought the pH level of the stream to near acceptable levels, as this stream was already heavily contaminated from a number of other sources. Cleanup of these three incidents required the express permission of local or state environmental agencies.

Corrosivity

We have covered at length tanks and containers that hold corrosives, so you should already be aware of some of the concerns in dealing with corrosives. This is a term that is applied to both acids and bases and is used to describe a material that has the potential to corrode or eat away skin or metal. We deal with corrosives every day: Our body is naturally acidic, and we use many corrosive materials in our everyday lives. Acids are sometimes referred to as corrosives, while bases are also known as alkalis or caustics. The accurate way to describe a corrosive is to identify the material's pH, which provides some measure of corrosiveness. pH is an abbreviation for a number of terms: potential of hydrogen, power of hydrogen, or percentage of hydrogen ions. It is used to designate the corrosive nature of a material. Acids have hydronium ions, and bases have hydroxide ions. It is the percent (ratio) of these ions that makes up the pH number. Materials having a pH of 0–6.9 are considered acids, and materials with a pH of 7.1–14 are considered bases. A material having a pH of 7 is considered neutral. The pH scale is a logarithmic scale, meaning the movement from 0 to 1 is an increase by a factor of 10. The movement from 0 to 2 is an increase of a factor of 100. One issue with corrosives is the concentration of the corrosive, which determines the amount of harm that can be caused. As an example, Pepsi™ has a pH of 2 but is not hazardous because the corrosive phosphoric acid is not concentrated. The bottom line is that if the phosphoric acid were concentrated, it would have more potential to cause harm. The impact on humans by a corrosive material is best evaluated by a HAZMAT technician. The first responder should avoid any potential contact with corrosive materials in an emergency response setting.

When dealing with a corrosive response, one of the common methods of mitigating the release is to neutralize the corrosive. One thought is that water can be used to dilute, thereby neutralizing the spill. This presents two big issues: one, the mixing of water (chemical reaction) and, two, the runoff from the reaction. Corrosives and water can be a dangerous combination. One should never add water to a corrosive as it presents great risk; there may be some spattering and heat generation.

If you had 1 gallon (3.8 liters) of an acid with a pH of 0, you would need 10 gallons (38 liters) of water to move the material to a pH of 1. To move it to a pH of 2 you would need 100 gallons (380 liters) of water, and to change the pH to 6 it would require 1,000,000 gallons (3,785,412 liters) of water, all for a 1-gallon (3.8-liter) spill. Chemically neutralizing a corrosive spill is the better choice, but even that can present some issues. When neutralizing a strong acid, you should use a weak base to perform the neutralization. A street method of calculation for neutralization states you would need more than 8,800 pounds (3,629 kg) of potash to bring 1,000 gallons (3,800 liters) of 50 percent sulfuric acid to neutral, which is more realistic for you to control than 1,000,000 gallons (3,785,412 liters) of runoff. These examples are for your information only; the neutralization of corrosives is a technician-level skill and can be very dangerous.

If your skin and eyes are exposed to a corrosive material, they should be immediately flushed with large quantities of water. This flushing should continue for at least 20 minutes uninterrupted. Some corrosive materials will cause immediate blindness and skin burns, and water should be used to prevent further injury. **Table 2-13** lists the pH of some common compounds.

TABLE 2-13

pH of Common Materials			
Material	**pH**	**Material**	**pH**
Water[1]	7	Sulfuric acid	1
Stomach acid	2	Gasoline	7
Orange juice[2]	2	Hydrochloric acid	0
Drain cleaner[3]	14	Pepsi™[4]	2
Potassium hydroxide	14	Household ammonia[5]	12

[1]Tap water is usually a 7, while rainwater in the Northeast can be 3–6.
[2]Citric acid is the main ingredient.
[3]Main ingredient is sodium hydroxide (lye).
[4]Phosphoric acid is the main ingredient.
[5]Household ammonia is a 5% solution of ammonium hydroxide and water.

Chemical Reactivity

When mixed, chemicals will have one of three types of reactions: exothermic, endothermic, or no reaction.

NOTE The most common chemical reaction is exothermic, meaning the release of heat.

For example, fire is a rapid oxidation reaction (exothermic) accompanied by heat and light. When most chemicals mix and provide an exothermic reaction, there usually is not a lot of light but substantial heat may be produced. Mixing one tablespoon each vinegar (acetic acid) and ammonia (base) at the same temperature will produce an immediate rise in temperature of 10°F (12°C). When handling oleum (concentrated sulfuric acid) and applying water to a spill, the resulting mixture will bubble, fume, boil, and heat to over 300°F (149°C) the instant the water hits the acid. An **endothermic** reaction is one in which the heat generated by the reaction is absorbed, and no heat is released. For example, a cold pack used for first aid purposes is comprised of two chemicals. There is a solid and a liquid that when mixed together create an endothermic reaction, which becomes cold.

Flash Point

Flash point is described as the temperature of the liquid that, when an ignition source is introduced into the vapor/air mixture above the liquid during heating, will produce a flash fire. This resulting flash fire will ignite just the vapors and self-extinguish once those vapors are burned up. The liquid itself does not burn; only the mixture of vapors and air is ignited. Following closely behind flash point is the fire point of a liquid. The **fire point** is the temperature of a liquid, that produces vapors when heated, and that will ignite the liquid and sustain burning. In the lab scientists can replicate flash points and fire points; on the street, however, they are usually one and the same. **Table 2-14** lists the flash point of some common chemicals.

Auto-Ignition Temperature

Also sometimes referred to as **ignition temperature,** this is the temperature that the material will ignite on its own without an ignition source. Ignition temperatures are much higher than flash points. Once the ignition temperature is reached, the material will burn. **Self-accelerating decomposition temperature (SADT)** is another term, essentially the same as auto-ignition temperature. SADT is the temperature at which a material will spontaneously decompose. This is almost always an exothermic reaction which starts slowly and accelerates violently. A runaway polymerization is an example. The temperature to initiate polymerization is less than the boiling point of a substance. If literature says "decompose" for the boiling point, suspect a SADT potential. Both the ignition temperature and SADT are temperature levels to avoid.

Flammable Range

Two levels define **flammable range,** lower explosive limit and upper explosive limit. The flammable range is the difference between the two extremes. To have a fire or explosion you need an ignition source, a fuel (vapors), and air. The proper vapor-to-air ratio must exist to produce ignition. The **lower explosive (flammable) limit (LEL)** is the lowest amount of vapor mixed with air that can provide the proper mixture for a fire/explosion. The air monitor used to detect combustible gases is designed to read this level. The **upper explosive (flammable) limit (UEL)** is the highest amount of vapor mixed with air that will sustain a fire or explosion. Each year there are any number of natural gas explosions throughout the United States, and when we examine the mixture required we find that the LEL of natural gas is 5 percent. In other words, when natural gas vapors are mixed with 95 percent air with an ignition source present, an explosion or fire is possible, as shown in **Figure 2-106.** The upper explosive limit for natural gas is 15 percent, which combines with 85 percent air to have a fire or explosion. The flammable range for natural gas is 5 to 15 percent,

TABLE 2-14

Flash Point of Some Common Materials			
Material	**Flash point**	**Material**	**Flash point**
Gasoline	−45°F (−43°C)	Diesel fuel	100°F (38°C)
Isopropyl alcohol	53°F (11.6°C)	Motor oil	300–450°F (149–232°C)
Acetone	−4°F (−20°C)	Xylene	90°F (32°C)

Figure 2-106 Flammable range. To have a fire or explosion the lower explosive limit must be reached. Each gas has a flammable range in which there can be a fire or explosion. Below the LEL or above the UEL, there cannot be a fire.

and a fire or explosion can occur at any point in that range, as long as there is enough air to complete the mixture. A combination of 4 percent methane and 96 percent air could not sustain a fire or explosion; nor would a mix of 16 percent natural gas and 84 percent air. Acetylene, which is a common gas, has a LEL of 2.5 percent and UEL of 100 percent, which means that when you reach 2.5 percent you have the potential for a fire. Less than the LEL is called *too lean,* and amounts in excess of the UEL are called *too rich.* Finding levels less than the LEL are much safer than finding levels above the UEL.

SAFETY When confronted with a situation that has a level higher than the UEL, whenever the area is ventilated you must pass through the flammable range, an extremely dangerous situation.

Ventilation should be carried on using non–spark-inducing devices, and great care should be taken to minimize any potential electrical arcs, such as using light switches or doorbells.

Containers and Properties

The interrelationship with chemical and physical properties and the containers that hold hazardous materials is important. When there are chemical releases or incidents involving containers, knowing how the materials may react is important for responder safety. The lower the boiling point that a chemical has means that the container is going to be under more pressure, and if involved in a fire situation the heat will cause an increase in pressure within the container. Materials that react or that are corrosive can also create pressure within a container, or if the corrosive is placed in the wrong container can cause the container to fail. When containers are under pressure there is a good chance that the venting or rupture of the container will be violent. Understanding flash point and flammable range is important as the response to potentially flammable materials is common. The lower the flash point, the greater the fire risk, and the more likely the container will be under some pressure, as the vapor pressure is generally more forceful as the flash point decreases.

RADIATION

We are subjected to radiation exposure every day, in various forms. Our bodies have radioactive substances in our makeup. We eat various forms of radiation sources, and we breathe in radiation without any harm every day. Our exposure to these everyday radiation sources far exceeds those that would be found if we worked at a nuclear power plant. Through television, medical tests, and elevation we are subjected to levels of radiation that under normal circumstances cause us no harm. Some common radiation sources are listed in **Table 2-15.** To understand how radiation can hurt us, we must have an understanding of radioactivity. The basis of radioactivity is the makeup of the basic atom, specifically the nucleus. Having knowledge of radiation is important to understand not only how radiation affects the human body, but also how radiation monitors work.

An atom is comprised of electrons, neutrons, and protons. Protons and neutrons reside in the nucleus in center of the atom, while electrons orbit the nucleus. Protons have a positive charge and determine the element or type of atom. Neutrons are the same size as protons, but are neutral. Each element, with a given number of protons, can have several forms, or **isotopes,** which are determined by the number of neutrons in the nucleus. The chemical properties of each isotope of an element are the same—you cannot chemically or physically identify one from the other. If there are too few or too many neutrons, the nucleus becomes unstable. **Radioisotopes,** isotopes whose nuclei are unstable, are radioactive and emit radiation to become more stable. This emission of radiation is known as **radioactive decay,** and usually is in the form of gamma radiation, but may also be alpha, beta, or neutron radiation as shown in **Table 2-16.** An unstable material may become stable after one or two decays, but others may take many decay cycles. If a radioisotope decays by emission of an alpha or beta particle, the number of protons in the nucleus is changed, and the radioisotope becomes a different element. Uranium is the base for the development of **radon,** a common radioactive gas found in homes. Eventually the radon will decay into lead, as shown in **Figure 2-107.** Each radioisotope has a constant **half-life.** Half-life is defined as the amount of time for half of a given radioactive source to decay. The **activity** of a source of radioactive material is a measure of the number of decays per second that occurs within the source. Source activity is measured in **becquerel (Bq)** or **curies (Ci).** One becquerel is equal to one radioactive decay per second. One curie is equal to 37 gigabecquerels, or 37 billion disintegrations per second, in the same way that one kilometer is equal to 1000 meters. The physical size of a radioactive source is not an indicator of radioactive strength or activity. One curie is also the activity of exactly 1 gram of radium-222.

TABLE 2-15

		Common Radioactive Sources		
Name	**Energy Emitted**	**Half-Life**	**Use**	**Decays to Forr**
Polonium-214	α	164 microseconds	Dust-removing brushes	Lead-210
Lead-214	β and γ	27 min	Industrial	Bismuth-214
Cobalt-60	β and γ	5.3 years	X-ray machines, industrial applications	Nickel-60
Cesium-137	β	30 years	Medical field, industrial imaging, check source	Barium-137
Radon-222	α	3.8 days	Naturally occurring gas	Polonium-218
Uranium-238	α	4.5 billion	Ore	Thorium-234
Plutonium-239	α	24,100 years	Nuclear weapons	Uranium-235
Americium-241	α	433 years	Smoke detectors	Neptunium-237

TABLE 2-16

Common Radionuclides			
Name	**Alpha**	**Beta**	**Gamma**
American-241	√		√
Cesium-137		√	√
Cobalt-60		√	√
Iodine-129 and-131		√	√
Plutonium	√	√	√
Radium	√		√
Radon	√		
Strontium-90		√	
Technetium-99		√	√
Tritium		√	
Thorium	√		√
Uranium	√		√

Types of Radiation

Radiation comprises two basic categories: ionizing radiation and nonionizng radiation. Some examples of nonionizing radiation include radio waves, microwaves, infrared visible light, and ultraviolet light. Alpha, beta, gamma, and X-rays are forms of ionizing radiation, and some characteristics of each are provided in **Figure 2-108.** There are two subcategories of ionizing radiation, one with energy and weight, and the other comprising just energy. Alpha and beta are known as particulates and have weight and energy. Gamma radiation has just energy and no weight.

ALPHA (α)

Alpha radiation consists of two neutrons and two protons, which carry a 2+ charge, and is identical to a helium nucleus. Alpha is a particulate and has the ability to move only a few feet in air. Its primary hazard is through inhalation or ingestion. Street clothing or other protective clothing provides ample protection against alpha radiation.

BETA (β)

Beta particles are electrons (–) or positrons (+) and weigh 1/1836 the weight of a proton. Beta radiation is

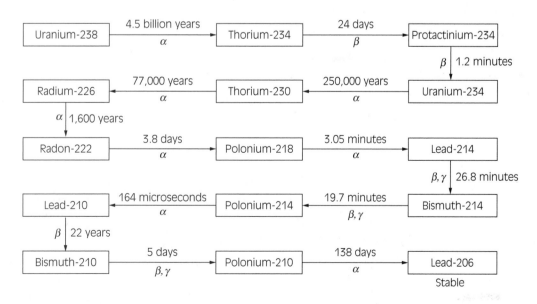

Figure 2-107 Common radiation sources.

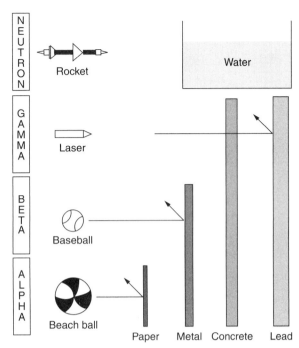

Figure 2-108 Characteristics of ionizing radiation.

one of two forms, low and high energy, but both are particulates. The low energy is comparable to alpha but has the ability to move a little farther in air and cause more damage. High energy beta moves even farther and can cause greater harm. Beta can move several feet and higher levels of protective clothing are required, but structural firefighter's clothing offers some protection.

GAMMA (γ)

Gamma rays come from the energy changes in the nucleus of an atom. Gamma is not a particulate, but is airborne energy described as wavelike radiation that comes from within the source. These waves are sometimes called electromagnetic waves or electromagnetic radiation. Gamma can move a considerable distance. Lead offers some protection. It is high energy and can cause internal damage to the body without causing external damage.

NEUTRON (n)

Neutron radiation is not common, but it is the basis for nuclear power plants and nuclear weapons. Neutrons are ejected from the nuclei of atoms during fission, when an atom splits into two smaller atoms. When fission occurs there is a release of energy in the form of heat, gamma radiation, and several neutrons. Fission can be spontaneous, in the case of californium-252, or induced, in the case of uranium-235 in nuclear power plants. Isotopes that can be induced to fission are called **fissile.**

Neutrons can travel great distances, but are stopped by several feet of water, concrete, or hydrogen-rich

materials. Neutron radiation is dangerous because it readily transfers its energy to water, and on average the human body is about 68 to 75 percent water. One unusual aspect of neutron radiation is its ability to activate nonradioactive isotopes, in other words, make them radioactive. The nucleus of the targeted atom can capture a neutron, which makes it a different isotope. If the nucleus of the new isotope is unstable, the isotope will be radioactive. For example, when natural gold, gold-197, is bombarded with neutrons, it can be activated and become gold-198, which emits beta and gamma radiation. Activation is the reason that the materials used in a nuclear reactor become radioactive and must be disposed of as nuclear waste. Activation is accomplished only by neutron radiation.

X-RAYS

X-rays are comparable to gamma radiation waves, as they are wavelike, but X-rays are only dangerous when the X-ray machine is energized.

Toxic Products of Combustion

This is the one area in which responders suffer considerable chemical exposures. Any time you are in smoke or breath smoke, your body is being bombarded with toxic chemicals. A basic lesson for responders is that fire produces many toxic chemicals, such as carbon monoxide, carbon dioxide, hydrogen cyanide, hydrochloric acid, phosgene, and others. The worst chemical accident a responder can respond to is a house, car, and dumpster fire. Even brush fires are not exempt from the toxic products of combustion, as the field may have been sprayed with pesticides, herbicides, or other chemicals. Even burning wool or hay produces extremely toxic gases. Due to this constant exposure to these and other materials, it is important to wear all of your protective clothing, especially your SCBA.

SUMMARY

The ability to recognize and identify the potential for hazardous materials is important for the first responder. The numbers of tank trucks, tank cars, and containers can easily overwhelm the beginning student. At every incident responders must complete the recognition and identification process, whether it be the location, placards, containers, or physical senses. It could even be your sixth sense that alerts you to a potential problem; when that occurs, proceed with caution. An important lesson for responders is that you don't need to commit everything to memory, but you must know where you can access that information. It is not expected that each community have a railcar expert in its response organization. What is expected is that each agency have a contact person available around the clock to obtain that resource.

It is possible that every material on this Earth has the ability to cause harm in some fashion, but chemical and physical properties plays a factor in the potential for harm. Materials that do not have a vapor pressure present little risk to responders unless touched or eaten. Materials with high vapor pressures present great risk to responders and to the community and should be treated with caution. Vapor pressure is only one of the terms that you as a responder should understand, as a variety of chemical properties may come into play. Many responders have a fear of radiation, but this fear comes from a lack of education and experience. Understanding the harms from radiation is an important safety consideration. Your local HAZMAT responders are a good resource and should be contacted early in an incident and whenever you need assistance.

KEY TERMS

Aboveground storage tanks (AST) Tanks that are stored above the ground in a horizontal or vertical position. Smaller quantities of fuels are often stored in this fashion.

Activity A measure of the number of decays per second that occur with the source.

Becquerel (Bq) A measure of radioactivity; the metric version of Curies.

BLEVE Boiling liquid expanding vapor explosion, which occurs when a pressurized tank is heated and the pressure inside the tank exceeds the capacity of the relief valves to handle the pressure. When pressurized tanks rupture, they can travel significant distances, and if holding flammable materials, they will usually ignite, resulting in a large fireball.

Boiling point Temperature to which liquids must be heated to turn to a gas.

Building Officials Conference Association (BOCA) A group that establishes minimum building and fire safety standards.

Bulk tank Large transportable tank comparable to but larger than a tote.

Chemtrec The Chemical Transportation Emergency Center, which provides technical assistance and guidance in the event of a chemical emergency. This network of chemical manufacturers provides emergency information and response teams if necessary.

Cryogenic gas Any gas that exists as a liquid at a very cold temperature, always below −150°F (−101°C).

Curies (Ci) The measure of activity level for radiation sources.

Deflagrates Rapid burning, which can be considered a slow explosion, but travels at a lesser speed than a detonation.

Department of Transportation (DOT) Federal agency that issues regulations regarding the transportation of hazardous materials.

Emergency Response Guidebook (ERG) Reference book provided by the DOT to assist first responders in making decisions at a transportation-related chemical incident.

Endothermic reaction Chemical reaction in which heat is absorbed and the resulting mixture is cold.

Excepted packaging Used to transport low risk radioactive materials; the packaging is exempted from DOT regulations.

Exothermic reaction Chemical reaction that releases heat, as when two chemicals are mixed and the resulting mixture is hot.

External floating roof tank Tank in which the roof is exposed to the outside and covers the liquid within the tank. The roof floats on the top of the liquid, which does not allow for vapor build-up.

Fissile Isotopes that can be induced into fission, which creates a release of energy.

Flammable range The correct mixture of flammable gas and air in which there can be a fire; for most gases there is a certain span (range) that will allow for a fire to ignite.

Flash point Temperature of a liquid, which if an ignition source is present, will ignite only the vapors being produced by the liquid, creating a flash fire.

Frangible disk A pressure-relieving device that ruptures to vent the excess pressure. Once opened, the disk remains open; it does not close after the pressure is released.

Freezing point Temperature at which liquids become solids.

Gas State of matter that describes the material in a form that moves freely about and is difficult to control. Steam is an example.

Half-life The amount of time for a given radiation source to decay, which means it is emitting radioactivity.

Ignition temperature Temperature of a solid, liquid, or gas that will ignite on its own without an ignition source.

Intermodal containers Constructed to be transported by highway, rail, or ship. Intermodal containers exist for solids, liquids, and gases.

Internal floating roof tank Tank with a roof that floats on the surface of the stored liquid; this tank also has a cover on its top to protect the top of the floating roof top.

Isotope Material that has a different form due to the number of neutrons that are in the nucleus.

Leaking Underground Storage Tank (LUST) Term to describe this potential health and environmental hazard.

Liquid State of matter that implies fluidity, which means a chemical can move as water would. There are varying states of being a liquid from moving very quickly to very slowly. Water is an example.

Low Specific Activity (LSA) Indicates that a material is emitting low levels of radiation.

Lower explosive limit (LEL) The lower part of the flammable range, or the minimum required to have a fire or explosion.

Material Safety Data Sheet (MSDS) An information sheet that provides chemical-specific safety information.

Melting point Temperature at which solids become liquids.

Nuclear Regulatory Agency (NRC) The government regulatory agency responsible for oversight of nuclear materials.

Ordinary tank Horizontal or vertical tank that usually contains combustible or other less hazardous chemicals. Flammable materials and other hazardous chemicals may be stored in smaller quantities in these types of tanks.

Polymerization A runaway chain reaction that can be violent if contained. Once started it cannot be stopped until the chemical reaction has run its course.

Radioactive decay As a material emits radiation, it decays, changing its form. Uranium eventually becomes lead after it decays.

Radioisotope Radioactive form of an element, it is unstable due to the number of neutrons.

Radon Radioactive gas that is emitted from the earth, sometimes collecting in basement areas.

Relief valve Device designed to vent pressure in a tank, so that the tank itself does not rupture due to an increase in pressure. In most cases these devices are spring loaded so that when the pressure decreases, the valve closes to keep the chemical inside the tank.

Reportable quantity (RQ) A term used by the EPA and DOT to describe a quantity of chemicals that may require some type of action, such as reporting an inventory or reporting an accident involving a certain amount of chemicals.

Sea containers Shipping boxes designed to be stacked on a ship and then placed onto a truck or railcar; also called sea boxes.

Self-accelerating decomposition temperature (SADT) The temperature at which a material will spontaneously decompose. A runaway polymerization is an example. The temperature to initiate polymerization is less than the boiling point of a substance.

Solid State of matter that describes chemicals that may exist in chunks, blocks, chips, crystals, powders, dusts, and other types. Ice is an example.

Specific gravity Value that determines if the liquid will sink or float in water. Water is given a value of 1, and chemicals with a specific gravity of less than 1 will float on water. Those chemicals with a specific gravity of greater than 1 will sink in water.

Specification (spec) plates Label on trucks and tanks that lists the type of vehicle, capacity, construction, and testing information.

Spent nuclear fuel Radioactive fuel that was used in a nuclear reactor.

States of matter Term applied to chemicals to describe the forms in which they may exist: solids, liquids, or gases.

Sublimation Ability of a solid to go to the gas phase without being liquid.

Surface Contaminated Object (SCO) Material that may be contaminated with radioactive waste; is usually low hazard.

Tote Large tank, usually 250 to 500 gallons (946–1,893 liters), constructed to be transported to a facility and dropped for use.

Type A container Container designed to hold low risk radioactive material, offers moderate protection against accidents.

Type B container Container designed to hold high risk radioactive materials, is a hardened transport case capable of withstanding severe accidents.

Underground storage tanks (UST) Tanks, commonly for gasoline and other fuels, that are buried under the ground.

Upper explosive limit (UEL) Upper part of the flammable range; above the UEL, fire or an explosion cannot occur as there is too much fuel and not enough oxygen.

Vapor density Value that determines if the gas will sink or rise in air. Air is given a value of 1, and chemicals with a vapor density less than 1 will rise in air. Those with a density of greater than 1 will sink and stay low to the ground.

Vapor pressure Amount of force pushing vapors from a liquid; the higher the force, the more vapors (gas) are being put into the air.

Violent tank rupture (VTR) This concept is the same as a BLEVE, but there is not a characteristic fireball and explosion.

REVIEW QUESTIONS

1. What are the nine hazard classes as defined by DOT?
2. An explosive placard is what color?

3. What three things on a placard indicate potential hazards?

4. An MC 306/DOT 406 tank truck commonly carries what product?

5. A DOT 406 tank truck has a characteristic shape from the rear. What is it?

6. An MC 331 carries what type of gases?

7. What does the blue section of the NFPA 704 system refer to?

8. What are the locations of emergency shutoffs on an MC 331?

9. What are the four basic clues to recognition and identification?

10. A truck with dangerous placards is carrying what product(s)?

11. **Are placards reqiured for** Table 1 materials? What is the quanity that requires placarding?

12. What does a material listed on the shipping papers as Packing group I indicate?

13. A tractor-trailer carrying class 1.1 materials is well involved in fire. What are your initial tactics regarding this fire?

14. In the NFPA 704 system, what does the use of ALK mean?

15. When propane tanks are involved in a fire, what is a potential consequence?

16. What types of radiation are particles?

17. Which one type of radiation has the ability to activate other materials and make them radioactive?

18. What impact does half-life have with regard to a radiation source?

19. Which tank truck is commonly referred to as a chemical hauler?

20. How many conpartments does an MC 312 have?

ENDNOTES

[1]The requirements for the subdivisions are presented in general, and there may be other requirements that need to be met for a material to be assigned to a subdivision. Refer to 49 CFR 170-180.

[2]The notation that a DOT-406 is a gasoline tanker is a general reference, and a DOT-406 can hold any number of commodities, not just gasoline. A DOT-406 may even carry food-grade products or other chemicals.

Suggested Readings

Bevelacqua, Armando, *Hazardous Materials Chemistry,* Delmar Cengage Learning. Clifton Park, NY, 2001.

Bevelacqua, Armando, and Richard Stilp, *Hazardous Materials Field Guide*, Delmar Cengage Learning. Clifton Park, NY. 1998.

Bevelacqua, Armando S., and Richard H. Stilp. *Terrorism Handbook for Operational Responders*, 2nd ed., Delmar Cengage Learning. Clifton Park, NY, 2004.

Hawley, Chris, *Hazardous Materials Incidents*, 2nd ed., Delmar Cengage Learning, Clifton Park, NY, 2004.

Hawley, Chris, *Hazardous Materials Air Monitoring & Detection Devices*. Delmar Cengage Learning, Clifton Park, NY, 2007.

Henry, Timothy V., *Decontamination for Hazardous Materials Emergencies*, Delmar Cengage Learning. Clifton Park, NY, 1999.

Hildebrand, Michal S., and Gregory Noll, *Composite Propane Cylinders,* White Paper, Propane Education & Research Council, Washington, DC, 2007.

Lesak, David, *Hazardous Materials Strategies and Tactics*, Prentice Hall, Upper Saddle River, NJ, 1998.

Noll, Gregory, Michael Hildebrand, and James Yvorra. *Hazardous Materials Managing the Incident*, 3rd ed., Red Hat Publishing, Chester, MD, 2005.

Schnepp, Rob, and Paul Gantt. *Hazardous Materials: Regulations, Response & Site Operations*, Delmar Cengage Learning. Clifton Park, NY, 1999.

Stilp, Richard, and Armando Bevelacqua. *Emergency Medical Response to Hazardous Materials Incidents*, Delmar Cengage Learning. Clifton Park, NY, 1997.

INFORMATION RESOURCES

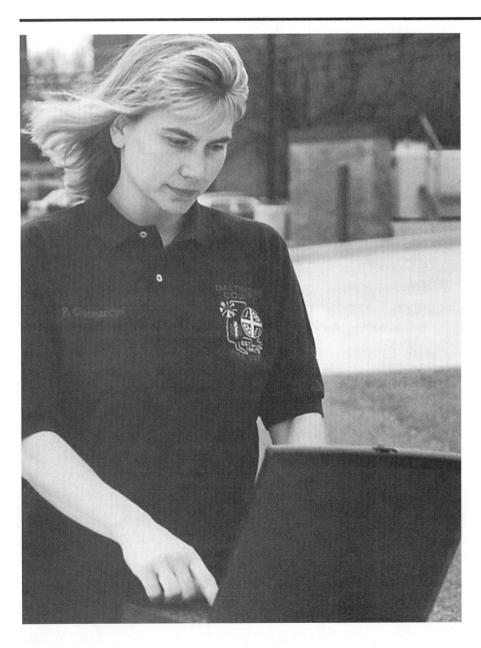

OUTLINE

▌ Objectives
▌ Introduction
▌ Emergency Response Guidebook
▌ Material Safety Data Sheets
▌ Shipping Papers
▌ Facility Documents
▌ Computer Resources
▌ Chemtrec
▌ Reference and Information Texts
▌ Industrial Technical Assistance
▌ Summary
▌ Key Terms
▌ Review Questions

STREET STORY

I had been on the job for several years. There was nothing unusual about the call when it first came in—nothing remarkable that would lead to any heightened sense of awareness. "Engine 29, check out the smell of smoke in the building at 211 Mason Street." It would be a simple food on the stove call or something to that effect.

When we arrived on the scene, the local police department had already made an initial entry to the facility (which turned out to be a restaurant) and reported a slight haze with a strange pungent odor but no visible fire. We made entry into the vacant facility and immediately noticed the haze and pungent odor, but because it was a restaurant, we thought it was simply something to do with the food preparation process. As we made our way through the facility, we began to notice a strange taste in our mouths and several of those present began to experience irritation to the eyes and throat, yet still nothing clicked. It was, after all, simply a restaurant, perhaps some burnt wiring or bad cooking, no big deal. As we continued our investigation, several of those present began to complain of tightness in the chest, headaches, and nausea. I too was feeling strange but did not want to throw in the towel, so another responder and I continued on.

Then suddenly as the haze began to clear via the doors and windows that had been opened for ventilation, we began to notice the dead insects, an abundance of dead insects and rodents throughout the facility. And now on the radio we heard the call for an ambulance as one of the responders who had been on the scene for some time began to experience more than simply mild irritations. We began to realize that the haze was in fact some type of aerosolized insecticide/pesticide and that we had now been exposed and contaminated. Now and only now did we begin to comprehend the error of our ways and implement the proper response and protocols.

As the weeks followed, I replayed the incident over and over and came to realize all of the mistakes that we had made. I understood how imperative it was to seek the proper training to ensure that I never put myself or those with me in this position again. Two of the responders were unable to return to work for a period of time (in one case several months). They had to be treated for the effects of organophosphates. Both are fine today but much has been written about the long-term effects of exposure to organophosphates, and one can only wonder what course these individuals' health could have taken had they been further exposed. And I do wonder, all too often.

Speaking only to my errors, there was no attempt to obtain the proper information from the appropriate resources or the owners. (After we realized we had no fire, we should have consulted the owners before further entry, because there was no life hazard.) Nor was subsequent contact made with other tenants in this row structure who may have been able to provide information or take basic actions to ensure the health and safety of those involved. The failure to properly research and gather information on this call could have resulted in much more severe health problems for all involved had the concentrations been higher or ventilation unsuccessful. But the fact that two responders were unnecessarily exposed still remains with me today.

—Street Story by Tom Creamer, Special Operations Coordinator, City of Worcester Fire Department, Worcester, Massachusetts

OBJECTIVES

After completing this section the student should be able to identify or explain:

■ The terms used on Material Safety Data Sheets (O)

■ The likely location of MSDSs (O)

■ Standard information available on an MSDS (O)

■ The use of the *Emergency Response Guidebook* (ERG) (O)

■ The type of assistance that can be provided by Chemtrec (O)

■ The methods that can be used to contact a shipper or emergency contact (O)

■ Other resources that may be available in the community (O)

INTRODUCTION

CAUTION Specific and current chemical information for the first responder is essential to protect lives and property.

Chemical information is available through a variety of sources, including those carried on emergency apparatus or information sources that can be reached via telephone, fax, or computer. The shipper and the facility are required to maintain certain documents that will assist the first responder in determining the chemical hazards they may face. Knowing what information is available and how to interpret this information is a valuable tool for your protection. This chapter is divided into the most common areas of information from various sources.

EMERGENCY RESPONSE GUIDEBOOK

The *Emergency Response Guidebook* (ERG), shown in **Figure 3-1**, is a well-known book for emergency responders. It is produced by the United States Department of Transportation (DOT). One copy is made available for every emergency response apparatus in the country. The guide is commonly referred to as the *DOT book* or the *orange book*. Usually updated about every 3 years, the book contains information regarding the most commonly transported

Figure 3-1 The DOT *Emergency Response Guidebook* is published and distributed to every emergency apparatus in the United States. It provides chemical emergency response information valuable for the first responder.

chemicals as regulated by the DOT. It is intended as a guidebook for first responders during the initial phase of a hazardous materials incident. It provides information regarding the greatest potential hazards of responding to these materials. It is one of the only books that provides specific evacuation recommendations. Although an excellent book for first responders, it does have limitations for the more advanced responder, who requires more specific chemical information.

The book consists of these major sections:

■ Placard information

■ ADR/RID marking system information

■ Silhouettes of tank trucks and train cars

■ Listing by DOT identification number

■ Alphabetical listing by shipping name

■ Response guides

■ Table of initial isolation and protective action distances

■ List of dangerous water-reactive materials

On the inside cover the book has an example of a shipping paper used in truck transportation, and it

<u>RESIST RUSHING IN !</u>
<u>APPROACH INCIDENT FROM UPWIND</u>
<u>STAY CLEAR OF ALL SPILLS, VAPORS, FUMES AND SMOKE</u>

HOW TO USE THIS GUIDEBOOK DURING AN INCIDENT INVOLVING DANGEROUS GOODS

ONE IDENTIFY THE MATERIAL BY FINDING ANY **ONE** OF THE FOLLOWING:

THE 4-DIGIT ID NUMBER ON A PLACARD OR ORANGE PANEL

THE 4-DIGIT ID NUMBER (after UN/NA) ON A SHIPPING DOCUMENT OR PACKAGE

THE NAME OF THE MATERIAL ON A SHIPPING DOCUMENT, PLACARD OR PACKAGE

IF AN **ID NUMBER** OR THE **NAME OF THE MATERIAL** CANNOT BE FOUND, SKIP TO THE NOTES BELOW.

TWO LOOK UP THE MATERIAL'S 3-DIGIT GUIDE NUMBER IN EITHER:

THE ID NUMBER INDEX..(the yellow-bordered pages of the guidebook)

THE NAME OF MATERIAL INDEX..(the blue-bordered pages of the guidebook)

If the guide number is supplemented with the letter "P", it indicates that the material may undergo violent polymerization if subjected to heat or contamination.

If the index entry is highlighted (in either yellow or blue), it is a TIH (Toxic Inhalation Hazard) material, a chemical warfare agent or a Dangerous Water Reactive Material (produces toxic gas upon contact with water). **LOOK FOR THE ID NUMBER AND NAME OF THE MATERIAL** IN THE TABLE OF INITIAL ISOLATION AND PROTECTIVE ACTION DISTANCES (the green-bordered pages). Then, if necessary, **BEGIN PROTECTIVE ACTIONS IMMEDIATELY** (see Protective Actions on page 298). If protective action is not required, use the information jointly with the 3-digit guide.

USE GUIDE 112 FOR ALL EXPLOSIVES EXCEPT FOR EXPLOSIVES 1.4 (EXPLOSIVES C) WHERE GUIDE 114 IS TO BE CONSULTED.

THREE TURN TO THE NUMBERED GUIDE (the orange-bordered pages) **AND READ CAREFULLY.**

NOTES **IF A NUMBERED GUIDE CANNOT BE OBTAINED BY FOLLOWING THE ABOVE STEPS**, AND A PLACARD CAN BE SEEN, LOCATE THE PLACARD IN THE TABLE OF PLACARDS (pages 16-17), THEN GO TO THE 3-DIGIT GUIDE SHOWN NEXT TO THE SAMPLE PLACARD.

IF A REFERENCE TO A GUIDE CANNOT BE FOUND AND THIS INCIDENT IS BELIEVED TO INVOLVE DANGEROUS GOODS, TURN TO **GUIDE 111** NOW, AND USE IT UNTIL ADDITIONAL INFORMATION BECOMES AVAILABLE. If the shipping document lists an emergency response telephone number, call that number. If the shipping document is not available, or no emergency response telephone number is listed, IMMEDIATELY CALL the appropriate **emergency response agency listed on the inside back cover of this guidebook.** Provide as much information as possible, such as the name of the carrier (trucking company or railroad) and vehicle number. AS A LAST RESORT, CONSULT THE TABLE OF RAIL CAR AND ROAD TRAILER IDENTIFICATION CHART (pages 18-19). IF THE CONTAINER CAN BE IDENTIFIED, REMEMBER THAT THE INFORMATION ASSOCIATED WITH THESE CONTAINERS IS FOR THE WORST CASE POSSIBLE.

Figure 3-2 The first page of the ERG is a step listing of how to use the guide. When not sure how to proceed, the responder should turn to this page.

provides information on how the shipping paper should be written. It also provides an example of a placard with an identification number panel. The first page, which is shown in **Figure 3-2,** is an essential page for your safety; it outlines all of the actions the first responder should take. It provides a decision tree process that advances step by step through the incident. It also has an important listing of the guides for explosives. The next few pages provide information on safety precautions and contacts to call for assistance. First responders should already know

where the closest local assistance will be coming from and how to request state assistance if required. The DOT book provides a contact number for federal assistance; the standard is to proceed by requesting local, state, and finally federal assistance. The contact numbers and agencies are specific to Canada, United States, and Mexico. It is important for the first responder to read the DOT book prior to an incident as it provides a large amount of background material that could not be effectively read during an emergency.

HAZARD CLASSIFICATION SYSTEM

The hazard class of dangerous goods is indicated either by its class (or division) number or name. For a placard corresponding to the primary hazard class of a material, the hazard class or division number must be displayed in the lower corner of the placard. However, no hazard class or division number may be displayed on a placard representing the subsidiary hazard of a material. For other than Class 7 or the OXYGEN placard, text indicating a hazard (for example, "CORROSIVE") is not required. Text is shown only in the U.S. The hazard class or division number must appear on the shipping document after each shipping name.

Class 1 - Explosives

Division 1.1	Explosives with a mass explosion hazard
Division 1.2	Explosives with a projection hazard
Division 1.3	Explosives with predominantly a fire hazard
Division 1.4	Explosives with no significant blast hazard
Division 1.5	Very insensitive explosives with a mass explosion hazard
Division 1.6	Extremely insensitive articles

Class 2 - Gases

Division 2.1	Flammable gases
Division 2.2	Non-flammable, non-toxic* gases
Division 2.3	Toxic* gases

Class 3 - Flammable liquids (and Combustible liquids [U.S.])

Class 4 - Flammable solids; Spontaneously combustible materials; and Dangerous when wet materials/Water-reactive substances

Division 4.1	Flammable solids
Division 4.2	Spontaneously combustible materials
Division 4.3	Water-reactive substances/Dangerous when wet materials

Class 5 - Oxidizing substances and Organic peroxides

Division 5.1	Oxidizing substances
Division 5.2	Organic peroxides

Class 6 - Toxic* substances and Infectious substances

Division 6.1	Toxic*substances
Division 6.2	Infectious substances

Class 7 - Radioactive materials

Class 8 - Corrosive substances

Class 9 - Miscellaneous hazardous materials/Products, Substances or Organisms

* The words "poison" or "poisonous" are synonymous with the word "toxic".

Figure 3-3 The DOT lists nine classes of hazardous materials in the ERG. Occasionally the only information available is the hazard class.

The book also provides a listing of the hazard class system, which is provided in **Figure 3-3,** and offers a reference point for the hazard classes that may be listed on placards or shipping papers. This listing precedes the placard section, which is valuable if the only information you have available is the placard. The placard section, which is shown in **Figure 3-4,** provides information as how to proceed at an incident where the only information is a placard. All of the possible current placards are listed

TABLE OF PLACARDS AND INITIAL RESPONSE GUIDE TO USE ON-SCENE

USE THIS TABLE ONLY IF MATERIALS CANNOT BE SPECIFICALLY IDENTIFIED BY USING THE SHIPPING DOCUMENT, NUMBERED PLACARD, OR ORANGE PANEL NUMBER

Figure 3-4 If the only information available is a placard, the ERG can provide assistance. Listed on each placard is a guide number that provides response information based on the placard information.

ROAD TRAILER IDENTIFICATION CHART*

DOT406, TC406 Non-pressure Liquid Tank (MC306, TC306) **(131)**

MC338, TC338 Cryogenic Liquid Tank (MC306, TC306) **(117)**

DOT407, TC407 Low pressure Chemical Tank (MC307, TC307) **(137)**

Compressed Gas/ Tube Trailer **(117)**

DOT412, TC412 Corrosive Liquid Tank (MC312, TC312) **(137)**

Dry Bulk Cargo Trailer **(134)**

MC331, TC331 High pressure Tank **(117)**

Mixed Cargo **(111)**

DOT407, TC407 Vacuum Loaded Tank DOT412, TC412, (TC350) **(137)**

CAUTION: This chart depicts only the most general shapes of road trailers. Emergency response personnel must be aware that there are many variations of road trailers, not illustrated above, that are used for shipping chemical products. The suggested guides are for the most hazardous products that may be transported in these trailer types.

* **The recommended guides should be considered as last resort if product cannot be identified by any other means.**

Figure 3-5 Truck silhouettes are provided with the appropriate guide page for each of the truck styles.

with their pictures, and the reader is referred to the accompanying guide page. Much like the placard section, included in the ERG are silhouettes of tank trucks shown in **Figure 3-5,** and rail cars shown in **Figure 3-6.** Some international shipments use additional hazard markings as shown in **Figure 3-7.** The yellow section, which is shown in **Figure 3-8,** is a numerical listing by the identification number or ID number. This number is assigned by the DOT to identify a material being shipped. This number takes two forms—a North America (NA) number or a United Nations (UN) number—but is always a four-digit

RAIL CAR IDENTIFICATION CHART*

Hopper Car Dry Bulk (140)

Box Car Mixed Cargo (111)

Pressure tank car Compressed liquified gases (117)

Non-pressure tank car Liquids (131)

REPORTING MARKS & CAR NUMBER
LOAD LIMIT (POUNDS OR KG)
EMPTY WEIGHT OF CAR
PLACARD HOLDER
TANK TEST & SAFETY VALVE TEST INFORMATION
CAR SPECIFICATION
COMMODITY NAME *
TC PERMIT NUMBER

REPORTING MARKS & CAR NUMBER
CAPACITY IN GALLONS OR LITERS
PLACARD HOLDER *

CAUTION: Emergency response personnel must be aware that rail tank cars vary widely in construction, fittings and purpose. Tank cars could transport products that may be solids, liquids or gases. The products may be under pressure. It is essential that products be identified by consulting shipping documents or train consist or contacting dispatch centers before emergency response is initiated.

The information stenciled on the sides or ends of tank cars, as illustrated above, may be used to identify the product utilizing:

a. the commodity name shown; or

b. the other information shown, especially reporting marks and car number which, when supplied to a dispatch center, will facilitate the identification of the product.

* **The recommended guides should be considered as last resort if product cannot be identified by any other means.**

Figure 3-6 Railcar silhouettes are provided with the appropriate guide page for each of the rail styles.

HAZARD IDENTIFICATION CODES
DISPLAYED ON SOME INTERMODAL CONTAINERS

Hazard identification codes, referred to as "hazard identification numbers" under European and some South American regulations, may be found in the top half of an orange panel on some intermodal bulk containers. The 4-digit identification number is in the bottom half of the orange panel.

The hazard identification code in the top half of the orange panel consists of two or three digits. In general, the digits indicate the following hazards:

2 - EMISSION OF GAS DUE TO PRESSURE OR CHEMICAL REACTION

3 - FLAMMABILITY OF LIQUIDS (VAPORS) AND GASES OR SELF-HEATING LIQUID

4 - FLAMMABILITY OF SOLIDS OR SELF-HEATING SOLID

5 - OXIDIZING (FIRE-INTENSIFYING) EFFECT

6 - TOXICITY OR RISK OF INFECTION

7 - RADIOACTIVITY

8 - CORROSIVITY

9 - MISCELLANEOUS DANGEROUS SUBSTANCE

- Doubling of a digit indicates an intensification of that particular hazard (i.e. 33, 66, 88).

- Where the hazard associated with a material can be adequately indicated by a single digit, the digit is followed by a zero (i.e. 30, 40, 50).

- A hazard identification code prefixed by the letter "X" indicates that the material will react dangerously with water (i.e. X88).

- When 9 appears as a 2nd or 3rd digit, this may present a risk of spontaneous violent reaction.

Figure 3-7 Intermodal containers will have an orange panel near the placard like that shown in the figure. The four-digit number is the DOT ID number, and the one- or two-digit number that is a classification code, providing information on the basic hazards.

number. Although one would think that this number would specifically identify a material, in some cases it may not. Depending on the material it may be lumped into a general category, such as Hazardous Waste, Liquid, or Not Otherwise Specified (N.O.S.), but in most cases it will identify a specific substance. This section lists the ID number, shipping name, and guide reference number. If the material is highlighted yellow, as is shown in **Figure 3-9,** then responders must refer to the guide page listed and to the table of initial isolation and protective action distances, located at the back of the book in the green section.

Abbreviations used in the DOT ERG include:

■ N.O.S.—Not otherwise specified

■ P.I.H.—Poison inhalation hazard

■ P—Polymerization hazard

■ SCA—Surface contaminated articles (radiation)

■ LSA—Low specific activity (radiation)

ID No.	Guide No.	Name of Material	ID No.	Guide No.	Name of Material
1027	115	Cyclopropane	1043	125	Fertilizer, ammoniating solution, with free Ammonia
1027	115	Cyclopropane, liquefied	1044	126	Fire extinguishers with compressed gas
1028	126	Dichlorodifluoromethane			
1028	126	Refrigerant gas R-12	1044	126	Fire extinguishers with liquefied gas
1029	126	Dichlorofluoromethane			
1029	126	Refrigerant gas R-21	1045	124	Fluorine
1030	115	1,1-Difluoroethane	1045	124	Fluorine, compressed
1030	115	Difluoroethane	1046	121	Helium
1030	115	Refrigerant gas R-152a	1046	121	Helium, compressed
1032	118	Dimethylamine, anhydrous	1048	125	Hydrogen bromide, anhydrous
1033	115	Dimethyl ether	1049	115	Hydrogen
1035	115	Ethane	1049	115	Hydrogen, compressed
1035	115	Ethane, compressed	1050	125	Hydrogen chloride, anhydrous
1036	118	Ethylamine	1051	117	AC
1037	115	Ethyl chloride	1051	117	Hydrocyanic acid, aqueous solutions, with more than 20% Hydrogen cyanide
1038	115	Ethylene, refrigerated liquid (cryogenic liquid)			
1039	115	Ethyl methyl ether	1051	117	Hydrocyanic acid, liquefied
1039	115	Methyl ethyl ether	1051	117	Hydrogen cyanide, anhydrous, stabilized
1040	119P	Ethylene oxide			
1040	119P	Ethylene oxide with Nitrogen	1051	117	Hydrogen cyanide, stabilized
1041	115	Carbon dioxide and Ethylene oxide mixture, with more than 9% but not more than 87% Ethylene oxide	1052	125	Hydrogen fluoride, anhydrous
			1053	117	Hydrogen sulfide
			1053	117	Hydrogen sulfide, liquefied
1041	115	Carbon dioxide and Ethylene oxide mixtures, with more than 6% Ethylene oxide	1053	117	Hydrogen sulphide
			1053	117	Hydrogen sulphide, liquefied
1041	115	Ethylene oxide and Carbon dioxide mixture, with more than 9% but not more than 87% Ethylene oxide	1055	115	Isobutylene
			1056	121	Krypton
			1056	121	Krypton, compressed
			1057	115	Lighter refills (cigarettes) (flammable gas)
1041	115	Ethylene oxide and Carbon dioxide mixtures, with more than 6 % Ethylene oxide	1057	115	Lighters (cigarettes) (flammable gas)
			1058	120	Liquefied gas (nonflammable)

Figure 3-8 The yellow pages are a numeric listing of the chemicals that the DOT regulates. This listing is by four-digit United Nations/North America identification number and also provides the shipping name and the emergency response guide page information.

ID No.	Guide No.	Name of Material
2474	157	Thiophosgene
2475	157	Vanadium trichloride
2477	131	Methyl isothiocyanate
2478	155	Isocyanate solution, flammable, poisonous, n.o.s.
2478	155	Isocyanate solution, flammable, toxic, n.o.s.
2478	155	Isocyanate solutions, n.o.s.
2478	155	Isocyanates, flammable, poisonous, n.o.s.
2478	155	Isocyanates, flammable, toxic, n.o.s.
2478	155	Isocyanates, n.o.s.
2480	155	Methyl isocyanate
2481	155	Ethyl isocyanate
2482	155	n-Propyl isocyanate
2483	155	Isopropyl isocyanate
2484	155	tert-Butyl isocyanate
2485	155	n-Butyl isocyanate
2486	155	Isobutyl isocyanate
2487	155	Phenyl isocyanate
2488	155	Cyclohexyl isocyanate
2490	153	Dichloroisopropyl ether
2491	153	Ethanolamine
2491	153	Ethanolamine, solution
2491	153	Monoethanolamine
2493	132	Hexamethyleneimine
2495	144	Iodine pentafluoride
2496	156	Propionic anhydride
2498	129	1,2,3,6-Tetrahydrobenzaldehyde
2501	152	1-Aziridinyl phosphine oxide (Tris)
2501	152	Tri-(1-aziridinyl)phosphine oxide, solution

Figure 3-9 If the listing in the yellow pages or the blue pages is highlighted, as is the listing of ID number 2481, ethyl isocyanate, then the responder must turn to the orange guide 155 and to the green section. The green section is the table of initial isolation and protective action distances.

■ PG I, II, or III—Packing group I, II, or III

■ O.R.M.—Other regulated material (listed as ORM A-E)

■ LC_{50}—Lethal concentration to 50 percent of the population being tested, such as rats

■ LD_{50}—Lethal dose to 50 percent of the population being tested

The blue section of the book mirrors the yellow section except that it is alphabetical by shipping name, as shown in **Figure 3-10.** These shipping names are assigned by DOT and for the most part are identical to the chemical name, although they may vary in some cases. Chemicals may go by several names, including chemical name, synonym, trade name, and shipping name.

SAFETY Some chemicals can have more than 60 synonyms, which can create a confusing situation.

Table 3-1 provides one example. This section lists the ID number, shipping name, and guide reference number. If the material is highlighted blue, then responders should refer to the guide page listed and to the table of initial isolation and protective action distances, located at the back of the book in the green section.

The middle of the book, the orange section, is the actual guide pages, as shown in **Figure 3-11.** They are listed from guide 111 to guide 172, a total of 61 guides for the more than 4,000 chemicals listed by DOT. This generalization makes this book more valuable to initial responders and less valuable to an advanced responder, such as a HAZMAT technician. A HAZMAT technician needs more specific information, which the ERG does not provide. Each guide is on two pages and is divided into three major sections: potential hazards, public safety, and emergency response. The guide may also provide some additional information about specific chemicals. It is important for the responder to read the entire guide and the table of initial isolation and evacuation distances, if referred to that section. The potential hazards section lists the predominant hazard on the top line. If fire is the major concern, then it will be listed on top; if health is the major concern, then it will be listed first. The information regarding the health effects provides the route of exposure and any major symptoms of exposure. It will detail other potential health and environmental concerns with fire products and runoff. The fire section is not as detailed as the health section but does provide some assistance in identifying potential hazards. It provides information related to the flammability of the product and will identify if the vapors will stay low to the ground or will rise in air. Depending on the product, it may also describe the material's ability to float on water. Products shipped at an elevated temperature will have a notation as will those materials that have the ability to polymerize. One of the problems with the ERG is that gasoline, which is highly flammable, is under the same guide that is provided for diesel fuel, which to the fire service is combustible. Although

Name of Material	Guide No.	ID No.
Allyl alcohol	131	1098
Allylamine	131	2334
Allyl bromide	131	1099
Allyl chloride	131	1100
Allyl chlorocarbonate	155	1722
Allyl chloroformate	155	1722
Allyl ethyl ether	131	2335
Allyl formate	131	2336
Allyl glycidyl ether	129	2219
Allyl iodide	132	1723
Allyl isothiocyanate, inhibited	155	1545
Allyl isothiocyanate, stabilized	155	1545
Allyltrichlorosilane, stabilized	155	1724
Aluminum, molten	169	9260
Aluminum alkyl halides	135	3052
Aluminum alkyl halides, liquid	135	3052
Aluminum alkyl halides, solid	135	3052
Aluminum alkyl halides, solid	135	3461
Aluminum alkyl hydrides	138	3076
Aluminum alkyls	135	3051
Aluminum borohydride	135	2870
Aluminum borohydride in devices	135	2870
Aluminum bromide, anhydrous	137	1725
Aluminum bromide, solution	154	2580
Aluminum carbide	138	1394
Aluminum chloride, anhydrous	137	1726
Aluminum chloride, solution	154	2581
Aluminum dross	138	3170
Aluminum ferrosilicon powder	139	1395
Aluminum hydride	138	2463
Aluminum nitrate	140	1438
Aluminum phosphide	139	1397

Name of Material	Guide No.	ID No.
Aluminum phosphide pesticide	157	3048
Aluminum powder, coated	170	1309
Aluminum powder, pyrophoric	135	1383
Aluminum powder, uncoated	138	1396
Aluminum processing by-products	138	3170
Aluminum remelting by-products	138	3170
Aluminum resinate	133	2715
Aluminum silicon powder, uncoated	138	1398
Aluminum smelting by-products	138	3170
Amines, flammable, corrosive, n.o.s.	132	2733
Amines, liquid, corrosive, flammable, n.o.s.	132	2734
Amines, liquid, corrosive, n.o.s.	153	2735
Amines, solid, corrosive, n.o.s.	154	3259
2-Amino-4-chlorophenol	151	2673
2-Amino-5-diethylaminopentane	153	2946
2-Amino-4,6-dinitrophenol, wetted with not less than 20% water	113	3317
2-(2-Aminoethoxy)ethanol	154	3055
N-Aminoethylpiperazine	153	2815
Aminophenols	152	2512
Aminopyridines	153	2671
Ammonia, anhydrous	125	1005
Ammonia, anhydrous, liquefied	125	1005
Ammonia, solution, with more than 10% but not more than 35% Ammonia	154	2672
Ammonia, solution, with more than 35% but not more than 50% Ammonia	125	2073
Ammonia solution, with more than 50% Ammonia	125	1005

Figure 3-10 The blue pages are alphabetical by shipping name. These listings also provide the UN/NA number and the emergency response guide page.

the DOT and the NFPA use 141°F (60.5°C) as the difference between flammable and combustible, some reference sources within the fire service still use 100°F (385°C) as the difference. There is a major difference between handling a gasoline spill and a diesel fuel spill under normal conditions. By using the guide as intended, the reader has no way of differentiating between the two products. Although

one could say that taking this route is safer, as it assumes the worst-case scenario, there are times when this action is inappropriate and can lead to other problems, such as an unnecessary evacuation for a diesel fuel spill.

The public safety section provides information for initial public protection options and key issues for the safety of the responders.

TABLE 3-1

Chemical Names

Many chemicals have several different names: chemical name, synonym, trade name, or manufacturer name. As an example, the chemical known by the DOT as chlordane liquid is also called the following names:

Aspon-Chlordane	Octachlorodihydrodicyclopentadiene
Belt	
CD 68	1,2,4,5,6,7,8,8-Octachloro-2,3,3a,4,7,7a-Hexahydro-4,7-Methanoindene
Chloordaan	1,2,4,5,6,7,8,8-Octachloro-3a,4,7,7a-Hexahydro-4,7-Methylene indane
Chlordan	Octachloro-4,7-Methanohydroindane
g-Chlordan	Octachloro-4,7-Methanotetrahydroindane
Chlorindan	1,2,4,5,6,7,8,8-Octachloro-4,7-Methano-3a,4,7,7a-Tetrahydroindane
Chlor kil	1,2,4,5,6,7,8,8-Octachloro-3a,4,7,7a-Tetrahydro-4,7-Methanoindane
Chlorodane	1,2,4,5,6,7,8,8-Octachloro-4,7,7a-Tetrahydro-4,7-Methanoindane
Chlortox	1,2,4,5,6,7,8,8-Octachlor-3a,4,7,7a-Tetrahydro-4,7-endo-Methano-Indan (German)
Clordan	Octa-Klor
Clorodane	Oktaterr
Cortilan-Neuq	Ortho-Klor
Dichlorochlordene	1,2,4,5,6,7,8,8-Ottochloro-3A,4,7,7A-Tetraidro-4,7-endo-Methno-indano (Italian)
Dowchlor	RCRA Waste Number U036
ENT 9,932	SD 5532
ENT 25,552-X	Shell SD-5532
HCS 3260	Synklor
Kypchlor	Tat Chlor 4
M 140	
NCI-C00099	Topiclor
Niran	
1,2,4,5,6,7,8,8-Octachloor-3a,4,7,7a-Tetrathydro-4,7-endo-Methano-Indaan (Dutch)	Toxichlor
1,2,4,5,6,7,10,10-Octachloro-4,7,8,9-Tetrahydro-4,7-Methyleneindane	Velsicol 1068

Source: Excerpted from *Dangerous Properties of Industrial Materials*, 10th ed.

NOTE Only the public safety section (orange pages) of the ERG provides recommendations as to how far to isolate the scene and some initial strategic goals for the first arriving company.

This section and the table of initial isolation and protective action distances, which is shown in **Figure 3-12,** make this book a necessary tool in the mitigation of a chemical release. The green section provides detailed isolation and evacuation information.

Figure 3-11 The response guide in the orange section provides the responder with basic information regarding health, fire, and public safety issues. Each part needs to be read completely before taking any action.

This information relates to extremely hazardous substances, ones that can present a significant risk to the community. The orange section lists isolation distances for the emergency responders, and the green section is for the community. It also recommends that responders contact the emergency contact listed on the shipping papers or, if this is not available, one of the contacts listed on the inside back cover. It provides some tactical objectives for handling radioactive substances and discusses additional considerations regarding these types of incidents.

The public safety section also lists personal protective equipment (PPE) recommendations and provides four basic PPE scenarios. It promotes the use of positive pressure self-contained breathing apparatus (SCBA), which is always a good recommendation when responding to chemical releases. The next level in severity recommends that chemical protective equipment is required and that turnout gear provides limited protection. In some cases the ERG states that turnout gear provides limited protection, without any recommendation for other PPE. We will discuss PPE in Chapter 4, and many of these same issues will be

covered in that section. A final recommendation is that in some cases chemical protective equipment is suggested, but turnout gear is acceptable for fire situations.

This section provides information regarding initial evacuation distances for large spills and fires. If the material is listed in the table of initial isolation and protective actions distances, the reader will be referred to this section located at the back of the book in the green section. The isolation distances listed in the DOT book vary from a minimum of 30 feet (9 meters) to a maximum of 800 feet (243 meters). The minimums are derived for materials that are solid or present little risk to the responder. The higher the toxicity or the risk, the greater the isolation distance. The average distance is 330 feet (101 meters), which is what should be used for the establishment of an isolation zone for an unknown material. That distance in an urban setting can present some significant problems. As more information becomes available, this isolation distance can increase or decrease as needed. When dealing with explosives, the minimum distance should be 800 feet (243 meters) for the initial isolation distance. The isolation distances for

TABLE OF INITIAL ISOLATION AND PROTECTIVE ACTION DISTANCES

ID No.	NAME OF MATERIAL	SMALL SPILLS (From a small package or small leak from a large package)						LARGE SPILLS (From a large package or from many small packages)					
		First ISOLATE in all Directions		Then PROTECT persons Downwind during-				First ISOLATE in all Directions		Then PROTECT persons Downwind during-			
		Meters	(Feet)	DAY Kilometers (Miles)		NIGHT Kilometers (Miles)		Meters	(Feet)	DAY Kilometers (Miles)		NIGHT Kilometers (Miles)	
2204 2204	Carbonyl sulfide Carbonyl sulphide	30 m	(100 ft)	0.1 km	(0.1 mi)	0.6 km	(0.4 mi)	300 m	(1000 ft)	3.0 km	(1.9 mi)	8.1 km	(5.0 mi)
2232 2232	Chloroacetaldehyde 2-Chloroethanal	30 m	(100 ft)	0.2 km	(0.1 mi)	0.3 km	(0.2 mi)	90 m	(300 ft)	0.8 km	(0.5 mi)	1.6 km	(1.0 mi)
2334	Allylamine	30 m	(100 ft)	0.1 km	(0.1 mi)	0.5 km	(0.3 mi)	120 m	(400 ft)	1.1 km	(0.7 mi)	2.5 km	(1.5 mi)
2337	Phenyl mercaptan	30 m	(100 ft)	0.1 km	(0.1 mi)	0.1 km	(0.1 mi)	60 m	(200 ft)	0.4 km	(0.2 mi)	0.6 km	(0.4 mi)
2382 2382	1,2-Dimethylhydrazine Dimethylhydrazine, symmetrical	30 m	(100 ft)	0.1 km	(0.1 mi)	0.2 km	(0.1 mi)	60 m	(200 ft)	0.6 km	(0.4 mi)	1.2 km	(0.8 mi)
2407	Isopropyl chloroformate	30 m	(100 ft)	0.1 km	(0.1 mi)	0.3 km	(0.2 mi)	90 m	(300 ft)	0.7 km	(0.5 mi)	1.5 km	(0.9 mi)
2417 2417	Carbonyl fluoride Carbonyl fluoride, compressed	30 m	(100 ft)	0.2 km	(0.1 mi)	1.1 km	(0.7 mi)	90 m	(300 ft)	1.0 km	(0.6 mi)	3.6 km	(2.3 mi)
2418 2418	Sulfur tetrafluoride Sulphur tetrafluoride	60 m	(200 ft)	0.7 km	(0.4 mi)	3.2 km	(2.0 mi)	500 m	(1600 ft)	4.7 km	(2.9 mi)	10.6 km	(6.6 mi)
2420	Hexafluoroacetone	30 m	(100 ft)	0.3 km	(0.2 mi)	1.3 km	(0.8 mi)	800 m	(2500 ft)	7.2 km	(4.5 mi)	11.0+ km	(7.0+ mi)
2421	Nitrogen trioxide	30 m	(100 ft)	0.1 km	(0.1 mi)	0.5 km	(0.3 mi)	60 m	(200 ft)	0.4 km	(0.3 mi)	1.9 km	(1.2 mi)
2437	Methylphenyldichlorosilane (when spilled in water)	30 m	(100 ft)	0.1 km	(0.1 mi)	0.1 km	(0.1 mi)	30 m	(100 ft)	0.3 km	(0.2 mi)	1.1 km	(0.7 mi)
2438	Trimethylacetyl chloride	30 m	(100 ft)	0.1 km	(0.1 mi)	0.2 km	(0.1 mi)	60 m	(200 ft)	0.5 km	(0.3 mi)	0.8 km	(0.5 mi)
2442	Trichloroacetyl chloride	30 m	(100 ft)	0.2 km	(0.2 mi)	0.8 km	(0.5 mi)	120 m	(400 ft)	1.2 km	(0.8 mi)	2.2 km	(1.4 mi)
2474	Thiophosgene	90 m	(300 ft)	0.8 km	(0.5 mi)	2.4 km	(1.5 mi)	360 m	(1200 ft)	3.6 km	(2.3 mi)	6.8 km	(4.2 mi)
2477	Methyl isothiocyanate	30 m	(100 ft)	0.1 km	(0.1 mi)	0.2 km	(0.1 mi)	60 m	(200 ft)	0.5 km	(0.3 mi)	1.0 km	(0.7 mi)
2480	Methyl isocyanate	60 m	(200 ft)	0.5 km	(0.3 mi)	1.9 km	(1.2 mi)	600 m	(1800 ft)	5.4 km	(3.3 mi)	11.0+ km	(7.0+ mi)

Figure 3-12 The table of initial isolation and protective action distances provides recommended isolation distances for the materials that were highlighted in the yellow or blue sections. The table is divided into small spills and large spills.

downwind are expanded from the initial isolation distances and vary from 0.1 mile to 7 miles (0.16–11 km) in the green section.

The emergency response section provides information regarding fires, spills, and first aid. The fire section is divided between small fires and large fires and suggests potential tactics that can be used for both. For both types of fires, the guide lists what type of extinguishing agent may be needed: water, foam, dry chemical, Halon, or CO_2. Specific materials may also list necessary special agents. Listed under the large fire section are some tactical considerations and recommendations for specific substances. If the material is carried by tank truck or railcar, the ERG also provides some additional information on fighting those types of fires. This section provides some helpful hints that apply in many cases involving fires and transport containers:

■ Fight fire from a distance using unstaffed monitors.

■ Withdraw immediately in case of rising sound from venting safety device or if the tank begins to discolor.

■ Cool containers with flooding quantities of water until well after the fire is out.

The spill or leak section lists some general tactical objectives and also provides specific information on some substances. This section may lead responders into an action that may be beyond their level of training if not used correctly. As an example, most of the guides recommend that readers stop a leak if they can do it without risk. The intention is that a responder trained to the operations level could close a remote shutoff valve. Instead, responders might incorrectly follow the recommendation and try and stop the leak no matter the method, a tactic that requires a HAZMAT technician to accomplish.

Another area of concern is the suggestion to use water spray to knock down vapors. If necessary to save lives, using a water spray is an acceptable tactic.

Figure 3-13 A large spill of ethyl isocyanate (ID 2481) is listed as having an isolation distance of 3,000 feet in all directions with protection for persons downwind for 7 miles.

SAFETY The indiscriminate use of water to knock down "vapors" can create many other problems, such as spreading the chemical, causing a violent reaction, creating runoff issues, as well as increasing the size of the incident.

These problems include increasing the spill size, creating runoff problems, setting off possible reactions, and posing an additional hazard to other responders working to mitigate the leak. If necessary, as in the case to knock down anhydrous ammonia vapors that may affect a neighborhood, it is an acceptable tactic. Flowing water on a leaking container that holds liquified compressed gases, or a cryogenic material could create an ice formation that could prevent the sealing of the leak. Putting water on a material that is violently water reactive would not be a recommended practice.

The first aid section provides basic medical treatment and other information and offers basic decontamination recommendations for chemical burns. The information is elementary; if confronted with a patient, the responder should contact a local hospital or Chemtrec for further assistance.

The table of initial isolation and protective action distances is the green section in the back of the book,

as shown in **Figure 3-12.** It provides specific isolation and evacuation distances for the materials highlighted in the yellow or blue section. An example of how to use these isolation distances is shown in **Figure 3-13** using ethyl isocyanate (ID 2481). This section is further subdivided between small spills and large spills, and both are divided between day and night distances. The criteria used to establish these distances include the following information: the DOT incident database (HMIS), typical package size, typical flow rate from a ruptured package, and the release rate of vapors from a spill. The DOT chose the average day as being warm and sunny with a temperature of 95°F (35°C). The agency conducted a 5-year study of the weather in sixty-one cities to establish a pattern. It was during this study that it was determined that materials will travel farther at night than during the day. The DOT also chose to use an evacuation distance that in 90 percent of incidents would be too large. In this scenario, in 10 percent of the incidents the distance would be too small. The distances are a guide for the first 30 minutes of an incident and use a typical day as defined by the DOT. The ERG is a guide to get an evacuation started; as soon as possible, air monitoring must be established to definitively establish evacuation distances. In addition, an emergency response software package

TABLE OF WATER-REACTIVE MATERIALS WHICH PRODUCE TOXIC GASES

Materials Which Produce Large Amounts of Toxic-by-Inhalation (TIH) Gas(es) When Spilled in Water

ID No.	Guide No.	Name of Material	TIH Gas(es) Produced	
1716	156	Acetyl bromide	HBr	
1717	155	Acetyl chloride	HCl	
1724	155	Allyltrichlorosilane, stabilized	HCl	
1725	137	Aluminum bromide, anhydrous	HBr	
1726	137	Aluminum chloride, anhydrous	HCl	
1728	155	Amyltrichlorosilane	HCl	
1732	157	Antimony pentafluoride	HF	
1745	144	Bromine pentafluoride	HF	Br$_2$
1746	144	Bromine trifluoride	HF	Br$_2$
1747	155	Butyltrichlorosilane	HCl	
1752	156	Chloroacetyl chloride	HCl	
1754	137	Chlorosulfonic acid	HCl	
1754	137	Chlorosulfonic acid and Sulfur trioxide mixture	HCl	
1754	137	Chlorosulphonic acid	HCl	
1754	137	Chlorosulphonic acid and Sulphur trioxide mixture	HCl	
1754	137	Sulfur trioxide and Chlorosulfonic acid	HCl	
1754	137	Sulphur trioxide and Chlorosulphonic acid	HCl	
1758	137	Chromium oxychloride	HCl	
1763	156	Cyclohexyltrichlorosilane	HCl	
1766	156	Dichlorophenyltrichlorosilane	HCl	
1767	155	Diethyldichlorosilane	HCl	
1769	156	Diphenyldichlorosilane	HCl	
1771	156	Dodecyltrichlorosilane	HCl	
1777	137	Fluorosulfonic acid	HF	

Chemical Symbols for TIH Gases:

Br$_2$	Bromine	HF	Hydrogen fluoride	PH$_3$	Phosphine
Cl$_2$	Chlorine	HI	Hydrogen iodide	SO$_2$	Sulfur dioxide
HBr	Hydrogen bromide	H$_2$S	Hydrogen sulfide	SO$_2$	Sulphur dioxide
HCl	Hydrogen chloride	H$_2$S	Hydrogen sulphide	SO$_3$	Sulfur trioxide
HCN	Hydrogen cyanide	NH$_3$	Ammonia	SO$_3$	Sulphur trioxide

Figure 3-14 Water-reactive listings found in the back of the ERG.

known as CAMEO uses a vapor cloud modeling program known as ALOHA. This modeling program is generally referred to as a plume dispersion model and can assist with evaluating potential downwind evacuations, using realtime weather and data.

The DOT defines a small spill as a leaking container smaller than a 55-gallon (208-liter) drum or a leak from a small cylinder, while a large spill is coming from a container larger than 55 gallons (208 liters) or a large

leak from a cylinder. Leaks from several small containers may be considered a large leak. A small leak from a large container may also be considered a small leak.

The last section included in the green section, which is shown in **Figure 3-14,** is the list of dangerous water-reactive materials. This section provides the evacuation distances for these materials if they contact water; the distance varies from 0.3 mile (0.5 km) to 6 miles (9.7 km). One helpful item that this section provides is the chemical that a material forms when contacting water, something that is unusual for this type of text.

The last pages are definitions, glossary, and explanations. The back section also has several pages on the Use of Chemical/Biological Agents in terrorist or criminal events. The inside back cover provides a listing of additional emergency contacts for the United States, Canada, and Mexico. Further discussion regarding these agencies is found later in this chapter.

Some general reminders when using the DOT ERG:

- It is an excellent source for evacuation and isolation distances.
- It is an excellent source for water-reactive hazards.
- It is an excellent first responder document, one that will keep responders safe.
- It is a great starting point for the first thirty minutes of the incident.
- The specific chemical information is limited, and in some cases too general for a HAZMAT team, although it is well suited for the first responder level.
- Specific identification of the product is essential for the safety of responders and citizens.

MATERIAL SAFETY DATA SHEETS

Material Safety Data Sheets (MSDSs) are a result of the hazard communication standard (HCS), which is OSHA regulation 29 CFR 1910.1200, effective 1994. As we noted in Chapter 1, this regulation requires employers who use chemicals above the household quantity to have MSDSs.

Employers are also required to have a hazard communication plan, label all chemical containers, and provide training to the employees on an annual basis. This training is based upon the chemical hazards the employees will face in the workplace. MSDSs are required to provide a variety of information. The amount, quality, and order of information are determined by the manufacturer. The original intent of MSDSs was the protection of the employees working at the facility, not emergency responders. Although many of the sheets do have applicability and serve a useful purpose, in some cases they may not have the necessary emergency response information. The information that may be found on an MSDS is presented in **Table 3-2.**

The quality of information varies from MSDS to MSDS and from manufacturer to manufacturer. The early MSDSs had a lot of good information, but as litigation has increased, the typical MSDS now provides a worst-case scenario and recommends an extremely conservative approach to handling the chemical. Given the choice of using the DOT ERG or a MSDS, responders should rely more on the technical information on the MSDS, but should never rely on just one source of information. HAZMAT responders will usually follow a rule of three with regard to information sources. They will check the information and compare it with at least two other sources of information, and in many cases the MSDS will be one of the sources. In June 1999 the EPA issued an alert with regard to the quality of information that MSDSs contained. In the accompanying text box (pages 104 and 105) we have provided most of that bulletin. The DOT ERG places many chemicals into a few general categories; although it may be conservative, the MSDS is specific to the chemical. It is important through the preplanning process that responders review the chemicals at a facility with a representative from

CHEMICAL ABSTRACT SERVICE REGISTRY NUMBERS

With all of the synonyms and trade names, it can be very confusing trying to identify a chemical. One method is through the use of the Chemical Abstract Service (CAS) registry number, which specifically identifies a chemical. The CAS number is assigned to a chemical compound; no matter the name of the material, if it is the same compound as the original chemical, then it is assigned the same CAS number. Using chlordane as the example, even though we have listed forty-eight different names for the compound chlordane in **Table 3-1,** they all have the same CAS number.

The majority of chemical reference sources use the CAS number, and in some cases the CAS number is used to sort MSDSs. The CAS number is a series of three numbers separated by two dashes, such as xxx-xx-xxx. Early numbers indicate that the substance was provided a number many years ago, and the numbers issued today are quite lengthy. When looking for chemical information, the CAS number is very important, as it can be used as a cross-reference source. Shipping papers and the DOT ERG do not use the CAS number, and the most common location for a CAS number is the MSDS or occasionally the label.

TABLE 3-2

Elements of a MSDS	
Information	**Note**
Chemical product and company identification	The name the chemical company uses to identify the product is near the top of the MSDS. Information related to the manufacturer, such as the address and other contact information, is provided near the top of the first sheet. The MSDS sheet is required to list a 24-hour emergency contact number, although in most cases this number is Chemtrec.
Chemical composition	The ingredients of the chemical are listed here. If the mixture is a trade secret, there is a provision to exclude this information. If the information is needed for medical reasons, the manufacturer is to provide this information to a physician. In some cases the true identity of the material may never be known, and only the hazards may be provided.
Hazards identification	In this section an emergency overview is provided, for both an employer and emergency responders. This section is lengthy and includes potential health effects, first aid, and firefighting measures. The exposure levels are typically included with the health effects, but may be listed separately. New to MSDS will be a section on accidental releases, which will be beneficial to emergency responders. Handling and storage considerations and engineering controls, including proper personal protective equipment, are also provided in this section.
Physical and chemical properties	Specific chemical information, such as boiling points, vapor pressures, and flash points, are included in this section.
Stability and reactivity	Some chemicals become unstable after a period of time or storage conditions, or may react with other chemicals. Any information regarding this type of information is provided here.
Toxicology information	The long-term health effects and other concerns regarding acute and chronic exposures are provided in this section.
Ecological information	Information regarding any potential environmental effects are listed here.
Disposal considerations	Although in many cases this section states "follow local regulations," this is the section that outlines any regulatory requirements for disposal.
Transport information	Information regarding the DOT regulations is listed here; typically the hazard class and UN identification number are provided in this section.
Regulatory information	This information is listed in this section.

the facility. Working through the MSDS and the ERG can help determine prior to the incident which information is most useful to responders. Although many attempts have been made to modify the format of MSDSs and to improve the quality of information, the basic format has remained the same since its inception. There are a couple of recommended formats, and a new format exists for the European chemical industry. It can be anticipated that the United States will need to conform to this format. The information provided in **Table 3-2** is from this format. One of the improvements with this new MSDS format is a section specifically on emergency response procedures and considerations for responders. A sample MSDS is provided in **Figure 3-15.**

GASES AND EQUIPMENT GROUP
MATERIAL SAFETY
DATA SHEET

SECTION 1: PRODUCT IDENTIFICATION

PRODUCT NAME: Argon, compressed
CHEMICAL NAME: Argon   **FORMULA:** Ar
SYNONYMS: Argon gas, Gaseous argon, GAR
MANUFACTURER: Air Products and Chemicals, Inc.
7201 Hamilton Boulevard
Allentown, PA 18195 - 1501
PRODUCT INFORMATION: 1 - 800 - 752 - 1597
MSDS NUMBER: 1004   **EFFECTIVE DATE:** August 1997   **REVISION:** 5

SECTION 2: COMPOSITION - INFORMATION ON INGREDIENTS

Argon is sold as pure product > 99%.
CAS NUMBER: 7440-37-1
EXPOSURE LIMITS:
   OSHA: Not established.
   ACGIH: Simple asphyxiant
   NIOSH: Not established.

SECTION 3: HAZARDS IDENTIFICATION

EMERGENCY OVERVIEW

Argon is a nontoxic, odorless, colorless, nonflammable gas stored in cylinders at high pressure. It can cause rapid suffocation when concentrations are sufficient to reduce oxygen levels below 19.5%. It is heavier than air and may concentrate in low areas. Self Contained Breathing Apparatus (SCBA) may be required.

EMERGENCY TELEPHONE NUMBERS
800 - 523 - 9374 in Continental U.S., Canada and Puerto Rico
610 - 481 - 7711 outside U.S.

POTENTIAL HEALTH EFFECTS INFORMATION
   INHALATION: Simple asphyxiant. Argon is nontoxic, but may cause suffocation by displacing the oxygen in air. Lack of sufficient oxygen can cause serious injury or death.
   EYE CONTACT: No adverse effect.
   SKIN CONTACT: No adverse effect.
   CARCINOGENIC POTENTIAL: Argon is not listed as a carcinogen or potential carcinogen by NTP, IARC, or OSHA Subpart Z.
EXPOSURE INFORMATION
   ROUTE OF ENTRY: Inhalation

TARGET ORGANS: None
EFFECT: Asphyxiation (suffocation)
MEDICAL CONDITIONS AGGRAVATED BY OVEREXPOSURE: None
   SYMPTOMS: Exposure to an oxygen deficient atmosphere (<19.5%) may cause dizziness, drowsiness, nausea, vomiting, excess salivation, diminished mental alertness, loss of consciousness and death. Exposure to atmospheres containing 8-10% or less oxygen will bring about unconsciousness without warning and so quickly that the individuals cannot help themselves.

SECTION 4: FIRST AID

INHALATION: Persons suffering from lack of oxygen should be moved to fresh air. If victim is not breathing, administer artificial respiration. If breathing is difficult, administer oxygen. Obtain prompt medical attention.
EYE CONTACT: Not applicable.
SKIN CONTACT: Not applicable.

SECTION 5: FIRE AND EXPLOSION

FLASH POINT	AUTOIGNITION TEMP	FLAMMABLE LIMIT
N/A	Nonflammable	Nonflammable

EXTINGUISHING MEDIA: Argon is nonflammable and does not support combustion. Use extinguishing media appropriate for the surrounding fire.
HAZARDOUS COMBUSTION PRODUCTS: None
SPECIAL FIRE FIGHTING INSTRUCTIONS: Argon is a simple asphyxiant. If possible, remove argon cylinders from fire area or cool with water. Self contained breathing apparatus may be required for rescue workers.
UNUSUAL FIRE AND EXPLOSION HAZARDS: Upon exposure to intense heat or flame cylinder will vent rapidly and or rupture violently. Most cylinders are designed to vent contents when exposed to elevated temperatures. Pressure in a container can build up due to heat and it may rupture if pressure relief devices should fail to function.

SECTION 6: ACCIDENTAL RELEASE MEASURES

Evacuate all personnel from affected area. Increase ventilation to release area and monitor oxygen level. Use appropriate protective equipment (SCBA). If leak is from container or its valve, call the Air Products emergency telephone number. If leak is in user's system close cylinder valve and vent pressure before attempting repairs.

SECTION 7: HANDLING AND STORAGE

STORAGE: Cylinders should be stored upright in a well-ventilated, secure area, protected from the weather. Storage area temperatures should not exceed 125° F (52° C) and area should be free of combustible materials. Storage should be away from heavily traveled areas and emergency exits. Avoid areas where salt or other corrosive materials are present. Valve protection caps and valve outlet seals should remain on cylinders not connected for use. Separate full from empty cylinders. Avoid excessive inventory and storage time. Use a first-in-first-out system. Keep good inventory records.
HANDLING: Do not drag, roll, or slide cylinder. Use a suitable handtruck designed for cylinder movement. Never attempt to lift a cylinder by its cap. Secure cylinders at all times while in use. Use a pressure reducing regulator or separate control valve to safely discharge gas from cylinder. Use a check valve to prevent reverse flow into cylinder. Do not overheat cylinder to increase pressure or discharge rate. If user experiences any difficulty operating cylinder valve, discontinue use and contact supplier. Never insert an object (e.g., wrench, screwdriver, pry bar, etc.) into valve cap openings. Doing so may damage valve causing a leak to occur. Use an adjustable strap-wrench to remove over-tight or rusted caps.
Argon is compatible with all common materials of construction. Pressure requirements must be considered when selecting materials and designing systems.
SPECIAL REQUIREMENTS: Always store and handle compressed gases in accordance with Compressed Gas Association, Inc. (ph.703-412-0900) pamphlet CGA P-1, *Safe Handling of Compressed Gases in Containers*. Local regulations may require specific equipment for storage or use.

CAUTION: Compressed gas cylinders shall not be refilled except by qualified producers of compressed gases. Shipment of a compressed gas cylinder which has not been filled by the owner or with the owner's written consent is a violation of federal law (49 CFR 173.301).

SECTION 8: PERSONAL PROTECTION / EXPOSURE CONTROL

ENGINEERING CONTROLS: Provide good ventilation and/or local exhaust to prevent accumulation of high concentrations of gas. Oxygen levels in work area should be monitored to ensure they do not fall below 19.5%.
RESPIRATORY PROTECTION
   GENERAL USE: None required.
   EMERGENCY: Use SCBA or positive pressure air line with mask and escape pack in areas where oxygen concentration is < 19.5%. Air purifying respirators will not provide protection.
OTHER PROTECTIVE EQUIPMENT: Safety shoes and leather work gloves are recommended when handling cylinders.

SECTION 9: PHYSICAL AND CHEMICAL PROPERTIES

APPEARANCE: Colorless gas.   **ODOR:** Odorless .
MOLECULAR WEIGHT: 39.95   **BOILING POINT:** -302.2°F (-185.9°C)
SPECIFIC GRAVITY (Air =1): At 70°F (21.1°C) and 1 Atm: 1.38   **SPECIFIC VOLUME:** 9.7 ft³/lb. (0.606 m³/kg)
FREEZING POINT/MELTING POINT: -308.9°F (-189.4°C)   **VAPOR PRESSURE:** Not applicable @70°F
GAS DENSITY: At 70°F (21.1°C) and 1 Atm: 0.103 lbs/ft³ (1.65 kg/m³)
SOLUBILITY IN WATER: Vol/Vol at 32 °F (0°C): 0.056

SECTION 10: STABILITY AND REACTIVITY

CHEMICAL STABILITY: Stable
CONDITIONS TO AVOID: None
INCOMPATIBILITY: None
HAZARDOUS DECOMPOSITION PRODUCTS: None
HAZARDOUS POLYMERIZATION: Will not occur.

SECTION 11: TOXICOLOGICAL INFORMATION

Argon is a simple asphyxiant.

SECTION 12: ECOLOGICAL INFORMATION

The atmosphere contains approximately 1% argon. No adverse ecological effects are expected. Argon does not contain any Class I or Class II ozone depleting chemicals. Argon is not listed as a marine pollutant by DOT (49 CFR 171).

SECTION 13: DISPOSAL

UNUSED PRODUCT / EMPTY CONTAINER: Return cylinder and unused product to supplier with the cylinder valve tightly closed and the valve caps in place. Do not attempt to dispose of residual or unused quantities.
DISPOSAL: For emergency disposal, secure the cylinder and slowly discharge gas to the atmosphere in a well ventilated area or outdoors.

Figure 3-15 Material Safety Data Sheet.

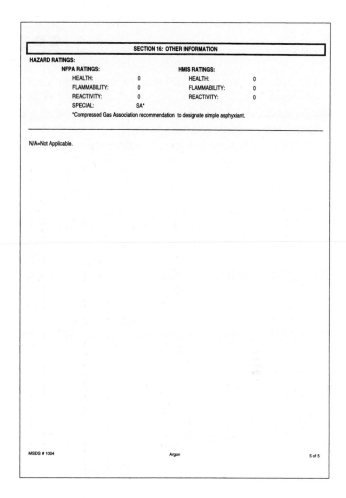

Figure 3-15 Material Safety Data Sheet. *(continued)*

SHIPPING PAPERS

In addition to the use of placards when chemicals are transported, the carrier is required to provide shipping papers for the cargo. The shipping papers in general provide the following information:

▪ Shipping company

▪ Destination of the packages

▪ Emergency contact information

▪ The number and weight of the packages

▪ The proper shipping name of the material and the hazard class of the materials

▪ Special notations for hazardous materials

If the vehicle is not carrying hazardous materials, there is no requirement for any specific information, which at times can be confusing for first responders. The shipping papers that accompany a vehicle carrying hazardous materials may include a packing group number (PG) listed as I, II, or III. The lower the number, the more dangerous the chemical is; usually some special requirements accompany the shipping of the materials. Other information may include a reportable quantity (RQ) for a hazardous material.

Some chemicals have a threshold of reporting when spilled, much like in storage situations and SARA section 304.

The RQ will be listed on the shipping papers, and if the material is spilled and exceeds the quantity listed, the driver must report the spill to the **National Response Center (NRC).** The driver/operator of the vehicle is supposed to keep the shipping papers with the vehicle at all times and should be able to provide them upon request. Problems occur in accidents where there are multiple shipping papers from a variety of pickups. If the driver has kept all of the previous shipping papers from other days, it can be confusing sorting out what the actual cargo is. If the driver provides information regarding the cargo and the shipping papers agree with that information, responders should confirm the results with the shipping company via Chemtrec. If all three are in agreement as to what is being shipped, it is probably accurate, but this is not guaranteed. A driver who picked up some illegal waste will not put it on a shipping paper or placard the vehicle; in fact, the driver probably will not admit to having it. It is obvious that the shipping company will not know about it, so until visually confirmed with appropriate

EPA ALERT

Responders should always use more than one source of information. The EPA issued a safety alert in June 1999 on the use of MSDSs, and the majority of that bulletin is reproduced here.

PROBLEM: A critical consideration when choosing a response strategy is the safety of emergency responders. Adequate information about on-site chemicals can make a big difference when choosing a safe response strategy. This information must include: name, toxicity, physical and chemical characteristics, fire and reactivity hazards, emergency response procedures, spill control, and protective equipment. Generally, responders rely primarily on MSDSs maintained at the facility. However, MSDSs may not provide sufficient information to effectively and safely respond to accidental releases. This *Alert* is designed to increase awareness of MSDS limitations, so that first responders can take proper precautions, and identify additional sources of chemical information, which could help prevent death or injury.

Accidents and How MSDSs Relate

In May 1997, a massive explosion and fire occurred at an agricultural chemical packaging facility in eastern Arkansas. Prior to the explosion, employees observed smoke in a back warehouse and evacuated. The facility called local responders and asked for help to control smoldering inside a pesticide container. The local fire department rapidly responded and reviewed the smoldering product's MSDS. The MSDS lacked information on decomposition temperatures or explosion hazards. The firefighters decided to investigate the building. While they were approaching, a violent explosion occurred. Fragments from a collapsing cinder block wall killed three firefighters and seriously injured a fourth. In April 1995, an explosion and fire at a manufacturing facility in Lodi, New Jersey, caused the death of five responders. The explosion occurred while the company was blending aluminum powder, sodium hydrosulfite, and other ingredients. Even though the material was water reactive, the MSDS for the product advised the use of a "water spray . . . to extinguish fire." The recommendation in the MSDS for "small fires" was to flood with water; however, "small fire" was not defined, the amount of water necessary was not specified, and

no information dealt with how to respond to large fires (which can occur during blending processes).

The MSDS only described the hazards associated with the product. In this case, responders needed information on the hazards associated with the reactivity during the blending process (which was significantly different from the product). Emergency responders should note that the chemical information provided on an MSDS usually presents the hazards associated with that particular product. Once the product is placed in a process some factors may change, resulting in the increase/decrease, or elimination of hazards. These factors may include reactions with other chemicals and changes in temperature, pressure, and physical/chemical characteristics.

MSDSs in the Workplace

In 1988, the Occupational Safety and Health Administration (OSHA) required facilities storing or using hazardous chemicals to comply with the Hazard Communication Standard. This standard requires employers to provide employees with an MSDS for every hazardous chemical present on site, and to train those employees to properly recognize the hazards of the chemicals and to handle them safely. MSDSs normally provide information on the physical/chemical characteristics and first aid procedures. This information is valuable for employees to safely work with the chemical. However, the content for MSDSs on emergency response procedures, fire, and reactive hazards may be insufficient for local responder use in an emergency situation. Vagueness, technical jargon, understandability, product vs. process concerns, and missing information on an MSDS may increase the risk to emergency responders. MSDSs are provided by manufacturers, importers, and/or distributors. MSDS chemical hazard information can vary substantially depending on the provider. Sometimes this discrepancy is due to different testing procedures. However, whoever prepared the MSDS is responsible for assuring the accuracy of the hazard information. The following chart summarizes information from various MSDSs for the chemical *azinphos methyl,* and it illustrates how different sources can provide varied and conflicting information.

(continued)

EPA ALERT *(continued)*

Comparison of MSDS Data for Azinphos Methyl (AZM)

	MSDS–A	MSDS–B	MSDS–C	MSDS–D	CAMEO
Hazard Rating	Health—2 Flammable—0 Reactivity—0	None listed.	Health—3 Flammable—2 Reactivity—2	Health—4 Flammable—0 Reactivity—0	Health—3 (extremely hazardous) Flammable—2 (ignites when moderately heated) Reactivity—2 (violent chemical change possible)
Reactivity Hazards	Stable under normal conditions. Hazardous. Polymerization will not occur.	Depends on characteristics of dust, decomposes under influence of acids and bases.	Stable material. Unstable above 100°F sustained temperature. Hazardous. Polymerization will not occur.	Releases toxic, corrosive, flammable, or explosive gases. Polymerization will not occur.	Will decompose.
Incompatibility	High temperatures, oxidizers, alkaline substances.	Acids and bases.	Heat, moisture.	Heat, flames, sparks, and other ignition sources.	Heat, UV light.
Fire Hazards	Vapors from fire are hazardous.	Combustible, gives off irritating or toxic fumes (or gases) in a fire.	Decomposes above 130°F with gas evolution and dense smoke. Dust explosion hazard for large dust cloud.	Containers may rupture or explode if exposed to heat.	Decomposes giving off ammonia, hydrogen, and CO.

levels of PPE, responders can never be certain about a cargo's contents.

Mode of Transportation

In highway transportation the shipping papers are called a *bill of lading,* or most commonly just shipping papers. The papers are supposed to be with the driver within arms reach, most commonly in a pouch on the drivers door. On tank trucks there are occasions when a duplicate set of papers is in a tube attached near the landing gear on the driver's side. In most cases the driver will leave the papers in the truck, and they usually need to be retrieved. If drivers had many pickup stops, they will have multiple papers, and some companies put each individual item on a ticket. The hazardous materials are sometimes color coded, as in the case of United Parcel Service (UPS), which uses red tabs to identify HAZMAT packages.

For rail the shipping papers are called the **consist** or **waybill** and are in the control of the engineer or the conductor, if the train has one. There may be two sets of papers, one listing the contents of each car (waybill) and one by car number (consist), starting with the first car past the engine as number 1 or in some cases with the last car as number 1. In a derailment this numbering system usually goes out the window, but in some cases it may be useful. Most rail cars are well identified with car numbers and, as long as upright, can usually be identified. Rail uses a number known as a **Standard Transportation Commodity Code (STCC),** usually referred to as a "stick" number. It is a seven-digit number, and if it starts with 49, the material is a hazardous material. Out of all the modes of transportation, in rail incidents the engineer is least likely to surrender the papers and will want to accompany the papers anywhere they go.

The regulations governing rail transport require the railroad to maintain procession of the papers. A system called *Operation Respond* is a computer program that was originally designed to track rail shipments of hazardous materials. The system now tracks both freight and passenger trains, as well as some trucking company transports. The system has grown to include all the major railroads and some large trucking firms and would be valuable if rail traffic is in your area.

On a ship the papers are called the **Dangerous Cargo Manifest (DCM).** These papers are in the control of the ship's captain or the master who is second in command of the ship. No matter what the location of the ship and the crew complement, there is always someone in charge of the ship. There may be a considerable crew on board a ship, but the crew may only have one person who speaks English. The most difficult time to get information regarding a cargo is when the ship is loading and unloading, especially for container ships. Although they may have the papers on board for all the containers, it can be some time before crew members actually learn where each container is and what the actual cargo is. One of the major problems is learning what is in each individual container, and this information may be limited or not available at all. Luckily each container is well marked with identification information, and after some time the source can be traced to get information.

NOTE One of the biggest hazards from ships comes from the fuel on board. A cruise ship can have up to a million gallons of fuel on board. When dealing with ships, special training is required, as they are extremely hazardous to operate on, and responders can easily get lost and/or fall into deep areas of the ship. Another little known fact is that many ships carrying freight can also take a small number of passengers. There is a separate passenger and crew area, and the maximum number of passengers, excluding the crew, is twenty.

In air the shipping papers are called **Air Bills** and are in the control of the pilot, usually stored in the cockpit. The type of aircraft (passenger or cargo) will dictate the type and amount of materials that can be sent via the air. The most common aircraft involved with chemical spills are cargo aircraft, and in many

cases exact identification is time consuming and difficult. In the event of a crash, information would be difficult to obtain.

FACILITY DOCUMENTS

Each facility that has chemicals that are not consumer quantities is required to have MSDSs. In SARA Title III, facilities that are covered by these requirements are also required to submit a Tier 2 form, a listing of all chemicals on site that have MSDS, and a site plan. Facilities using extremely hazardous substances (EHS) are also required to have an emergency plan. Facilities may also be covered by other regulations and may be required to have other plans in place. Each facility should be able to provide upon request, relatively quickly, and without hassle a MSDS for a given material. Many facilities will have a binder of MSDSs left at the gate or with the security guard. Twenty-four–hour staff should have full access to information and have emergency contact numbers in the event of an emergency. The size of the facility will determine the amount of staff available to track this process. The SARA reports and information are updated annually and must be submitted by March 1 of each year. The MSDSs may not be revised annually, but the facility must ensure their accuracy. Responders should meet with the facility to review these documents as they view the facility.

COMPUTER RESOURCES

Many of the chemical information texts are also available on a computer, like that shown in **Figure 3-16,** typically on a CD-ROM disk or online. Many first responders

Figure 3-16 Most HAZMAT teams and many fire departments have chemical information on computer and can quickly and easily access this information.

Figure 3-17 Some computer programs such as Aloha can generate plume models based on local data and conditions.

use **Computer Aided Management for Emergency Operations (CAMEO)** for chemical information. CAMEO is a combination software package that combines chemical response information with emergency planning capability. The information within CAMEO is easily accessible and can be used by first responders. CAMEO also has the ability to provide a vapor cloud model, known as plume dispersion, as shown in **Figure 3-17.** When responders input local data and leak information, a program known as Aerial Location of Hazardous Atmospheres (ALOHA) can determine the worst-case scenario for the vapor cloud travel.

CDs may also be used to store MSDS data, and many companies, especially those that are chemical distributors, have their MSDSs on the Internet. Several colleges and other large companies put their MSDSs on the Internet for public access. A simple Internet search of "Material Safety Data Sheets" provides several thousand results of possible MSDS locations. Other programs provide chemical information, but in most cases this software is specifically designed for a HAZMAT team and may be above the first responder level. One of the extreme advantages of computer software is the ability to search quickly for a chemical by its name and synonyms.

CHEMTREC

The Chemical Transportation Emergency Center (Chemtrec) is an information service provided by the American Chemistry Council. A group of chemical manufacturers established the association for several purposes, but one of the outgrowths is the Chemtrec service. This free service to emergency responders is funded by the chemical manufacturers through an annual fee. Although other services provide chemical information, Chemtrec is the largest and most experienced information service. When a company joins the Chemtrec system, it provides MSDS sheets to Chemtrec along with emergency contact information. Think of Chemtrec as a large phonebook: if you have a company name or chemical, Chemtrec can provide a contact name and number. Many shippers use Chemtrec as their emergency contact point. If Chemtrec does not have an MSDS on file, it can consult other chemical information databases. One of the big advantages is the ability to contact the manufacturer directly via conference call to get accurate information right from the product specialist. If Chemtrec does not have a specific manufacturer's name, it can connect you with other specialists who may be able to provide technical assistance. Chemtrec is well connected when providing information about chemical injuries and exposures; the service can provide medical information and a contact for further information. One thing that Chemtrec does not do is make regulatory notifications; that is the responsibility of the shipper. When calling Chemtrec, responders should have the following information available:

- Caller's name and phone number—most responders provide their dispatch's phone number as the on-scene cell phone may be tied up.
- Name of shipper or manufacturer—if this information is not known, then Chemtrec will contact manufacturer after manufacturer until the identity is established.
- Shipping paper information—such as truck or railcar number, the carrier name, and the consignee (receiver) name.
- Type of incident and local conditions.

The Canadian equivalent is called Canutec, for Canadian Transportation Emergency Center, and provides the same services as Chemtrec. In Mexico the Emergency Transportation System for the chemical industry is known as SETIQ, and it provides the same service as Chemtrec and Canutec. All three services' phone numbers and other emergency contact numbers are provided in the DOT ERG.

REFERENCE AND INFORMATION TEXTS

A lot of information texts from a variety of sources provide chemical information. At the end of this chapter is a listing of the common texts used by HAZMAT teams. One recommendation is that every piece of apparatus have several reference sources. If your community has a HAZMAT team, you probably have additional sources of information, as contacting the HAZMAT team by radio or phone is quick and easy. If you are in an area where the HAZMAT team is not immediately available or has considerable travel times, then it would be prudent to have additional sources of information readily available. The DOT ERG is fairly easy to use, but many other reference sources can be complicated and may require some knowledge of chemistry terms, many of which may be above the first responder level. With some study of the preliminary information that is contained in each text, most responders can readily access the information they need. The other difference is that these texts are chemical specific and the information is for that material only. Reference texts also tend to be slanted toward the group that develops the text. These books are generally used by HAZMAT technicians, but would be recommended for first responders as well. Each book has a different perspective, and there is not one that is the "only" book to use. The *NIOSH Pocket Guide to Chemical Hazards,* as shown in **Figure 3-18,** comes the closest to being the "one" book used by HAZMAT technicians. When operating at a HAZMAT release, it is best to check more than one source of information and always confirm the information with other sources.

The National Institute of Occupational Safety and Health (NIOSH) is the research arm to OSHA. It does not issue regulations, but recommends regulatory issues to OSHA. It does provide a considerable amount of research resources to the fire service and a number of sources of information. One book that is commonly used by firefighters is the *NIOSH Pocket Guide to Chemical Hazards,* which provides information on the commonly used industrial chemicals. This pocket guide provides information on exposure levels, chemical and physical properties, and protective equipment recommendations. The pocket guide is one of the few that provides ionization potentials (IP), which are used to determine if a photoionization detector will determine if a chemical is present. Another common book is the *Chemical Hazards Risk Information System (CHRIS),* published by the U.S. Coast Guard. It provides information on chemical hazards, chemical and physical properties, and water spill information. The *CHRIS* book is actually a three-volume set, and it has a great synonym index. Between the *NIOSH Pocket Guide* and the *CHRIS* book most of the common chemicals are covered.

The book that provides the most detail, covering more than 23,000 chemicals, is *Sax's Dangerous Properties of Industrial Materials.* This book is invaluable to HAZMAT teams and has considerable toxicological information, as well as chemical and physical properties. It covers trade names, synonyms, CAS numbers, and DOT numbers in the index, which is its own volume.

There are a number of reference texts that are available to first responders and HAZMAT teams. Many are listed in the Suggested Readings section at the end of the chapter. The books covered here are the basic texts and only represent a small number of texts that should be used to determine chemical response information.

A variety of texts is recommended as you can never anticipate the type of material that you may run across. If you are in an area where the HAZMAT team has an extended response time, several persons in your department should be trained to the technician level, so they can provide technical assistance to the incident commander. To function as a HAZMAT team you would need several persons trained, but one could provide chemical-specific information and act as a liaison after the arrival of the HAZMAT team.

INDUSTRIAL TECHNICAL ASSISTANCE

Within each community there is usually a technical specialist in a given field. As an example, in a facility that ships and receives railcars, there is usually someone within that facility who is knowledgeable regarding railcars. In the event of an emergency in that community, the railcar specialist's expertise would be a welcome addition to the response team. Many areas of the country, including Baltimore, Houston, and Baton Rouge, have industrial mutual aid groups designed to assist each other and the community in the event of a chemical release. In

Figure 3-18 Many texts provide chemical information. No single text covers all chemicals and responders should use several references to make sure the information is available and correct.

Baltimore, the South Baltimore Industrial Mutual Aid Plan (SBIMAP) provides technical specialists who assist on the scene and specialized equipment throughout a three-state area. SBIMAP has been in existence for more than 23 years, and two HAZMAT teams use the services of a chemist each time they respond to an incident.

Industrial facilities usually have employees responsible for safety and health, and those professionals are usually a good technical resource. Within the facility there may be chemists or chemical engineers who can provide chemical information. When dealing with chemical exposures and toxicology, many facilities have industrial hygienists who work with those issues daily. As these people exist in your community, you need to search them out prior to a large incident. Determine their location and availability. When approached, many of these professionals are more than willing to assist their community.

SUMMARY

NOTE Accurate and quick information is essential to every HAZMAT incident.

The more information that can be obtained, the easier it will be to assess and respond to the incident. First responders should be starting the information process by trying to obtain as much information as possible, so they can make safe and accurate tactical decisions. Using the recognition and identification skills, a considerable amount of information can be obtained. This information combined with the use of reference sources can help provide a significant amount of information. If waiting for a HAZMAT team, the more information that can be obtained prior to the arrival of the team, the easier and less stressful its task will be. As essential as this information is, first responders should not take any risks attempting to get this information, such as entering a hazardous area to get shipping papers.

KEY TERMS

ADR/RID Abbreviations used by the international shipping community and established by a European Agreement. The U.S. DOT is a voting member of this shipping body which governs Regulations Concerning the International Transport of Dangerous Goods by Rail (RID) and Road (ADR)

Air bill Term used to describe the shipping papers used in air transportation.

Chemtrec The Chemical Transportation Emergency Center, which provides technical assistance and guidance in the event of a chemical emergency. This network of chemical manufacturers provides emergency information and response teams if necessary.

Computer Aided Management for Emergency Operations (CAMEO) Computer program that combines a chemical information database with emergency planning software; commonly used by HAZMAT teams to determine chemical information.

Consist Shipping papers that list the cargo of a train; the listing is by railcar.

Dangerous Cargo Manifest (DCM) Shipping papers for a ship, which list the hazardous materials on board.

Ionization Potential (IP) The ability of a gas or vapor to be ionized; most commonly used to determine if a photoionization detector can detect a gas or vapor.

National Response Center (NRC) Location a spiller is required to call to report a spill if it is in excess of the reportable quantity.

Photoionization Detector (PID) A detection device that can detect toxic gases or vapors; commonly used by HAZMAT teams.

Standard Transportation Commodity Code (STCC) Number assigned to chemicals that travel by rail.

Waybill Term that may be used in conjunction with consist, but is a description of what is on a specific railcar.

REVIEW QUESTIONS

1. What are three common methods of obtaining chemical information?
2. The DOT ERG is intended to be useful for how long at an incident?
3. Given a four-digit UN identification number, which resource provides quick initial information?
4. What mode of transportation uses STCC numbers?
5. Are MSDSs required in a home for household quantities of hazardous materials?
6. Can Chemtrec be used for medical information?
7. Which section of the DOT book is used if you have a shipping name?
8. Which reference book provides information related to suggested isolation distances?
9. What advantage does an industrial contact provide at a chemical spill?
10. What cargo probably will not be listed on the shipping papers?

Suggested Readings

American Association of Railroads, *American Association of Railroads Hazardous Materials Action Guides,* AAR, Washington, DC, 2006. Provides basic to advanced response information; one of the few guides that actually provides cleanup and mitigation strategies; lists common materials transported by rail.

American Association of Railroads, *Emergency Handling of Hazardous Materials in Surface Transportation,* AAR, Washington, DC, 2005. Covers materials commonly transported on rail, which applies to highway incidents as well.

American Conference of Governmental Industrial Hygienists, *ACGIH Threshold Limit Values,* 2007 ed., ACGIH, Cincinnati, OH. Provides the threshold limit values (TLV) for all of the chemicals that have been studied by the ACGIH; targets common industrial chemicals.

Ash, Michael and Irene (editors), *Gardener's Chemical Synonym and Trade Names,* 11th ed., Gower, 1999. Listing of common synonyms for industrial materials and household products; provided that some chemicals have more than 60 different synonyms, this text can be invaluable; no response information.

Brethrick, L., *Brethricks Handbook of Reactive Substances,* 6th ed., Butterworths, Boston, MA, 2000. One of the few sources of information on what happens when chemicals combine; not slanted toward any grouping of chemicals.

Chemical Hazard Risk Information System (CHRIS), U.S. DOT/U.S. Coast Guard, Washington, DC, 2001. http://www.chrismanual.com. Provides detailed chemical information and is one of the best reference texts, and is free.

Compressed Gas Association, *Handbook of Compressed Gases,* 4th ed., Van Nostrand Reinhold, New York, 1999. Provides extensive detail on common compressed gases; includes detailed information on each type of cylinder and common gases.

Crop Protection Handbook, formerly *Farm Chemicals Handbook,* Meister Publishing, Willoughby, OH, 2003. Definitive source for pesticide, herbicide, and insecticide information; no other text is as up-to-date or as in-depth.

Fosberg, K., and S. Z. Mansdorf, *Quick Selection Guide to Chemical Protective Clothing,* 4th ed., Van Nostrand Reinhold, New York, 2003. Provides chemical compatibility information on several hundred chemicals; listings are for the most common types of protective clothing.

Lewis, Richard J., Sr., *Dangerous Properties of Industrial Materials,* 11th ed., John Wiley & Sons, New York, 2005. Is one of the best and most detailed reference sources. It is a three-volume text that has almost every material that one would respond to. Has an extensive cross-reference and information on 26,000 chemicals.

NIOSH Pocket Guide to Chemical Hazards, National Institute of Occupational Safety and Health, Washington, DC., 2005. http://www.cdc.gov/niosh/npg. Covers most occupational chemicals and most chemicals that responders encounter. Is an excellent source of information, and has some detailed chemical and physical properties. Both hard copy and electronic versions are available. The electronic version is free, and single copies of the text are also free.

Additional Resources

Bevelacqua, Armando, and Richard Stilp, *Hazardous Materials Field Guide,* Delmar Cengage Learning, Clifton Park, NY, 1998.

Bevelacqua, Armando, and Richard Stilp, *Terrorism Handbook for Operational Responders*, Delmar Cengage Learning, Clifton Park, NY, 1998.

Callan, Michael, *Street Smart HazMat Response*, Red Hat Publishing, Chester, MD, 2001.

Hawley, Chris, *Hazardous Materials Incidents,* 2nd ed., Delmar Cengage Learning, Clifton Park, NY, 2004.

Hawley, Christopher, *Hazardous Materials Air Monitoring and Detection Devices*, Delmar Cengage Learning, Clifton Park, NY, 2007.

Hawley, Chris, Gregory G. Noll, and Michael S. Hildebrand, *Special Operations for Terrorism and HazMat Crimes*, Red Hat Publishing, Chester, MD, 2002.

Lesak, David, *Hazardous Materials Strategies and Tactics*, Prentice Hall, Upper Saddle River, NJ, 1998.

Noll, Gregory G., Michael S. Hildebrand, and James Yvorra, *Hazardous Materials Managing the Incident*, 3rd ed., Red Hat Publishing, Chester, MD, 2005.

Schnepp, Rob, and Paul Gantt, *Hazardous Materials: Regulations, Response, and Site Operations*, Delmar Cengage Learning, Clifton Park, NY, 1999.

PROTECTION

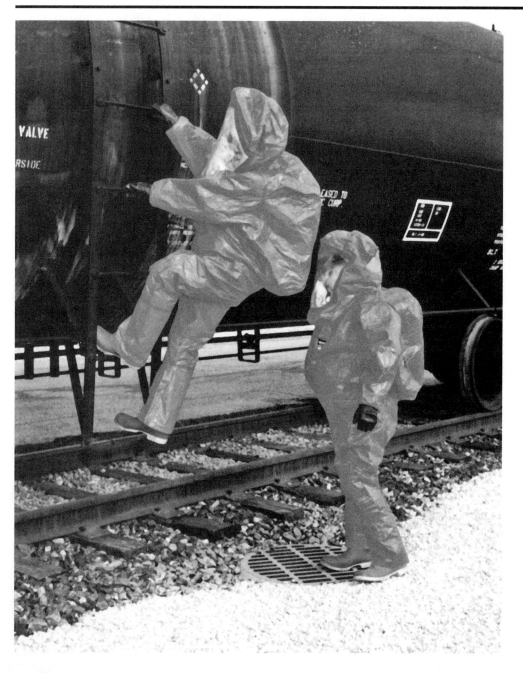

CHAPTER 4

OUTLINE

- **Objectives**
- **Introduction**
- **Health Hazards**
- **Exposure Levels**
- **Types of PPE**
- **Summary**
- **Key Terms**
- **Review Questions**

VALVE

RSIDE

STREET STORY

Chemical protective clothing has come a long way since I was introduced to chemical hazards one evening in 1974. I had responded to a dumpster fire early one morning. We arrived to see a thick red-brown cloud issuing from the waste container. Unfortunately, we did not recognize that we had a hazardous materials situation, not a fire. Undaunted, we went about our tasks, and after half an hour, we eventually extinguished the fire. However, we all had burning eyes, nose, throat, and severe irritation to our wrists and necks. We called for the assistant chief on duty. He suggested our ailments were from the diesel fumes issuing from the fire truck. We thought otherwise but went back to the station, finished our tour, and went home.

The next morning I was visited by a man from the chemical company who wanted to make sure we were all right. He explained that the dumpster contained discarded nitric acid bottles and that some had still been full. They had started the fire. He was more concerned about our exposure, however, since we had not worn chemical-protective clothing or even SCBA. (Hey, the fire was outside! Remember those days?) He went on to explain that nitric acid had a "delayed reaction to any acute exposure." This, he told me, would have affected the lungs, which might have developed fluid or edema, creating chemical pneumonia. In addition, my throat might have swollen and made it difficult to breathe. I remember saying to him, "You're probably glad I answered the door!" He agreed that it was a relief.

Over the years I have looked back at that incident, and I know it is what got me into HAZMAT. It was the driving force behind my decision to sign up for one of the first National Fire Academy Chemistry of Hazardous Materials courses. I know now that chemical protective clothing or, at the very least, SCBA should always be worn in any smoke, vapor, or cloud situation. Hindsight is the best sight, I guess.

A final note for anyone who might be concerned about me. Do not worry. Everything works out in the end. A few months later I went to an acetylene tank fire. I was the acting officer and had decided to extinguish the fire with a sodium bicarbonate extinguisher. Upon closer examination I noticed the fire was out. I turned to my engine crew to tell them to not discharge the extinguisher. Sadly, one overanxious fellow discharged the unit right into my face—and me, once again without an SCBA! With my mouth wide open, I got a lung full of baking soda. So we can all relax, because I am balanced—chemically neutral—again!

—Street Story by Mike Callan, President, Callan and Company, Middlefield, Connecticut

OBJECTIVES

After completing this chapter the student should be able to identify and describe:

- The causes of harm (A)
- The health hazards associated with chemical releases (A)
- The definitions of various chemical-related health terms (O)
- The various levels of PPE (O)
- The use of SCBA and other respiratory protection at chemical releases (O)
- The use of firefighting protective clothing at chemical releases (O)
- The signs and symptoms of heat stress (O)

INTRODUCTION

Personal protective equipment (PPE) takes many different shapes and versions; one example is provided in **Figure 4-1**. Even Firefighter protective clothing (FFPC) has many variations and ensembles. When dealing with hazardous materials, these configurations are endless. The failure to use PPE or using it improperly is asking for an injury and may have fatal consequences.

NOTE The use of PPE is essential to the health and safety of first responders.

Many responder injuries can be prevented by the proper use of PPE, most of which only take a few seconds to put in place. The best respiratory protection for the first responder above all else is the **self-contained breathing apparatus (SCBA),** which offers a substantial amount of respiratory protection and should be considered the minimum for chemical spills. Firefighters' turnout gear does offer some limited protection against some chemicals, but it is not intended to be used for chemical spill response. FFPC is not tested or approved for chemical spills. Although it may offer some protection in some chemical environments, FFPC should only be used for immediate life-threatening rescue situations where there will be limited time in the hazard area. Military researchers have tested the effectiveness of FFPC and SCBA in chemical warfare agent release. Agents tested included nerve agents and blister agents, such as sarin, which is a highly toxic nerve agent. What the military found was unusual in the fact that FFPC provided more protection than the chemical gear issued to soldiers. Against the chemicals that the military

Figure 4-1 PPE is used to protect the wearer from a variety of hazards, but no one type of PPE protects the wearer from all chemicals. This type of PPE is worn by bomb technicians to offer some protection in the event of an inadvertent detonation.

tested, the gear provides a high level of protection. As a matter of practice FFPC should not be used in a chemical environment, with the exception of performing rescues of live victims. As the need to be in the hazard area increases, so does the need for proper PPE. This chapter outlines these types of PPE, but specific hands-on training is required prior to the use of this PPE. In general HAZMAT technicians and specialists wear chemical protective clothing, but persons trained to the operations level may be required to don chemical protective clothing to perform decon operations or other patient-related activities, after specific PPE training is provided.

HEALTH HAZARDS

Chemical exposure poses a variety of health hazards, from a short-term irritant effect to an immediate life-threatening situation to a longer-term carcinogenic risk.

Toxicology

Toxicology is the study of poisons and their effects on the body. People who study the effects of poisons on the body are known as toxicologists. Although toxicologists are typically found in the medical community, many industrial facilities employ industrial hygienists who are responsible for the protection of the workers' health and safety. Their primary focus is on the chemical hazards that exist within the facility. These persons have extensive training in toxicology and chemical exposures and are great resources to the emergency services. As the world of toxic exposures can be complicated, information is needed quickly when dealing with emergency situations. A quick consultation with an industrial hygienist may make the incident go a whole lot easier.

Types of Exposures

Two types of exposures, **acute** and **chronic,** can have serious health effects. Another term that is associated with exposure is hazard. Being exposed to a chemical may present a risk. The hazard that a chemical presents provides an indication as to the potential harm. Typically, an acute exposure is one where the body is subjected to a large dose over a short period of time. This period of time can be from seconds to 72 hours. It should be noted that after an acute exposure the body, depending on the product and its hazards, may exhibit injuries ranging from very minimal to major and potentially leading to death.

Overall, the human body does well with short duration exposures and recovers from these exposures, but all chemical exposures should be avoided. Some acute exposures may have dangerous consequences, such as being exposed to large amounts of chlorine or ammonia for a long period of time. Your body reacts adversely to these substances and has an urgent desire to evacuate any area where these chemicals may be present, usually preventing any significant health problems with these types of materials. An example of an acute exposure is a nonsmoker trying a cigarette.

This one-time exposure for most people is not harmful; nor would it cause any long-term health effects. By comparison, a chronic exposure is a person who has smoked three packs of cigarettes a day for 20 years. This person has received a dose of cigarette smoke several times a day for a long period of time and is most likely to have health problems associated with this chronic exposure. Abnormalities occur in both cases: It is possible that a person who smokes one cigarette may develop lung problems from that one acute exposure, just as the person who chain smokes for 20 years might never develop health problems associated with the chronic exposure.

NOTE An acute exposure is a quick one-time exposure to a chemical.

SAFETY Hazardous materials incidents have the ability to impact humans, the environment, and property. In some cases, such as with dioxin, the risk to humans, the environment, and property can last for many years. Through immediate effects or long-term chronic effects hazardous materials can present a risk. Proper identification of potential hazards and the wearing of proper protective clothing are essential for responder safety.

Types of Hazards

In the realm of hazardous materials hazard is defined as the category of risk that can be inflicted by exposure or contamination with a chemical. There are several methods of identifying possible hazards at a chemical release. The most common in use today is the acronym TRACEMP, which stands for thermal, radiation, asphyxiation, chemical, etiological, mechanical, and psychological hazards. Each of the individual hazards has additional hazards that fit within that classification. Much like the **risk-based response** theory, the use of TRACEMP puts a chemical into a risk category, and tactical decisions can be based upon that classification. The subcategories within TRACEMP are as follows:

Thermal   Both heat and cold hazards fit into this category. If a flammable liquid ignites, it is classified as a thermal hazard. If liquefied oxygen contacts your skin, it would cause frostbite and a thermal (cold) burn. When liquefied gases are released into small spaces or rooms the temperature can drop dramatically, presenting hypothermia and cold stress concerns.

Radiation   Types of radiation include nonionizing and ionizing. Nonionizing types include microwave and infrared. Ionizing types such as alpha, beta, gamma, neutron, and X-rays fit into this category. Alpha and beta are particles, while gamma, neutron, and X-ray are forms of energy.

Asphyxiation   Both simple and chemical asphyxiants fit into this category.

Chemical   This category includes poisons and corrosives. The poisons may also be referred to as toxic. Some chemicals are highly toxic. There are specific levels of exposure that determine which category a chemical fits into. Also within this category are convulsants, irritants, sensitizers, and allergens. Reactions to a particular chemical vary person to person much in the way some people are allergic to bee stings while others are not. Chemicals have effects on some people while having no effect on others.

Etiological   Blood-borne pathogens and biological materials are in this category.

Mechanical   Although not chemical in nature, mechanical hazards exist within the hazard area of a chemical spill. There exists standard slip, trip, and fall hazards that one should always be concerned about. Another example of a mechanical hazard would be getting hit from blast particles, such as from a bomb or BLEVE. A drum falling on a responder is another example of a mechanical hazard.

Psychological   When exposed to traumatic or disaster situations some responders may suffer immediate, short-term, or long-term psychological stress. Post-traumatic stress syndrome (PTSS) is one example.

SAFETY   In situations where liquefied gases are released, their low temperatures will drop the temperature of the room. In some cases the drop can be dramatic. In one situation where anhydrous ammonia had been released, and there was liquid ammonia on the floor, the hazardous materials team's boots were frozen to the floor. Their SCBA regulators were starting to freeze and several team members had frostbitten hands. In addition to these cold hazards, the vaporization of the ammonia resulted in a vapor cloud that obstructed visibility; the crews were not able to see their team members an arm's length away, not to mention the impact of the fogged-up faceshields.

Categories of Health Hazards

Within the TRACEMP categories are several terms that responders should understand. MSDS sheets or industrial contacts may describe some chemicals as fitting into one or more of these categories. One of the most commonly used terms is **carcinogen,** which means cancer-causing potential. There are two classifications of carcinogens, known and suspected, with the majority falling into the latter category. According to the Chemical Abstracts Service (CAS), a division of the American Chemical Society (ACS), a group that tracks chemicals, there are 30 million chemicals in existence today. Each day the ACS adds 4,000 more chemicals to their listing. The National Toxicology Program under the Public Health Service of the U.S. Health and Human Services Administration issues an annual report on carcinogens. The eleventh annual report lists 54 chemicals that are known to cause cancer and 184 chemicals that are suspected of causing cancer. When dealing with chemical spill response, the risk of getting cancer always causes great fear. Firefighters are exposed to a large number of chemicals, many of them known

cancer-causing agents, but if firefighters wear their SCBA these exposures are unlikely to cause problems. For older firefighters who worked in earlier years when SCBA was not used as extensively as it is today, cancer is still a leading cause of death, and many retired firefighters have not had the chance to enjoy retirement due to an early death.

NOTE   The ACS estimates that there are 30 million chemicals on this earth; it is estimated that there are 54 known carcinogens and 184 chemicals suspected of causing cancer.

SAFETY   Wearing SCBA and PPE and avoiding other off-duty activities that involve cancer-causing materials can prevent the development of cancer in most persons.

One of the common mistakes made by responders is getting tunnel vision, or focusing on only one aspect of the emergency while neglecting others. This tunnel vision can cause responders' death and injuries. My personal experience with tunnel vision that may have long-term health effects involves a known cancer-causing substance.

I arrived at work in the afternoon to begin night shift. Upon arrival at the station I learned that the daylight shift was out on a HAZMAT call in an adjacent jurisdiction. As the other night shift crew arrived, we departed to relieve the day shift on the scene. When we arrived, we met with the incident commander who told us that the day shift was inside the house. They had been called to the house to investigate an unusual odor. We asked the incident commander what level of protective clothing the day shift was using. Just their uniforms, we were told. This is when tunnel vision really set in.

We decided to enter the building without speaking to any of the HAZMAT team members. When we got into the basement, we found the day shift and learned that the crew was searching for the source of an odor. There was a distinct odor that was slightly petroleum, and the day shift responders said they had been unable to find the source. We asked what they had done so far, and they departed for the station. As the day shift responders had said they had monitored, we did not bother getting any of the monitors. We were unable to find the source from within the house, so we decided to drill a hole in the basement and see if the source was coming from underneath the house. When the hole was finished, the odor got strong to the point of irritation. We decided to use an air monitor that the day shift had not used. The monitoring process took a while, and on the last test we found the presence of benzene. The amount of benzene in the air exceeded the legal limit several times over. At that point we had been in the basement for several hours with no respiratory protection.

Ground imaging equipment was brought in the next day and it located a drum buried near the house. The drum contained benzene, a known cancer-causing substance in humans. Our exposure to the benzene was extreme, and it is not known what the long-term health effects could be. The standard latency period for the development of cancer after a severe exposure to benzene is several years. The only thing I and colleagues who were there that night can do is wait and hope we don't develop cancer from this exposure. Tunnel vision got us, and I learned that I needed to evaluate the hazards no matter what the circumstances are when we arrive.

Another commonly used term, **irritant,** is fairly self-explanatory. An irritant is not corrosive but mimics the effect of a corrosive material. It can cause irritation of the eyes and possibly the respiratory tract. One notable characteristic of exposure to irritants is that the effects are easily reversed by exposure to fresh air. Mace or pepper spray are classified as irritants and may be called *incapacitating agents* by the military. **Sensitizer** is a term used to describe a chemical effect that is in reality an allergic reaction, typically by skin contact. In most cases an employee can work with a chemical for years and suffer no effects, and then one day suffer a severe reaction to the material. Skin reddening, hives, itching, and difficulty breathing are possible symptoms when dealing with a sensitizing agent. **Allergens** are chemicals that cause symptoms much like sensitizers but are the result of a reaction with an individual's immune system. In many cases someone is allergic to a substance—has been allergic to the substance for many years. In some cases allergies can develop later in life to materials that previously did not cause a reaction.

A **convulsant** chemical is one that has the ability to cause convulsions upon exposure, typically by ingestion. There are a number of drugs that can cause seizure-like activity. One group of chemicals that can cause such a reaction are organophosphate pesticides (OPP). When someone is exposed to an OPP the brain reacts to the chemicals and results in seizure-like activity. The grouping of chemical warfare agents, also known as nerve agents, is comparable to OPP compounds, and causes the same seizure-like activity.

NOTE Some people can become sensitive to a chemical after one exposure.

Some chemicals only affect one or more organs and are described as target organ hazards, or they may affect a body system, such as the central nervous system. The effects depend on the individual, dose, concentration, and length of exposure. Some of the target organ descriptions are provided in **Table 4-1.**

Radiation Hazards

Radiation has the ability to cause a number of health problems including death, which can occur within weeks. **Table 4-2** provides an overview of some example radiation sources. Exposure to radiation can cause radiation sickness or other illnesses, some of which are shown in **Table 4-3.** Coming in contact with or being close to a radiation source can cause significant burns. The level of radiation and the dose determine the damage to the body. The primary issue with radiation is the fact that it can kill without any warning, as it is colorless, odorless, and the danger cannot be seen. Detection devices can easily detect radiation levels and warn responders when they are entering a potentially dangerous environment.

Example radiation action levels are provided in **Table 4-4.** Understanding the level of danger can be confusing. Equally confusing is the process of calculating a radiation dose. There are three measurements that can be used to describe a radiation dose: absorbed dose, equivalent dose, and effective dose. The **absorbed dose** is a measure of energy transferred to a material by radiation. The absorbed dose is measured in units called gray (Gy) or rad (radiation absorbed dose). One gray is equal to 1 joule of energy absorbed by 1 kilogram of material, and is the SI (System International) unit for absorbed dose. One gray is also equal to 100 rad, an older unit. To understand the impact on humans we need to expand our thoughts on dose measurements and consider the effect of the absorbed energy on the body. The basic unit of equivalent dose in the United States is the roentgen (R), which is a value that is provided for the amount of ionization in air caused by X-rays or gamma radiation. Using the roentgen and applying the quality factor we can determine the dose and impact on humans, which is known as roentgen equivalent man, or rem. The value of 1 roentgen and 1 rem are equal to each other, and are used interchangeably. The rest of the world uses sieverts (Sv) to describe the equivalent dose to man. To further define how the dose impacts humans another factor can be added known as a weighting factor. Target organs have a higher factor than bone, for example. Radiation monitors measure in three scales: rem or sieverts (Sv), millirem (mrem) or millisieverts (mSv), and microrem (μR) or microsieverts (μSv). The dose of radiation is expressed by a time factor, typically an hour.

TABLE 4-1

Target Organs and Systems

Name	Target organ or system
Central nervous system	Can cause short-term or long-term effects. Commonly short-term (CNS) memory can be lost after an exposure to a CNS hazard material. Many hydrocarbons cause CNS effects, and some people purposely expose themselves to a CNS agent to receive a "high" from the exposure. In the long term, the brain cells are damaged permanently. May be referred to as neurotoxins; although they are not identical, they essentially cause the same effects.
Peripheral nervous system	Much like CNS chemicals, PNS chemicals affect the body's ability to move in a coordinated fashion. Exposure to a PNS chemical causes a disruption of the brain's ability to move messages to other body systems.
Hepatoxins	These types of materials affect the liver and, if the exposure is high enough, can cause severe liver damage.
Nephrotoxins	Affect the kidney.
Reproductive toxins	Affect the ability to reproduce or can cause birth defects. These types of toxins can stay within the body and complicate a pregnancy, even if the exposure occurred a considerable time before the pregnancy.
Mutagen	An exposure to a mutagen may not cause any effects to the people who received the exposure, but the effect could be transmitted to their offspring. Mutagens damage the genetic system and can cause mutations, which may become hereditary.
Teratogens	These materials can harm an unborn child. The effects do not happen on a cellular basis and would not be passed along from generation to generation.

TABLE 4-2

Common Radioactive Sources

Name	Energy Emitted	Half-Life	Use	Decays to Form
Polonium-214	α	164 microseconds	Dust-removing brushes	Lead-210
Lead-214	β and γ	27 min	Industrial	Bismuth-214
Cobalt-60	β and γ	5.3 years	X-ray machines, industrial applications	Nickel-60
Cesium-137	β	30 years	Medical field, industrial imaging, check source	Barium-137
Radon-222	α	3.8 days	Naturally occurring gas	Polonium-218
Uranium-238	α	4.5 billion years	Ore	Thorium-234
Plutonium-239	α	24,100 years	Nuclear weapons	Uranium-235
Americium-241	α	433 years	Smoke detectors	Neptunium-237

TABLE 4-3

Health Effects			
Exposure (REM)	**Exposure (Sieverts)**	**Health Effect**	**Time to Onset**
5–10	0.05–0.1	Changes in blood chemistry	
50	0.5	Nausea	Hours
50–100	0.5–1	Headache	
55	0.55	Fatigue	
70	0.7	Vomiting	
75	0.75	Hair loss	2–3 weeks
90	0.9	Diarrhea	
100	1	Hemorrhage	
100–200	1–2	Mild radiation poisoning	Within hours
200–300	2–3	Severe radiation poisoning	Within hours
400–600	4–6	Acute radiation poisoning, 60% fatal	Within an hour
600–1,000	6–10	Destruction of intestinal lining, internal bleeding, and death	15 minutes
1,000–5,000	10–50	Damage to central nervous system, loss of consciousness, and death	Less than 10 minutes
5,000–8,000	50–80	Immediate disorientation and coma, death	Under a minute
More than 8,000	More than 80	Immediate death	Immediate

Source: From EPA, DOD, and other sources.

TABLE 4-4

Radiation Action Levels and Assorted Doses		
Dose	**Cause or Type**	**Note**
Average Exposures to Radiation		
0.01 R annually	Chest X-ray	Per-year exposure
0.2 annually	Radon in the home	Per-year exposure
0.081 R annually	Living at high elevations (Denver)	Per-year exposure
1.4 R	Gastrointestinal series	
Action Levels		
1 mR/hr	Isolation zone (public protection level)	Recommended exposure limit for normal activities
5 R	Emergency response	For all activities
10 R	Emergency response	Protecting valuable property
25 R	Emergency response	Lifesaving or protection of large populations
>25R	Emergency response	Lifesaving or protection of large populations. Only on a voluntary basis for persons who are aware of the risk involved.

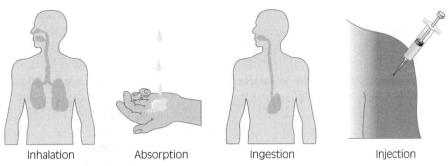

Figure 4-2 Routes of exposure.

Radiation Protection

Radiation doses should be kept "As Low As Reasonably Achievable." This is known in the nuclear industry as ALARA, and is the cornerstone of radiation safety. The three factors that can influence radiation dose are time, distance, and shielding, concepts that should be applied to all types of chemical exposures as well. To minimize their dose, one should stay near the radiation source for as little time as possible, stand as far away from the source as possible, and place as much shielding between them and the source as possible. Naturally occurring background radiation is in microrems (or microsieverts) per hour and so when using a radiation detection device an accurate device will be reading this background radiation. When levels above the normal background are encountered responders should use caution. Variations in background readings do vary depending on the location, elevation, terrain, building type, and a number of other factors. It is when the radiation dose changes scales, such as going from microrem to millirem, that there is a potential for health effects. Going from microrem to rem levels indicates severe danger. When encountering radiation levels that are significantly above background and have changed from micro to milli levels or rem levels, responders should move to an area that is reduced in radiation dose.

Routes of Exposure

Four primary routes of exposure—respiratory, absorption, ingestion, and injection—are listed in **Figure 4-2.** The route that is most commonly associated with causing ill health effects, both acute and chronic, is the respiratory system, as shown in **Figure 4-3.**

SAFETY In almost all cases the respiratory system requires some type of protection. SCBA will protect emergency services workers, but it must be donned and working.

The respiratory system can be affected by gases, vapors, and solid materials, such as dust and other particles. In many cases the chemicals may not have any effect on the respiratory system itself, but may enter the body through the respiratory system and harm other organs or body systems.

There are two categories of asphyxiants, simple and chemical. Although the result is usually the same, the cause of death may differ significantly.

NOTE Simple asphyxiants exclude the oxygen in the air and push it out of the area.

Simple Asphyxiant

Chemical Asphyxiant

Figure 4-3 Respiratory system route of exposure.

It is not a chemical reaction; something else is occupying the space of the oxygen. Normal oxygen levels are 20.9 percent in air. The human body starts to develop difficulty breathing at less than 19 percent and will start to have serious problems at less than 16.5 percent oxygen. Gases such as nitrogen, halon, and carbon dioxide (CO_2) will all move oxygen out of the area and cause people in the area to have difficulty breathing; if those gases exist in high enough concentrations, death may result.

NOTE Chemical asphyxiants work in a different manner. They cause a chemical reaction within the body and will not allow it to use the readily available oxygen.

The most common chemical asphyxiant is carbon monoxide (CO). Common locations for CO exposures are listed in Chapter 6 under "Carbon Monoxide Incidents." When CO is in the air in sufficient quantities, it enters the bloodstream through the lungs. It binds with the hemoglobin in the blood, forming carboxyhemoglobin. As the hemoglobin has an attraction for CO at about 225 times that of oxygen, it will not allow oxygen molecules to bind with the blood, which causes severe flulike symptoms, and with higher exposures, unconsciousness and even death.

The other common route of entry is skin absorption, as the skin is the body's largest organ. Although some chemicals can irritate the skin, this does not mean they are toxic by skin absorption.

The number of chemicals that are toxic by skin absorption is relatively low, but precautions should be taken to minimize skin contact—even eliminate it altogether—with chemicals. Skin contact with chemicals can cause burns, rashes, or drying of the skin. The only way to provide skin protection is to wear proper protective clothing that will not allow chemicals to get onto skin. FFPC will slow the process but will not prevent the eventual migration of the chemical to the skin. Effective decontamination is required to ensure that all of the chemical is cleaned from the skin. To best prevent chronic exposure to fire by-products, FFPC should be washed on a regular basis.

Another route of entry is ingestion, which may be more common than one would think.

CAUTION After fighting a fire, responders are typically covered in soot and other debris. Most firefighters do not decontaminate themselves prior to eating or drinking refreshments that may be available at the scene. Without proper cleaning they may ingest some of these products of combustion, none of which are healthy to eat. The use of SCBA generally prevents this route of exposure, at least until the rehab phase of the incident.

One other route of exposure is injection, although it is not considered to be a major route. Many emergency services workers get exposed to hazardous materials via this route; it may be the leading route behind inhalation. The most common material that responders are exposed to are body fluids, or what is referred to as blood-borne pathogens (BBP). Exposure to BBP can occur through contact with a patients' body fluids, many times through an inadvertent needle stick. Other methods of injection are through being near a high pressure line when it breaks, such as a hydraulic rescue tool fluid line that could inject hydraulic fluid into the body. Other than standard infectious control practices there is little protection for these types of exposures except to wear full PPE properly when working in and around situations that could expose you to these fluids. Some examples of the routes of exposure by hazard class are provided in **Table 4-5.**

Factors that affect the rate of exposure, no matter the route, are basic items such as temperature, pulse, and respiratory rate. The higher each of these items is, the more likely the chemical will have some effect. The damage that chemicals have is based on the equation: Effect = Dose + Concentration + Time. This equation relates to acute and chronic exposures. A small dose of a small concentration for a short period of time is not likely to have any adverse effect on a normal human. A large dose at a high concentration over a long period of time will in most cases have an effect. A person who is exposed to a chemical at normal vital signs may not suffer any ill effects. But if the person jogs around the block prior to exposure, negative effects are more likely.

NOTE The increased temperature of the body under exertion will allow for faster absorption into the body, and the accompanying increased respiratory rate will cause more chemicals to enter the respiratory system.

The increased pulse rate allows for the chemicals to be spread throughout the body faster. In some confined space incidents where chemicals may have played a factor in injuries and deaths, the fact that victims' vital signs are elevated above normal allows for an increased chemical exposure and plays a role in those injuries and deaths. In most cases the first victims may be unconscious and therefore their bodies' systems have slowed down; in some cases they may have gone down due to a lack of oxygen. When people recognize that coworkers have gone down, their vital signs are increased and, when trying to perform a rescue, they actually get exposed to a higher level of the chemical than those they are trying to rescue. This is one of the reasons that in some cases the rescuer dies and the original victim ends up surviving.

TABLE 4-5

Routes of Exposure/Harm by Hazard Class			
Hazard Class	Primary Route of Entry	Possible Route of Entry	Harm
Explosives	Inhalation, ingestion, and absorption	Injection	Thermal, etiological, mechanical, and psychological
Gases	Inhalation	Absorption	Thermal, asphyxiation, chemical, and psychological
Flammable liquids	Inhalation, ingestion, and absorption	Injection	Thermal, asphyxiation, chemical, and psychological
Flammable solids	Ingestion and absorption	Inhalation and injection	Thermal, chemical, and psychological
Oxidizers and organic peroxides	Inhalation, ingestion, and absorption	Injection	Thermal, asphyxiation, chemical, mechanical, and psychological
Toxic or infectious materials	Inhalation, ingestion, and injection	Absorption	Asphyxiation, chemical, etiological, and psychological
Radioactive	Inhalation, ingestion, and absorption	Injection	Thermal, radiation, and psychological
Corrosive	Inhalation, ingestion, and absorption	Injection	Chemical and psychological
Misc. dangerous goods	Inhalation, ingestion, and absorption		Thermal, radiation, asphyxiation, chemical, etiological, mechanical, and psychological

EXPOSURE LEVELS

In industry, monitoring for exposures is commonplace and is usually intended as a preventive measure, but in the emergency services it can be an afterthought, typically after an incident has occurred. There are several different types of exposure values issued by a variety of agencies, and some of those values can be very confusing. Exposure values are established for commonly used chemicals in a variety of situations. The key to preventing exposures is to monitor for hazardous materials and to wear appropriate PPE. The one key agency involved with exposure values is the Occupational Safety and Health Administration (OSHA). The exposure values OSHA sets must be followed by all industries, including the fire service, which is considered an industry.[1] Another organization that issues exposure values is the ACGIH, which advocates worker safety and conducts a lot of studies regarding chemical exposures. The National Institute of Occupational Safety and Health (NIOSH), a research arm of OSHA, issues recommendations for exposure levels. OSHA is the only agency that provides legally binding exposure values; all of the others are recommendations. In some OSHA

regulations the employer is required to use the lowest published exposure values, which in many cases are not OSHA's own values. When dealing with emergency situations, responders should always use air monitors to determine exposure levels and to follow the lowest published values. In most cases the HAZMAT team will be required to perform this function. In a response to carbon monoxide incidents, first responders may need to fulfill this role.

The exposure values are based on an average male and are for an industrial application.

NOTE The exposure values are typically based on an 8-hour day, 40-hour workweek, and 16-hour break between exposures.

The values that are issued for the various substances are typically conservative, as in a given population it is not atypical to find someone who is sensitive to a chemical at a lower value than the rest of the group. Each value has an extra margin for safety built in, ranging from 1 to 10,000 times the actual value. The exposure values are typically listed in parts per million (PPM) or are listed as milligrams

Figure 4-4 Explanation of units of measure.

per meter cubed (mg/m³); examples of these terms are provided in **Figure 4-4.** OSHA refers to this 8-hour exposure value as the **permissible exposure limit (PEL).** The ACGIH refers to this 8-hour exposure as the **threshold limit value (TLV).** Both are average exposures over the 8-hour period. A worker can be exposed to more than the PEL or TLV, as long as at the end of the day the exposure value is less than the PEL or TLV. In some cases these exposure values are called time-weighted averages (TWA), and you may see them expressed as the PEL-TWA or the TLV-TWA, which is an 8-hour-a-day average exposure. NIOSH issues **recommended exposure levels (REL),** which are recommendations for a 10-hour day, 40-hour workweek. These exposure values are outlined in graphic form in **Figure 4-5.**

Other values that may be listed for chemicals are the **ceiling levels,** generally referred to as PEL-C or TLV-C. These provide an amount that is the highest level an employee can be safely exposed to. When figuring an average, there are times when an employee may be exposed to chemicals at a level higher than the PEL or TLV, but there are also times when they will be exposed to less than those levels. As long as their exposure average is less than the PEL or TLV, the exposure level is acceptable. Certain chemicals will cause effects at levels that may be obtained through worker exposure and would not be considered safe, but the overall exposure would fall below the PEL or TLV. To avoid unsafe levels regulators and researchers may attach a ceiling level to an exposure value, and that is the highest level that the employee can be safely exposed to, no matter what the end average is.

Another exposure value is known as the **short-term exposure limit (STEL or ST).** This is a value assigned to a 15-minute exposure. An employee can be exposed at this level for 15 minutes and then is required to take an hour break from the exposure. The employee can do this four times a day without any adverse effects. NIOSH is also using an excursion value that is coupled with a time limit of 5 to 30 minutes typically. At this level an employee can enter an environment one time and not suffer any effects. The last value can be confusing as it is called the immediately dangerous to life or health (IDLH) value. One would think that being exposed to a chemical at the IDLH level would mean that death may be imminent. In reality this is an OSHA value that is the maximum airborne concentration that an individual could have the ability to escape and not suffer any adverse effects. The implied definition does not match the legal definition. At the IDLH level emergency responders need to be using SCBA as an absolute minimum, and if the chemical is toxic by skin absorption, then a fully encapsulated gas tight suit (Level A) must be used.

Other values that a responder may see are called **lethal doses (LD$_{50}$)** or **lethal concentrations (LC$_{50}$).** The LD is for solids and liquids, and the LC is for gases. In most cases animal studies provide these values, but there are some derived from human studies, suicides, and murders. The 50 attached to the LD or LC means the dose or concentration would be fatal to 50 percent of the exposed population. In the studies a certain number of test subjects were exposed to a low level of chemicals. After a period of time, the test subjects were studied for any adverse effects. Another group was exposed to a higher level of the chemical. When the subjects were exposed to a certain level and 50 percent of the test subjects died, this established the LD$_{50}$ or LC$_{50}$. With threats of terrorism emergency responders must use some military data, much of which is based on LD$_{50}$ or LC$_{50}$ type studies. The military values are generally expressed as **LCt$_{50}$** or lethal concentration to 50 percent of the population with the *t* representing time, usually expressed in minutes. The military also uses **ICt$_{50}$,** which is the incapacitating concentration to 50 percent population in a certain amount of time. **Table 4-6** lists exposure values for some hazardous chemicals.

Although the exposure values can be confusing, it is important to know what each of the values means, and emergency responders should be aware of the exposures they receive during incidents.

SAFETY Air monitoring is a key companion to these exposure levels, and it is the only way to ensure responder safety. Air monitoring for these toxic materials is a HAZMAT technician level skill.

STEL: Allows four exposures of 15 minutes in length, with a minimum of an hour break in-between. The ceiling is the highest exposure, no matter the time.

Figure 4-5   Exposure values.

TABLE 4-6

Exposure Values				
Chemical	**PEL**	**IDLH**	**LCt$_{50}$ (t = 3 min.)**	**ICt$_{50}$ (t = 3 min.)**
Sarin	0.000017 ppm	0.03 ppm	12 ppm	8 ppm
Mustard	0.0005 ppm	0.0005 ppm	231 ppm	21.5 ppm
Acetone	1000 ppm	2500 ppm	N/A	N/A
Acrolein	0.1 ppm	2 ppm	N/A	N/A
Ammonia	50 ppm (ST)	300 ppm	N/A	N/A
Chlorine	1 ppm	10 ppm	N/A	N/A
Ethion	0.02 ppm	N/A	N/A	N/A
Hydrogen sulfide	20 ppm (ceiling)	100 ppm	N/A	N/A
Carbon monoxide	50 ppm 35 ppm NIOSH	200 (ceiling)	N/A	N/A

It is generally recommended that emergency responders use the TLV or PEL as the point at which SCBA must be used. Emergency responders are faced with many chemical exposures, and at these levels safety can be ensured only if proper PPE is worn.

TYPES OF PPE

For firefighters the most common PPE is their firefighter protective clothing (FFPC). For law enforcement officers it is their body armor. The task determines the type of PPE, but when law enforcement officers need to enter an environment that presents a fire risk, then they need some type of PPE. The best type of PPE for this task is FFPC or some other fire-resistive clothing. Used in this text full FFPC is defined as helmet, hood, coat, pants, boots, gloves, PASS, and SCBA, as shown in **Figure 4-6.** For full protection, all of this equipment must be in use. All of the snaps, zippers, and closures must be used to offer optimal protection. FFPC is designed to offer protection against heat and water. The exact amount of protection depends on the type of gear used.

The vapor barrier used in the gear plays a vital role in protection. The vapor barrier used in the military study was a Crosstech^{TM} vapor barrier manufactured by W. L. Gore. This vapor barrier allows the body's moisture to flow through the barrier but offers a waterproof membrane that doesn't allow water to flow back onto the skin. An unintended benefit of this vapor barrier is that it keeps chemicals out as well.

The use of FFPC is controversial as it is not certified for chemical contact. It should not be used for chemical protection, with the exception of making rescues of live victims, which should be immediately followed by emergency decontamination in most cases. Chapter 5 discusses in more detail rescue issues involved with hazardous materials. Some of the latest gear is certified to protect against BBPs, but to be sure check for NFPA certification.

SAFETY The wearing of protective clothing can present significant risk, and both OSHA and the NFPA require the use of the buddy system when entering potentially hazardous environments. When entering IDLH atmospheres there is a requirement for at least two to enter together, and for a backup team to be in place. When one partner has to leave the hazard area, the other partner should always leave at the same time. There should not be a situation where responders are in a hazard area alone.

Self-Contained Breathing Apparatus

NOTE SCBA offers a protection factor of 10,000 from chemical exposure. In other words, a person wearing SCBA has a survivability rating 10,000 times greater than a person who is not wearing any SCBA. To offer this protection, the SCBA must be fitted properly and must be a positive pressure device.

If the SCBA fits poorly or is not a positive pressure device, the protection factor can be considerably less than 10,000. A positive pressure SCBA has an air flow in the face piece all the time and maintains a positive air pressure inside the mask to keep contaminants out. For firefighting or chemical spill response, it is imperative that the SCBA be positive pressure and that it be activated automatically without intervention of the wearer.

SAFETY The odds of survival at a hazardous materials incident are dramatically improved when wearing SCBA, and it should be considered the minimum when dealing with chemical spills.

Figure 4-6 Protective clothing, specifically the SCBA, offers a high level of protection. Each chemical situation is different, and the type of PPE takes careful consideration and risk assessment. Firefighters' protective clothing offers protection for materials that present a fire risk.

Although a variety of SCBA is available, the most common for chemical spill response is a 60-minute type, that on an average allows for a 20- to 30-minute work time for a HAZMAT team member. When determining the use of SCBA, one must consider the time it takes to enter the hazard area, working time, time to leave the area, and time to be decontaminated and undressed. In general, air supplies of 30 and 45 minutes are inadequate for spill response. Some responders to hazardous waste sites may use **supplied air respirators (SAR),** which have some advantages over SCBA. They do not have the weight of the SCBA, although the hose line necessitates some restrictions of movement. Some logistical issues are associated with the use of SARs, but for long-term incidents they are of assistance. Some response teams use **air purifying respirators (APR)** for minor spills, as shown in **Figure 4-7.** Although they offer some advantages and are common in industry, APRs are not commonly used within the emergency services. However, with terrorism and blood-borne pathogen issues becoming more commonplace, we will see an increased use of APRs and

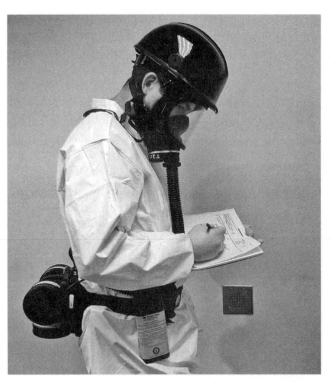

Figure 4-8 This responder is wearing a powered air purifying respirator (PAPR) which filters potentially contaminated air. This is a facepiece style PAPR. The filters (three) are on the back of the belt. The battery to operate the unit is on the hip.

powered air purifying respirators (PAPR). The PAPRs use multiple filters and a blower arrangement that blows a stream of filtered air into a mask. The mask is one of two styles, a hood or a face piece. The face piece style is shown in **Figure 4-8.** Many response teams are using PAPRs in place of APRs as they are cooler to wear. They do have issues with batteries and added weight as compared to APRs. They also require **fit testing** just like the SCBA to ensure that the respirator fits and will offer the wearer protection.

Responders who choose to use an APR at an incident must follow a decision flow chart. According to OSHA regulations, to use an APR or a PAPR these conditions must be met:

■ Oxygen must be at least 19.5 percent and less than 23.5 percent.
■ Level of airborne contamination must be known.
■ Filter cartridge must be appropriate for the chemical.
■ Chemical must have good warning properties.
■ Level must be below the IDLH level.

A variety of cartridges can be used with APRs, and when dealing with different chemicals, a large stock must be maintained. With the exception of waste sites, minor releases, and possibly some decontamination work, SCBA is a simpler option. The only

Figure 4-7 This HAZMAT responder wears an air purifying respirator (APR) to filter out possible contaminants. The APR does not offer as much protection as SCBA.

determining factor for SCBA is to ensure that the air bottle is full.

Problems with SCBA include extra weight, fatigue, lack of full visibility, lack of mobility, contribution to heat stress, need to refill air supply, etc. For this reason, many HAZMAT teams may choose the other types of respiratory protection. The type of SCBA used will determine the amount of weight added to the responder. In general, the longer the work time, the higher the weight of the apparatus. This added weight adds to the overall heat stress on the user. When using SCBA the wearer has limited vision, as the face piece does not allow for a full spectrum of vision. Add a chemical protective suit and this limited field of vision gets even smaller. The use of SCBA in some applications, such as confined spaces, limits responders as to how they can move about the area. People who may be slightly claustrophobic with SCBA must cope with an additional layer of stress placed upon them when donning chemical protective clothing. The users of SCBA and chemical protective clothing should be medically cleared to function with this type of PPE, and they should receive periodic medical exams, according to the applicable regulations.

Chemical Protective Clothing

The four basic levels of chemical protective clothing are broken down into further components. The levels are assigned letters to signify their protection levels. Both OSHA and the EPA use Level A, B, C, and D, with Level A offering the highest level of chemical protection. The NFPA has established standards for protective clothing, and a description of these is provided in the accompanying text box. Since the establishment of these levels, there have been many changes to PPE styles and types. When HAZMAT teams first started, they used reusable protective clothing that was used, cleaned, tested, and then reused. As a result of concerns over the integrity of these suits after each use, suit manufacturers decided to switch to disposable limited-use garments. Although both types of suits are available today, most teams use disposable suits. Some teams maintain at least one type of reusable garment, usually Teflon fabric suits, which are reused due to high cost.

Prior to use, all chemical protective clothing must be checked for compatibility with the chemical that is spilled.

CAUTION No one suit can be used with all chemicals.

Some NFPA-certified suits come close, but even they cannot be used with a couple of chemicals. Chemical compatibility is based on the **permeation** of

Figure 4-9 When conducting a permeation test, the fabric splits a test container, and a measurement device is used to see if the chemical goes through the fabric.

a chemical through the fabric of the suit. Permeation is the movement of a chemical through the fabric on the molecular level, as shown in **Figure 4-9.** The fabric is not damaged or visibly changed in any way. Imagine wrapping garlic in plastic wrap. Initially you don't smell anything, but after a few hours the scent of garlic has permeated the refrigerator as if you hadn't wrapped it at all, because the garlic vapors have permeated the plastic wrap. Compatibility charts are provided by the manufacturer to list the chemicals the fabric of protective clothing will withstand.

Level A Ensemble

Level A protective ensemble, as shown in **Figure 4-10,** provides the highest level of protection against chemical exposure.

NOTE For all the protection Level A gear affords responders, they remain at risk of suffering heat stress and physical and psychological stress from wearing the garment.

Level A ensembles may also be called an **encapsulated suit,** or a vapor-tight totally encapsulating suit. To be rated Level A, the suit must have vapor-tight attached gloves, attached vapor-tight boots, and a gastight zipper. Because the suit is designed to prevent

NFPA PROTECTIVE CLOTHING STANDARDS

Although the NFPA refers to Levels A–D throughout its documents, it uses a different naming system to describe the various levels of protective clothing. OSHA/EPA Level C protective clothing equates to the NFPA 1993 Standard for Support Function Protective Garments. The standard for OSHA/EPA Level B protective clothing fits into NFPA 1992 Liquid Splash Suits for Hazardous Chemical Emergencies. As with the OSHA/EPA levels, these suits can be both encapsulating or non-encapsulating. Level A suits are outlined in NFPA 1991 Vapor Protective Suits for Hazardous Chemical Emergencies. All three standards set construction requirements and require the suits to be tested for compliance with the standard. The NFPA 1991 certified suits cost about $1,000 more than a non-certified suit, and one manufacturer reports that certified suits make up only 20 percent of its Level A sales.

---

gases from penetrating the garment, it is liquid tight as well. Since materials, including air, cannot get in or out of the suit, a responder wearing it must use a SCBA the whole time. The suit does have relief valves that vent exhaled air after a certain pressure buildup. The requirement to use Level A ensemble within the HAZWOPER regulation is that the atmosphere must be at or above the IDLH value, and the chemical must be toxic by skin absorption. In some cases, the use of Level A would be recommended for chemicals that would not meet that definition. Level A would be advisable in incidents in which a responder may be covered with a hazardous chemical. A good example of this situation is responders working on a valve in the bottom of a corrosive tank, especially if the valve is above their heads as they are working on it. If the valve fails, it would dump the corrosive chemical all over the responders.

SAFETY When wearing a Level A ensemble or any other type of PPE, pre-hydration is necessary to prevent heat-related emergencies.

Figure 4-10 The Level A offers good protection from chemicals. It is a vapor-tight suit and resists permeation by most chemicals.

Dressing should take place in a cool quiet area, preferably in the shade. During emergency operations all members must be monitored for heat stress. Lack of full visibility and heat stress are major concerns when using Level A ensembles. The Level A ensemble typically consists of:

- Encapsulated suit with attached gastight gloves and boots
- Inner and outer gloves
- Hard hat (optional)
- Communication system (optional)
- Cooling system (optional)
- SCBA
- PBI/NOMEX® coveralls (optional)
- Overboots

A NFPA 1991 (Encapsulated Suit Specifications) certified Level A suit also offers some flash fire resistance. To assist in compliance newer suits use a blended fabric to offer chemical resistance and flash resistance. Older style suits use a flash suit overgarment made of aluminized PBI/KEVLAR® fabric. This flash protection is not for firefighting, but it offers 3 to 15 seconds of protection when exposed to a flash fire. A flash suit should also not be used when responding to an alkaline material such as anhydrous ammonia, which reacts with the aluminum and destroys the flash protection.

Level B Ensemble

Within the **Level B** family is a wide variety of suit types, as shown in **Figure 4-11.** Although the EPA and OSHA acknowledge two basic types, additional styles are available. The two basic types are coverall style and encapsulated, but even these have subvarieties. Encapsulated Level B ensembles are similar to the Level A ensemble but are not vapor

PROTECTIVE CLOTHING RATINGS

Permeation is based on an 8-hour day, or 480 minutes. When a responder's department's inventory has more than one suit fabric, it is best to choose a fabric with a value of 8 hours, which indicates that the fabric should show no breakthrough in 8 hours. After the fabric reaches 8 hours, the test is usually halted, unless specific times are indicated by the manufacturer. Most fabrics are tested against a battery of chemicals as specified by the American Society for Testing and Materials (ASTM) or the NFPA. These batteries of chemicals represent the majority of the chemical families that responders are most likely to encounter.

Two other areas of concern when dealing with PPE are degradation and penetration. The degradation of a suit involves its physical destruction, such as leaving a hole or damage. Penetration is the movement of chemicals through natural openings such as zippers, glove/suit interface points, and other areas of the suit; it does not involve damage but is an area where chemicals may enter the suit. Listed in the accompanying table are some examples of permeation data the manufacturer may provide. The suits (fabric type) are listed across the top of the table, and seven fabric types are provided in the table. Kappler Protective Fabrics provide a Level B/C suit selection called Chemical Protective Fabric (CPF) and a number 1 to 4. The Responder fabrics are usually configured in Level A suits, and the butyl and CPE fabrics are used for Level A and B suits. As you will note, no one fabric is good for all chemicals. The manufacturer tests its fabrics against hundreds of chemicals; the list provided here is only a small portion of the test results.

Chemical suit fabric	Kappler Zytron 200	Kappler Zytron 300	Kappler Zytron 500	Dupont Tychem 10,000	Dupont ThermoPro	Trelleborg Trelchem HPS type T
Ammonia gas	NT	39	>480	>480	>480	>480
Acetone	17	>480	>480	>480	>480	>480
Bromine				91		360
Carbon disulfide	2	>480	>480	>480	>480	>480
Chlorine	>480	>480	>480	>480	>480	>480
Sodium hydroxide	>480	>480	>480	>480	>480	>480
Toluene	6	>480	>480	>480	>480	>480

Excerpted from http://www.kappler.com and from http://www2.dupont.com/Personal_Protection/en_US/tech_info/literature/literature_tychem.html and http://www.trelleborg.com/protective/images/hps_t.pdf. All values are in minutes.

This table shows test results for 480 minutes (8 hours); if the suit provides protection for more than 480 minutes, it is expressed as >480 minutes. Note that the fabrics listed here provide good protection against the chemicals listed. Several suits provide adequate protection against ammonia. The Zytron 300 fabric offers 39 minutes of protection, while the Zytron 500 suit up offers >480 minutes. The remainder of the suits offer >480 minutes of protection. Each suit and therefore fabric generally have a problem with at least one chemical. Although the chart does not indicate such, the Zytron 300 has several chemicals that break through relatively quickly.

Keep in mind that these permeation tests put pure chemical directly against the fabric, a situation that is not normally found in the street. In most cases we do not deal with pure products. Nor are HAZMAT responders in the habit of coming in direct contact with the chemical. Another factor to consider is that most entry team times are 20 to 40 minutes in length, thereby limiting the contact time with the chemicals.

tight and do not have attached gloves. They will typically have attached booties, and the SCBA is worn on the inside. Some manufacturers provide glove ring assemblies that allow for the gloves to be preattached during storage. A variety of fabrics is available for Level B ensembles, and compatibility is important.

NOTE The Level B ensemble suit is the workhorse of HAZMAT teams, as it is the most common suit used.

The encapsulated Level B suit is sometimes referred to as a *Bubble B* or *B plus* suit. Although they

Figure 4-11 All three suits are Level B suits. From the left is a coverall style, an encapsulated suit, and a two piece. Level B is intended to offer splash protection and is not gas tight.

are lighter, these suits have some of the same heat stress issues as do the Level A ensembles. The encapsulated Level B ensemble usually does not have relief valves, but one to three holes covered by a flap help exhaust the exhaled air. Unlike a Level A, the Level B ensemble will not inflate or build pressure as this flap will not hold excess pressure. Chemicals can enter through this flap in unusual situations, whereas chemicals cannot enter through a Level A relief valve. Other styles of Level B include a two-piece garment consisting of a jacket and pants, usually bib-overall style. The coverall style Level B may have attached booties or a hood. The one item that makes the Level B different from the lower levels of PPE is the use of SCBA. Level A is a gastight suit, whereas Level B is intended for splash protection, with respiratory protection from a SCBA. A Level B suit ensemble consists of:

■ Level B suit
■ Hard hat (optional)
■ Inner/outer gloves
■ SCBA
■ Communication system (optional)
■ Outer boots
■ NOMEX®/PBI coveralls (optional)

Level C Ensemble

Level C, as shown in **Figure 4-12,** incorporates the use of an APR within the ensemble. For obvious reasons an APR cannot be used within an encapsulated

Figure 4-12 The difference between Level C and Level B is the respirator. When using an APR, the level is C; SCBA equates to Level B. Responders cannot use an encapsulated suit and an APR, as the wearer of the APR requires good oxygen to survive.

suit, but it can be used with the other styles. A Level C suit can be a coverall or a two-piece garment.

NOTE A Level C suit is used where splashes may occur but where respiratory hazards are minimal and are covered by the use of an APR.

When using an APR, an extensive listing of requirements must be met. A Level C ensemble consists of:

■ Level C suit
■ APR
■ Hard hat
■ Inner/outer gloves
■ Outer boots
■ NOMEX®/PBI coveralls

Level D Suits

A Level D suit, shown in **Figure 4-13,** is actually work clothing. It is used when respiratory protection is not required and splashes are not a concern. Level D provides no chemical protection but offers

Figure 4-13 Level D is standard work clothing and may include chemical gloves, goggles, and boots.

protection against other workplace hazards. Level D protection consists of:

■ Hard hat
■ Safety glasses
■ Work clothes
■ Safety shoes/boots
■ Chemical/work gloves

High Temperature Clothing

The two basic types of high temperature clothing are proximity and entry gear. A set of proximity gear is shown in **Figure 4-14.**

Entry suits are sometimes used in high temperature applications within industry, such as steel making. This high temperature gear is usually identified by its characteristic aluminized outer shell. This shell is usually attached to PBI/KEVLAR® mixture, which offers a higher heat resistance than normal FFPC. This proximity gear is named for its ability to allow the wearer in close proximity to the burning liquid. It offers protection for temperatures up to 300° to 400°F (149° to 204°C). Fire entry gear is designed to allow the wearer to enter a fully involved fire area for a period of 30 to 60 seconds over the life of the suit. Its original intent was to allow for the rescue of trapped victims, but it is now well established that the victims would not be alive if that garment were needed. Useful applications for this gear involve industrial incidents where high temperatures would be found, but these are rare and most departments no longer carry this type of gear. Fire entry gear can be used in temperatures ranging up to 2000°F (1093°C).

Figure 4-14 Proximity gear is used for flammable liquids firefighting, most commonly by airport firefighters and industrial firefighters. *(Photo by Tech Sgt. Brian E. Christiansen, USAF)*

Cold Temperature Clothing

When dealing with cryogenics, the responder must wear protective clothing that protects the wearer against very cold temperatures. Even propane from a cylinder, once released to the atmosphere, can reach well below 0°F (–17°C). By definition, cryogenics are at least –150°F (–101°C) and are usually colder. Standard protective clothing will not be effective against these types of materials. Responders should take precautions against cold stress and prevent hypothermia. Although no full protective clothing ensemble is designed specifically for cold materials, various gloves, which are shown in **Figure 4-15,** gauntlets, and aprons protect against extreme cold. Layering of the clothing can offer some additional protection. Unfortunately, in addition to the cold, cryogenics usually present other hazards, such as fire, corrosive, toxicity, or asphyxiation.

Figure 4-15 Cryogenic gloves are used when dealing with very cold liquids. They offer protection from a range of −150 to −450°F.

Limitations of PPE

Four basic limitations apply to protective clothing, including law enforcement body armor, SWAT gear, bomb suits, EMS infectious control gear, FFPC, and fully encapsulated Level A suits. Most of these issues come to mind when dealing with chemical protective clothing, but they should be considered when using any type of protective clothing. The four major issues are heat stress, mobility, visibility, and communications problems.

Heat Stress

SAFETY Stress is a leading killer of firefighters, and heat stress plays a major factor in many of these deaths.

FFPC, although getting lighter, places tremendous stress on the wearer. The type of vapor barrier also influences the amount of heat stress a person will encounter, and not all vapor barriers are created equal. W.L. Gore and Associates has developed a new vapor barrier that can be used in an emergency response environment. The clothing was designed for both firefighters and law enforcement. In addition to the firefighter's style, a law enforcement SWAT style coverall is also available. The clothing is known as GORE™ CHEMPAK™ Ultra Barrier Fabric. It is designed for a variety of urban search and rescue (USAR) and HAZMAT situations, as shown in **Figure 4-16.** Given

Figure 4-16 This type of PPE is very useful for USAR situations, and also offers chemical protection and is well suited for WMD response. (*Photo by W. L. Gore & Associates*)

the basic risk categories discussed above this protective clothing is flame resistant and chemical resistant, which are unique characteristics. It meets the NFPA 1994 Standard on Protective Clothing Ensemble for Chemical/Biological Terrorism Incidents, and the NFPA 1992 Standard on Liquid Splash Protective Ensembles and Clothing for Hazardous Materials Emergencies.

Among the greatest advantages of this clothing are its weight and mobility as it offers a high level of protection, but in a more form-fitting profile. For both firefighters and law enforcement personnel this presents an advantage over firefighters' protective clothing or Nomex™ SWAT clothes. Other than for active or structural firefighting this protective clothing has benefits, and offers protection against WMD agents. The gloves which are made of GORE™ CHEMPAK™ Ultra Barrier Fabric are some of the most tactile gloves made for hazardous situations and are unheard of in chemical protective garments.

DuPont(tm) has also developed a new type of protective clothing that has some unique capabilities as well. The main feature of the Tychem® ThermoPro garment is its chemical and flash fire resistance. In other clothing for HAZMAT response the combination of chemical and flash fire resistance was not an option; responders had to choose one protective stance and take some risk with the other. Now they can have both features with the ThermoPro suit. This garment meets the NFPA 1992 standard and the NFPA 2112 Standard on Flame Resistant Garments

Figure 4-17 Any type of PPE adds stress to the wearer even in cold weather. Hydration is the key to survival. The wearer of this bomb suit was subjected to significant heat stress due to the PPE.

for the Protection of Industrial Personnel against Flash Fire. This garment is also designed for law enforcement applications and WMD response. It is important for emergency responders to be able to recognize heat stress and its degrees of seriousness. Heat stress can lead to heat stroke, a condition that is almost always fatal. The temperature inside a Level A ensemble can easily exceed 100°F (38°C), even on a cold day. When using any type of PPE, responders should take care to pre-hydrate by drinking several glasses of water and take their time when doffing their PPE. Immediately removing their PPE can shock the body, as shown in **Figure 4-17,** and

DECISION FACTORS IN CHOOSING PPE

When choosing the best type of PPE, one should use the term *appropriate level of protective equipment.* There is not any one type of garment that protects the wearer against all hazards. A number of decision factors need to be taken into consideration, some of which carry more weight than others. The incident commander should rank the risk and the benefits to determine if the chosen PPE is appropriate. For example, many chemicals are flammable, and standard chemical protective clothing offers no protection against heat or flame. There are steps that can help reduce the flame-induced shrink-wrapping effect with chemical protective clothing, but none of them offer complete protection. One common misconception is that flash covers on Level A encapsulated suits will offer fire protection. They are designed to offer flash fire protection for a few seconds and are not effective in a free burning fire situation. The major issue with flash covers is that the cover is not gas tight, and it is

possible to have the flammable gas in-between the flash cover and the chemical suit. If the gases ignite, they may flash inside the cover. Tactically, it is best to remove the fire hazard through ventilation or other means and monitor continually to ensure that personnel are not working in a flammable atmosphere.

Some of the factors to consider are:

- **Risk category**—fire, corrosive, toxic, and/or radioactive
- **Explosive risk**
- **Vapor pressure of the chemical**—other properties are also considered
- **Contamination possibility**—to include amount and location, such as a whole body
- **Time on task**—how long is the exposure?
- **Type of task**—how hard are the team members going to work?
- **Temperature**—heat stress

cause serious health reactions. This is important to remember when removing PPE from HAZMAT responders, as they are usually zipped up inside a very warm environment and cannot control the undressing progress.

CAUTION Dehydration is a major hazard. When operating at a fire or a chemical spill, frequent hydration and rest breaks are important.

The progression of heat stress is dependent on the amount of work being performed and the physical ability of the responders. To head off heat-related emergencies, responders should hydrate **and** take frequent breaks, and take extra precautions until acclimated. How acclimated responders are to heat (or cold) is indicated by the first incident on an extremely hot (or cold) day. When an incident occurs on a day in which the weather conditions would not be considered normal, responders should pay extra attention to their activity level. Early recognition is important for the health and safety of the responders. The levels and their warning signs are:

Heat Cramps One of the initial signs of a heat-related problem is heat cramps, although some persons may never experience cramping prior to heat exhaustion. Cramping is usually in the stomach and extremities.

Heat Exhaustion This real first step to dehydration is fairly common. The root cause is excess sweating and a loss of body fluid. Symptoms include dizziness, headache, nausea, diarrhea, vomiting, and decreased urinary output. Heat exhaustion affects not only HAZMAT responders but firefighters as well. Responders should have had enough fluid that they feel a need to urinate. Untreated heat exhaustion may lead to heat stroke, which is a very serious and often fatal condition.

Heat Stroke This is a serious emergency and can be fatal to 80 percent of the patients. In simple terms, the body's ability to regulate its temperature has failed and is no longer functioning. Symptoms include unconsciousness, hot and dry skin, seizures, confusion, and disorientation. Being dehydrated when approaching heat stroke levels only complicates the process and speeds up the patient's deterioration. It is best to prevent heat stroke, as once the body shuts down, the result is usually fatal.

Cold Stress Liquefied gases or cryogenic gases present situations where cold stress could be a factor. When skin comes in contact with or in proximity to cold vapors or liquids it can be severely damaged. The symptoms of cold stress include discoloration of the skin, numbness, and pain. The more severe the damage the darker the skin will become. Skin severely damaged will be black from the exposure.

Mobility

Anytime you add a layer of protective clothing, you are reducing your mobility. Each subsequent level of protective clothing places additional stress on the wearer. Part of the risk-based response profile is to make sure responders wear appropriate levels of protective clothing for the hazard. In many cases the level of protection is chosen as a knee-jerk reaction to an event. Heat stress is the major concern with PPE, but the lack of mobility adds to physiological stress. If the responder is not comfortable with the protective clothing, the potential for injury increases. With Level A clothing the suit will slightly inflate with exhaled air, creating a large balloon-like effect, until the relief valve kicks in and releases some of the internal pressure. This ballooning effect makes it difficult to move about and can create vision problems. Envision crawling on top of a railcar for the first time. The top of a railcar is not designed for a person in a Level A suit to be on top of it. A very real danger to responders on top of a railcar is the great possibility they may fall off.

Visibility

Level A and B ensembles have a small window for vision in addition to the face piece on the SCBA. This double window effect causes problems. In many cases the suit face piece will fog up, causing severe vision problems. Sometimes the application of anti-fog spray alleviates this problem, but at other times nothing helps. Another vision problem occurs when a suit is larger than it should be; the responder's head may move but the face shield on the suit does not. In these cases responders must hold the face shield of the suit and turn it with their head to see. This adds to the other stresses of a dangerous situation. When responders cannot see their hands in front of their faces, the dangers increase.

Communications

The ability to communicate through a SCBA is problematic at best, and it takes a learned ear to understand many persons talking through the mask. Add a chemical suit on top of this, and communication is very difficult. In most cases two persons in a chemical suit are unable to use verbal communication with each other and must rely on hand signals. Most HAZMAT teams use radio communications to maintain contact with other responders. These communications may be ear mikes in which the responder places an ear piece to hear and talk through. Other systems use a bone mike, which is strapped to the top of the skull, to communicate. These systems work fairly well in most environments. One of the major problems with communications is that in some cases responders are working in a noisy environment such as a working chemical plant, which makes communications difficult.

SUMMARY

Having an understanding of the health effects and routes of exposure is important for responder safety. Most chemicals present chronic hazards, but acute hazards should be avoided as well. The routes of entry and type of harm that chemicals present vary and chemicals usually present multiple hazards. The use of protective clothing is important for the various hazards that responders may face. The absolute minimum respiratory protection should be fire resistive and SCBA, as this combination offers a high level of protection. If you cannot confirm the absence of hazardous materials through the use of air monitors, a basic level of protective clothing should be used. The use of chemical protective clothing is not an easy decision; many factors must be considered prior to using chemical protective clothing. Because of the ensuing heat stress that accompanies the wearing of chemical protective clothing, in some cases the protective clothing itself may be more dangerous than the chemical hazard. For this reason responders must use effective risk assessment to determine the true hazard and then dress for the hazard. Many responders have a fear of radiation, but it is no different than other hazardous materials in that one just has to understand the hazards. With radiation and with chemicals understanding dose is paramount to remaining safe. We are exposed to radiation every day, and just as with chemicals the dose makes the poison.

KEY TERMS

Absorbed dose A measure of the amount of radiation transferred to a material.

Activity The number of decays per second within a radiation source which can be equated to the strength of the source.

Acute Quick, one-time exposure to a chemical.

Air purifying respirators (APR) Respiratory protection that filters contaminants out of the air, using filter cartridges; requires the atmosphere have sufficient oxygen along with other regulatory requirements.

Allergen A material that causes a reaction to the immune system.

Becqurel (Bq) Unit of measure equivalent to one decay per second.

Carcinogen Material capable of causing cancer in humans.

Ceiling level Highest exposure a person can receive without suffering any ill effects; combined with the PEL, TLV, or REL, it establishes a maximum exposure.

Chronic Continual or repeated exposure to a hazardous material, over a period of time.

Convulsant A chemical that has the ability to cause seizure-like activity.

Curie Unit of measure equal to 37 billion disintegrations per second. One curie is the activity of 1 gram of radium-222, a radiation source.

Encapsulated suit Chemical suit that covers the responder, including the breathing apparatus; usually associated with gastight and liquid-tight Level A suits. Some Level B styles are fully encapsulated, but not gas or liquid tight.

Etiological Hazard that includes biological agents, viruses, and other disease-causing materials.

Fit testing Test that ensures that the respiratory protection fits the face and offers maximum protection.

Half-life A measurement of time, in which a radiation source's activity (strength) is cut in half.

Ict$_{50}$ Military term for incapacitating level over set time to 50 percent of the exposed population.

Irritant Material that is irritating to humans, but usually does not cause any long-term effects.

Isotope A variation of chemical with a different number of neutrons.

LC$_{50}$ See Lethal concentration

LCt$_{50}$ Military term for lethal concentration over set time to 50 percent of the population.

LD$_{50}$ See Lethal dose

Lethal concentration A term used to describe amount of vapors or gas that was inhaled to cause harm or death to the exposed test animals. A lethal concentration 50 percent means that the dose killed half the animals that were exposed to that level.

Lethal dose is a term used to describe a dose of a solid or a liquid that caused harm or death to test animals. A lethal dose 50 percent means that the dose provided killed half of the animals.

Level A protective clothing Fully encapsulated chemical protective clothing; gastight and liquid-tight gear offer protection against chemical attack.

Level B Level of protective clothing usually associated with splash protection; Level B requires the use of SCBA.

PAPR Powered air-purifying respirator, which is comparable to an APR, but has a fan to draw the air into the filtration system and into the mask.

Permeation Movement of chemicals through chemical protective clothing on a molecular level; does not cause visual damage to the clothing.

Permissible exposure limit (PEL) OSHA value that regulates the amount of a chemical that a person can be exposed to during an 8-hour day.

Radioactive decay The release of radiation, typically gamma, but may be alpha or beta.

Radioisotope An isotope whose nuclei are unstable and are radioactive.

Radon A common radioactive gas found in homes.

Recommended exposure level (REL) Exposure value established by NIOSH for a 10-hour day, 40-hour workweek. Is similar to the PEL and TLV.

Risk-based response An approach to responding to a chemical incident by categorizing a chemical into a fire, corrosive, or toxic risk. Using a risk-based approach can assist the responder in making tactical, evacuation, and PPE decisions.

Self-contained breathing apparatus (SCBA) Protective gear that safeguards responders' respiratory systems; considered essential for responding to chemical spills.

Sensitizer Chemical that after repeated exposures may cause an allergic effect for some people.

Short-term exposure limit (STEL) Fifteen-minute exposure to a chemical; requires a 1-hour break between exposures and is only allowed four times a day.

Supplied air respirators (SAR) Respiratory protection that provides a face mask, air hose connected to a large air supply, and an escape bottle; typically used for waste sites or confined spaces.

Threshold limit value (TLV) Exposure value that is similar to the PEL but is issued by the ACGIH. It is for an 8-hour day.

TRACEM Acronym for the types of hazards that may exist at a chemical incident: thermal, radiation, asphyxiation, chemical, etiological, and mechanical.

REVIEW QUESTIONS

1. What are the four common routes of exposure?
2. What route of entry is the easiest to protect against?
3. To what contaminant are emergency responders commonly exposed to through injection?
4. What type of exposure is 20 years of smoking considered?
5. Which type of exposure is a one-time event?
6. What six hazards potentially exist at a chemical release?
7. Which medical effect can become hereditary?
8. Nitrogen is what type of asphyxiant?
9. Which exposure value uses a 15-minute time limit?
10. What is the term *ceiling* used for in exposure values?
11. Which level of chemical protective clothing offers a high level of protection against chemicals that are toxic through skin absorption?
12. What is a major concern as the levels of protective clothing increase?
13. Which respiratory protection offers the highest level of protection?

ENDNOTE

[1]Not all emergency services workers are covered by OSHA; some are covered by their state occupational safety agency or are not covered by such regulations. This varies state to state, and you should consult your state occupational safety office.

Suggested Readings

Basics of Nuclear Radiation, Technical Note TN-176, RAE Systems, San Jose, CA, 2005.

Lesak, David, *Hazardous Materials: Strategies and Tactics,* Prentice Hall, Upper Saddle River, NJ, 1998.

Noll, Gregory, Michael Hildebrand, and James Yvorra, *Hazardous Materials: Managing the Incident,* Fire Protection Publications, Oklahoma University, 1995.

Radiation Basics, Technical Note, Health Physics Society, McLean, VA, 2005.

Schnepp, Rob, and Paul Gantt, *Hazardous Materials: Regulations, Response, and Site Operations,* Delmar Cengage Learning, Clifton Park, NY, 1999.

Smeby, L. Charles (editor), *Hazardous Materials Response Handbook,* 4th ed., National Fire Protection Association, Boston, MA, 2002.

Wahlström, Björn, *Understanding Radiation,* Medical Physics Publishing, Madison, WI, 1995.

PROTECTIVE ACTIONS

OUTLINE

▌ **Objectives**

▌ **Introduction**

▌ **Incident Management Systems**

▌ **Hazardous Materials Management Processes**

▌ **Site Management**

▌ **Common Incidents**

▌ **Decontamination**

▌ **Methods of Decontamination**

▌ **Summary**

▌ **Key Terms**

▌ **Review Questions**

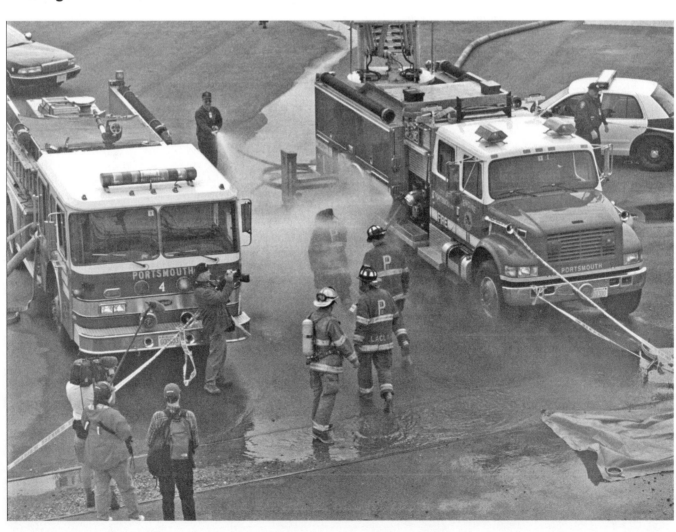

STREET STORY

It was the start of a normal day on the job, until later in the morning when an alarm for a chemical leak inside a beverage warehouse was sounded. The dispatch consisted of what we call a HAZMAT box: four engine companies, one truck company, a rescue squad, basic life support unit, HAZMAT company, and a commanding officer. I was working the HAZMAT company that day. While en route, a radio transmission by the first arriving company advised the communications center of a major ammonia leak inside the warehouse storage area. The first arriving company was taking protective actions. This area contained multiple storage of beverages and boxes within an enclosed and secured area that included the valves and piping for the anhydrous ammonia refrigeration system.

Upon our arrival, the warehouse had been evacuated and a strong odor of ammonia had already consumed the entire area surrounding the warehouse. Once we had performed a hazard risk assessment and ensured that the first responders had taken appropriate protective actions, we selected our level of protection. Myself and three hazardous materials technicians entered the release area to shut off the valve to the leaking pipe. After locating the release area we found the valve and made an attempt to close it. While closing the valve a sudden release of gaseous and liquid ammonia covered the personnel working at or around the valve. Visibility was taken from us almost instantaneously because of the gaseous release, and communications were lost between all four technicians. I was able to find my way out and noticed that my personnel were still in the release area. Prior to making another entry to locate my personnel, I noticed a white smoke coming from my chemical boots.

After further investigation I realized that the oil-based paint from the concrete floor was causing a chemical reaction under the soles of my boots. I reentered the release area, located my personnel, and immediately withdrew from the release area to the decontamination area. Once we were refreshed, a second entry attempt was made into the release area, where we were able to locate another sectional valve and stop the leak. The hardest part of the second attempt was removing our SCBA inside our suits to squeeze past piping and valves to get to the right one.

Before we left the scene, we finally determined that prior to our arrival a firefighter had entered the release area and closed the valve without notifying command and/or hazardous materials personnel. When hazardous materials personnel entered the release area thinking the valve was not closed, they actually reopened it, which caused the valve to freeze in the open position. In this event, a number of factors affected our response. First, arriving crews needed to address isolation and evacuation issues and the type of release, but one firefighter was endangered by not taking appropriate protective actions. No personnel were injured or exposed to the ammonia, but the incident proved to be very dangerous as a result of personnel freelancing and the lack of training present at an emergency scene.

—Street Story by Gregory Socks, Captain, Montgomery County, Maryland DFRS

OBJECTIVES

After completing this chapter, the reader should be able to identify and explain:

- ■ The various incident management systems (A)
- ■ The four methods of vapor cloud movement (O)
- ■ The methods used to determine isolation and evacuation decisions (O)
- ■ The determination and use of hot, warm, and cold zones (A)
- ■ The use of air monitors in the determination of zones (O)
- ■ Common incidents within each hazard class (O)
- ■ Methods of emergency decontamination (O)
- ■ The four types of decontamination (O)

INTRODUCTION

This chapter provides a myriad of topics for the responder and focuses on some general tactics that should be followed at an incident. The tactical considerations provided here are for general situations and may not apply to specific incidents. Each chemical spill is different, and for each spill there may be another way of handling that release.

NOTE The basic concept for first responders is one of isolation: Do not allow other people to become part of the incident, and protect those involved with the incident or those who may become part of the incident in a short time.

A lot of the information in this section may not apply to you specifically. As you are beginning your training, it is unlikely you will be making community evacuation decisions for the next couple of years. These are things to keep in mind in case you do have to make these decisions in the near future.

INCIDENT MANAGEMENT SYSTEMS

Several incident management systems (IMSs) are designed for use in managing emergencies. The three most commonly used systems are the Incident Command System (ICS), Incident Management System (IMS), and Firescope. The Firescope system originated in California to assist in the management of forest fires. Although originally designed to be used by fire response agencies, all of the ICS or IMS systems can be used by any agency. The systems are management systems that are used for emergencies. The basic system does not care what the emergency or disaster is; the system is a tool to manage the resources at that event. The system can be used for nonemergency situations, anywhere a group is being used to manage or work an event. All of the systems are comparable in the method of managing the incident, and the focus is to enable the incident commander to manage the incident effectively. The biggest difference in the systems is some terminology. OSHA requires the use of an IMS at a hazardous materials release but does not specify which system should be used. OSHA's major concern is that all the employees know the IMS and that it is written for review by employees.

The focus of the systems is to break the incident down into manageable pieces, with officers managing a small number of people. The desired number of personnel for one person to manage, known as the **span of control,** is four to six. The systems divide the incident into five major functions, with each function having additional tasks associated with it. Each task has a leader who reports up the chain of command to the incident commander. The major tasks are:

- ■ **Incident commander (IC)**—The person who is in charge of the incident and is responsible for all activities at the incident. The incident commander may also have staff support at the command post who are handling:
  - ■ Safety—Assists the incident commander with overall safety at the incident; has the authority to stop any unsafe activity without prior consultation of the incident commander. At a hazardous materials incident there are typically two safety officers, one for overall safety and one specifically assigned to the HAZMAT team to oversee specific issues.

NOTE OSHA requires the use of a safety officer and requires that the safety officer be knowledgeable about the tasks the responder will perform.

A person trained to the operations level is not adequate to be the safety officer for the HAZMAT operation. A HAZMAT safety officer should be trained to the technician or specialist level to make sure that the mitigation efforts are carried out effectively and safely.

- ■ Public Information—The person who will meet with the press and the public to provide updates and information from the incident commander.

- Liaison—There may be more than one person in this role, but the most common would be someone to interface with the police department or a police representative assigned to the command post. Other response agencies may be represented at the command post in the role of liaison.

■ **Operations**—This officer handles the tactical portion of the incident and manages the personnel who are mitigating the release. The person assigned to operations works closely with the incident commander and keeps the IC informed as to resource needs and the status of the event.

■ **Logistics**—The person responsible for obtaining and coordinating all the resources necessary to mitigate the incident. This may include just responder resources or may involve other local government or private sources of equipment. Once the equipment arrives, this person manages the equipment and provides it to operations.

■ **Finance**—The person responsible for tracking expenses involved with the incident, in addition to ensuring payment of any costs associated with the equipment that the logistics officer may obtain. In career organizations this person may track overtime and assist with the planning officer to minimize this impact. If the HAZMAT team bills the **responsible party,** the finance officer usually tracks the expenses incurred at the incident so that the responsible party can be billed for the incident.

■ **Planning**—The planning officer seeks to anticipate any needs of the incident commander. As many HAZMAT incidents have potential to last several hours to several days, this forecasting may be difficult. A good example would be when a planning officer anticipates that crews will be used for many hours and relief will be needed. Crews may also require food and refreshments, which the planning officer should anticipate. The planning officer does not request these items specifically, but runs the ideas through the incident commander who then requests these items.

The management of a hazardous materials incident can be very difficult even for a seasoned incident commander. A number of incident management strategies are available to the IC, some of which are outlined in Delmar's *Firefighter's Handbook.* HAZMAT specific systems are outlined in **Table 5-1.**

A fire department IC on a normal fire incident deals predominately within one agency. There may be occasion to discuss issues with the police department, and depending on the location, a police officer may be in the command area. The IC may also deal with other municipal agencies, such as the health department or social services. Other outside agencies, such as the Red Cross, may also be involved and work under the direction of the IC. Depending on the size of the community, the media may be a factor in the management of the incident. But in the overall scheme of the incident, ICs deal mostly with their agency and their own coworkers.

When involved in a chemical release, especially one of major proportion, considerably more agencies are involved. From the start in some communities, the HAZMAT team may not be associated with the local fire department; it may be from a county mutual aid group or from an adjoining community. Many incidents involve major road closures, which bring an increased presence of police officers; depending on the size of the road, these incidents may bring command level officers to the scene. If the incident involves a state road or highway or has the potential for impacting these roads, the state highway department may arrive at the incident. If these agencies exist, county or state environmental representatives or responders may arrive at the incident to assist. In some states the Department of Natural Resources (DNR) has jurisdiction in chemical spills and may respond to an incident.

On a large spill, such as the incident depicted in **Figure 5-1,** the media will be a much larger group and may be from outside your area. At a chemical release the incident usually involves the use of a cleanup contractor who will be arriving at the incident to assist in the cleanup effort. Depending on the incident, it is not uncommon to see insurance adjusters arriving within a few hours of the spill. On larger spills or spills that may endanger a waterway, the EPA may arrive to assist with the spill. Although this listing of agencies is nowhere near complete, it represents some of the agencies that may be represented at the incident, all coming to assist the IC with mitigation. In some cases ICs may have several alarms worth of equipment from their own agencies to manage along with equipment and personnel from other agencies. Until the incident is moved from the emergency phase to the nonemergency cleanup phase, the IC is usually still in charge.[1] In some cases the IC will have a more difficult time handling the "assistance" than actually managing the incident. When this occurs, public information officer (PIO) and the liaison officer will be of great assistance.

TABLE 5-1

Branch Positions

■ **Backup**—Assigned to rescue the entry team if necessary; dressed in the same suits as the entry team and fully prepared to make an entry, with the exception of being on air. The number of backup personnel needs to be a minimum of two, but can be expanded to more personnel depending on the incident. The backup team must be trained to the minimum of technician.

■ **Decontamination**—Person assigned to oversee the setup and operation of the decontamination area, sometimes referred to as the *decon area*, or the contamination reduction corridor. Other personnel will operate within the area performing the decon or PPE removal. Persons operating in this area are usually trained to the technician level, but persons trained to the operation level may also be operating in this area. If operations level personnel are used, they should have received specific training in this job function.

■ **Entry**—Two-person team at minimum who will enter the hazard area or hot zone. The type of chemical will determine the type of PPE that will be used. Prior to entry the IC must brief the personnel who will be working in the hazard area. Entry team members must be technicians or specialists.

■ **HAZMAT branch management**—Person responsible for handling the tactical objectives, as assigned by the IC. This person is in charge of the HAZMAT team and coordinates the efforts of this team; may be referred to as HAZMAT operations.

■ **HAZMAT branch safety**—Assistant to the overall safety officer, but is concerned mostly with HAZMAT specific issues. This HAZMAT safety officer is usually a technician or above and is responsible for HAZMAT specific concerns, such as PPE selection and use. The overall safety officer is concerned with all personnel at the incident and usually focuses on other hazards not related to the chemical, such as slip, trip, and fall. Slip, trip, and fall hazards are those that are caused by slippery floors or pavement, uneven terrain, or holes, and include falls from ladders or elevated areas. These are the most common type of workplace injuries.

■ **Information/research**—Person who provides the information regarding the chemical and physical properties to the HAZMAT branch officer and the IC. After completion of these duties, this person may then assume responsibility for documenting the incident.

■ **Reconnaissance**—Through the use of binoculars or possibly by entering the hazard area, this person or team determines incident severity and identification and gathers any additional information.

■ **Resources**—Team member(s) responsible for gathering and maintaining supplies required for the incident; may be referred to as logistics.

Figure 5-1 The media plays a role in a HAZMAT incident, from relaying evacuation instructions to minimizing hysteria. A good flow of public information is essential to keep the surrounding community at ease.

The management scenario painted in the accompanying text box does not mention victims, hospitals, or an evacuation, all of which further complicate the incident.

No matter what IMS is used, typically only a small component of the overall system is applied during routine fire operations. The finance branch probably will not be established during a house fire. It is likely that in a major chemical release this branch will be established and have several persons assigned to assist in this function. This applies to large and small communities with career, combination, or volunteer departments. In some fashion a liaison must be established with all of the responding agencies, and their specific roles must function within the IMS. Some specific HAZMAT branch functions may be used during an incident. These are listed in Table 5-1.

IC'S LEGAL RESPONSIBILITIES

At a hazardous materials incident the IC is responsible for a number of tasks. Regardless of the local standard operating procedure, the IC has many legal obligations under the HAZWOPER regulation. The IC is responsible for the overall actions at the incident, no matter who is performing the tasks. In most cases the HAZMAT team is the technical group that performs the mitigation of the incident. Depending on the interaction of the IC and the HAZMAT team, there may not be much verbal communication. If all goes "by the book," the IC makes the decisions, but the reality is that the technical group provides the suggested options. In some cases there may be only one option. If the incident ends up as the subject of a lawsuit, the IC must answer for the HAZMAT team's actions; thus the IC must stay informed of all the incident action plans and be given the opportunity to choose an appropriate response. At a chemical release the IC is also responsible for the pre-entry briefing, which is the time to inform entry crew members of the hazards that exist and what actions are expected of them. If there are no predesignated emergency signals, these must be decided prior to entry. The IC must use a system of monitoring the progress of the incident and make changes to the system as needed. The mitigation of an incident takes the cooperation of many persons and agencies, all working toward a common goal under the guidance of the IC.

HAZARDOUS MATERIALS MANAGEMENT PROCESSES

Several different management processes can be used for hazardous materials incidents, many of which have been in use for many years and offer proven methods of organizing an incident. All of the systems have been adapted from fire service systems to fit the needs of a chemical release.

The core of all these systems is basically the same, but they do differ in some areas. The core to all systems is the protection of life, property, and the environment. One IMS, the **8-Step Process**®, was devised by Mike Hildebrand, Greg Noll, and Jim Yvorra. Another system, designed by Dave Lesak, is the **GEDAPER**© process of hazardous materials management. Another system, developed early by Ludwig Benner, Jr., is the **DECIDE** process, which is outlined along with the other systems in **Table 5-2.** No matter which system a department selects, it is important to use a system that everyone understands and can use. An IMS must be in place and the group of responders who are responsible for HAZMAT must fit into that IMS, using one of these processes or a combination. In 2004 the Department of Homeland Security (DHS) mandated the use of the National Incident Management System (NIMS). The NIMS system was developed to merge the federal, state, and local response efforts to a disaster or terrorist attack. When developing NIMS the various incident management systems were combined into one. For use in this text we will use Incident Management System (IMS).

Isolation and Protection

NOTE When arriving at a suspected chemical release it is important to isolate the area from other persons who may inadvertently wander into a hazardous environment.

One of the most important tasks a responder trained to the awareness level can do is to isolate the area so that others do not become part of the problem. Methods of isolation can be as simple as barrier tape, such as the scene shown in **Figure 5-2,** to the use of law enforcement at traffic control points. Other methods such as traffic barriers, or even the use of emergency vehicles to block access, can be used. With the increased knowledge of chemical hazards, real or otherwise, it is important to control the incident quickly. The more people who enter the suspected hazard area, the more who may later need to be rescued or, depending on the situation, require decontamination.

The protection of the persons in a hazard area can best be accomplished by evacuation of the immediate area. This does not imply that you simply tell everyone to leave, as they may need to be decontaminated or, at a minimum, medically evaluated, depending on the situation. A plan must be established for holding these persons until a determination can be made as to their status. The HAZMAT team usually makes this determination. When dealing with people in the suspected hazard area, frequent communication with the HAZMAT team is important. The other use of protection at a chemical spill is usually associated with the adjacent community evacuation or sheltering in

TABLE 5-2

			HAZMAT Strategic
DECIDE	**8-Step Process©**	**GEDAPER©**	**Goals**
Detect the presence of hazardous materials	Site management and control	Gather information	Isolation
Estimate the likely harm	Identify the problem	Estimate potential course and harm	Evacuation
Choose a response objective	Hazard and risk identification	Determine strategic goals	Notification
Identify the action	Select personal protective clothing and equipment	Assess tactical options and resources	Product identification
Do the best you can	Information management and resource coordination	Plan and implement actions decided	Determination of appropriate personal protective equipment
Evaluate your progress	Implement response objectives	Evaluate	Decontamination
	Decontamination	Review	Spill and leak control
	Terminate the incident		Termination

The 8-Step Process© is excerpted from the text *Hazardous Materials: Managing the Incident*, by Greg Noll, Mike Hildebrand, and Jim Yvorra. The GEDAPER© process is excerpted from *Hazardous Materials: Strategies and Tactics*, by Dave Lesak.

place. Both of these issues are further discussed later in this chapter.

If the incident is suspected of involving criminal or terrorist activity, isolation is important not only for the chemical hazards but for evidence preservation as well. First responders at the awareness level should make efforts to keep people from leaving the area and keep note of persons or vehicles that are leaving the scene. They should isolate the area and ensure that others do not enter the hazard or crime scene area. First responders should establish a hazard area and restrict entry into the area. They should also notify law enforcement and the HAZMAT team of the conditions that they have found.

Rescue

When discussing isolation and protection, rescue is a topic that naturally follows. The rescue of victims from a suspected hazard area can be extremely controversial. The decision to make a rescue is a personal one, as it may involve substantial risk to the rescuer. *Local protocol and standard operating guidelines must be considered.*

Figure 5-2 A top priority should be to isolate the area to prevent other people from becoming involved with the incident.

SAFETY Emergency response is inherently risky. As much as we would like to eliminate all risk from our occupation, it is impossible. Our best hope is to safely manage that risk and use methods to identify that risk.

In reality, once responders arrive at the incident safe and sound, their chances for survival dramatically

CASE STUDY

The Stealth F117A fighter crash, which occurred on September 14, 1997, was an incident management challenge. The fighter crashed while performing at an air show in Baltimore County, Maryland. The fighter crashed just outside the National Guard airport, and the crash was witnessed by thousands of spectators and emergency responders. As soon as the fighter went down, crews from the airport and the county fire department responded. One of the major complications with this response was that the fighter crashed on a peninsula with only one road in and out. A large number of spectators left the air show and proceeded to the crash site. The pilot had safely ejected and was rescued by personnel from the Air Guard. The response included the Air Guard fire department, which responded with crash trucks, a first alarm assignment from the county, and a HAZMAT assignment.

The Air Guard crash trucks did an outstanding job of extinguishing the fire in and around the plane, while suppression crews were attacking multiple fires in structures. A bulk of the firefighting was done by crews who arrived from fire boats and fought the fires from the water side. The first arriving Battalion Chief Bill Purcell was faced with search and rescue, fire extinguishment in both the plane and a number of houses, EMS issues, and control of the site. Plane crashes are to be treated as crime scenes, so control of spectators and the area was essential. Because the road in was becoming clogged with spectators, the decision was made to dispatch a second alarm for additional assistance. The officers decided that extra help standing by near the scene would be better than needing help later only to find that it would be unable to make it through to the crash site.

Deputy Chief Gary Warren arrived prior to the second alarm and assumed command of the incident. The county's emergency management office was notified and was also on the scene early into the event. Many facets of the IMS were used, including operations, logistics, finance, and planning. During the investigation of the crash, residences along the peninsula were evacuated for several days. The isolation of the event initially was handled by the county

police department, and the large area required a large number of resources from the police. Early into the incident the police chief arrived to view the scene and review the activities.

Set up early on was a unified command post with the fire department assuming the role of incident commander. Other primary players in this unified command were the police department, Air Guard, Maryland Department of Environment, emergency management, and the county executive's office. The initial operations section was focused on fire control, but once the fire was knocked down, this was transitioned to the HAZMAT team to control the variety of hazards that existed. As more military personnel arrived, the military personnel changed from local air guardsmen to Air Force representatives. One of the keys to success with any incident is knowing all the players on a personal level. As this incident progressed, more and more personnel arrived who were outside of this personal level. This did have a significant impact on the event, as personnel were not familiar with each other and their agencies' policies and procedures.

As would be expected, this crash attracted considerable media attention, which taxed local resources. The Air Guard was established as the point of contact. There was substantial political attention, with many local politicians stopping by to visit the site. Considerable EMS issues that developed during the incident resulted in twenty-one patients being transported. Misinformation that did not follow the chain of command may have contributed to some of these issues. It is essential that all personnel from all agencies understand and follow the IMS. The crash required the resources and presence of the Air Guard, fire department, and emergency management office for a week, all under the original IMS. The management of the outside agencies was by far the most difficult part of the event and the most taxing. When the incident went beyond a local event with the fire department and Air Guard, it grew in complexity. Although this was a small plane crash with a $46 million loss, the event was effectively handled by the IMS.

increase. One of the most dangerous parts of firefighting is responding to the incident. Each year 20 to 30 percent of firefighter deaths are the result of responding to and from an incident.

With specific regard to FFPC, scientific data published in August 2003 provides ICs with some information as to the ability to make rescues. The Soldiers Biological and Chemical Command (SBCCOM) is now known as the Research, Development and

Engineering Command (RDECOM) located at the Aberdeen Proving Grounds in Maryland, performed several studies involving firefighters' protective clothing, chemical protective clothing, and detection devices. The lead study Risk Assessment of Using Firefighter Protective Ensemble (FFPE) with Self Contained Breathing Apparatus (SCBA) for Rescue Operations During a Terrorist Chemical Agent Incident is known as the 3/30 rule. Researchers found

THE IMPORTANCE OF ISOLATING THE SCENE

The first arriving responders can make a hazardous material release easier to manage if they begin the isolation process. In many areas a bus accident causes responders to act quickly to isolate the area and deny entry. In many cities the people on a bus that has been involved in an accident will receive a check in their names in exchange for a release from further damage claims against the bus company. Failure to control access points to the bus will result in a dramatic increase in the number of occupants, many seeking the payment from the bus company.

What holds true with the bus accident is now true with chemical releases. In many cases checks are being written or people plan to sue for possible damages, making isolation difficult. The mention of chemical exposure may result in people attempting to enter a potential hazard area. One of the most difficult areas to attempt to control is a mall or shopping center in which hundreds of people may be involved. A simple discharge of pepper spray will result in a number of real victims and usually generate a few more suspect victims. Everyone must be treated and possibly transported, which can create a strain on the EMS system. Early isolation and evacuation can help reduce the numbers of victims.

that firefighters' protective clothing can offer protection in situations where there are significant hazards to someone without protective clothing. They tested both nerve agents and blister agents, both of which present toxicity hazards; the nerve agent is extremely toxic. In situations where a nerve agent has been released and there are both live victims and dead victims, firefighters in full PPE and SCBA can enter this environment and make rescues with no or minimal effects. In a situation in which all the victims are dead, firefighters can enter this severely toxic environment for 3 minutes with no or minimal effects. It goes without saying that in these cases emergency decon is a must and the gear will probably be destroyed.

When involved in making rescues at hazardous materials incidents, responders must use the same type of clues that drive making rescues in fire situations. The type of occupancy, time of day, visible clues, audible clues, and eyewitness reports are only some of the clues to a viable rescue situation. Once on the scene, responders need to evaluate the incident. Is rescue necessary? This information should be confirmed as best as it can be, but in some cases this information cannot be verified. What are conditions at the incident? Is there a confirmed chemical spill? If there is a release, can a person not wearing protective clothing survive? In many cases the local responders are called to suspected chemical releases and determine through investigation whether or not chemical release has occurred. If the persons who require rescue are alive the actual risk to the responders making the rescue is slight when wearing full PPE. If on the other hand the first responders arrive at a local mall and are told twenty people are unconscious and appear to be dead in the hardware store, the situation is different. When the responders approach the store entrance and see the twenty people lying on the floor not moving, the responders may not want to enter that environment.

Other considerations when making a rescue decision include response and notification time. If people have been in the hazardous environment for 5 to 15 minutes, the response time of the department is critical to the decision-making process. The victims are not wearing any PPE and have been exposed to the material for a considerable amount of time. If they are alive when responders arrive, the risk is minimal to the rescuers. A method of emergency decontamination should be set up, and a backup crew should be standing by. The responding HAZMAT team should be consulted prior to entry to verify chemical information. There should be no delay in evacuating the victims, and a swoop and scoop technique may be the best course. Stokes baskets or other methods of quick evacuation should be employed. Do not take vitals, ask medical histories, or perform medical procedures. Move patients to a safe area. Responders should have established methods to notify the incident commander and other response personnel about critical conditions at the incident. Radios are the most common method, but hand signals may be required. Some form of relay system may need to be set up to ensure communication between the response crews and the incident commander, so that critical information can flow both ways.

SAFETY In order to understand the risks associated with the rescue of victims in a chemical warfare agent environment, responders should read the full report and understand the risks of operating in a dangerous environment. Within the report there are several caveats to several scenarios. The report can be downloaded at http://www.ecbc.army.mil/hld/cwirp/ffpe_scba_rescue_ops_download.htm.

SAFETY The rescue of victims is made using simple risk/benefit analysis; a lot of risk is taken when a lot is at stake. No risk is taken when the benefit is little.

Once responders and patients are out of the area, decontamination should be performed if required for the chemical. The patients and rescuers require decontamination and isolation from the remainder of the responders. Once decontaminated, EMS personnel can begin to work on the patients using appropriate levels of PPE. Depending on the condition of the patients, decontamination may be hurried and therefore not perfect. Secondary contamination is a concern for EMS personnel. If the patient has ingested a substance he or she may vomit, exposing personnel to potential contaminants in addition to the bloodborne (etiological) hazard. If the person has been contaminated with a material that is skin absorbable, the victim may off-gas the material or may release odors. After decontamination the rescue team should remove PPE and bag it. After evaluation by EMS providers and consultation with the HAZMAT team, the rescuers should be sent to rehab. Their PPE and SCBA will need to be evaluated and possibly sent for cleaning or replacement.

When handling hazardous materials incidents, encountering trapped victims is an unusual incident, but procedures should be in place to cover this contingency. The most likely scenario is a traffic accident in which persons are trapped and chemicals are involved. This scenario is common if we introduce the chemical gasoline to the picture. Firefighters are used to extricating trapped victims with gasoline leaking from one or more vehicles. Protection lines are established and typically a higher level of PPE may be used by rescue crews. The rescue crews limit the number of people in the hazard area, and quick extrication is usually performed. If other chemicals are involved, the scenario may change slightly: as HAZMAT companies arrive, personnel can be switched out for those with better chemical protection, or the HAZMAT company can work to control or eliminate the chemical hazard.

To make the decision to enter a hazardous environment takes training and experience. This decision should be made in direct consultation with the HAZMAT company. To make that decision some HAZMAT teams are using risk-based response. This system is designed to identify the risk that chemicals present. By the heavy use of air monitors, an unknown chemical can be characterized into one of four risk categories: fire, corrosive, toxic, or radioactive. When information is known about a chemical, it can still be

TABLE 5-3

Top Ten Chemicals Spilled
Sulfuric acid
Hydrochloric acid
Chlorine
Ammonia
Sodium hydroxide
Gasoline
Propane
Combustible liquids
Flammable liquids
Natural gas

These top ten (which are in random order) are derived from the EPA, OSHA, and DOT by examining the chemical release records published in a study as required by the 1990 Clean Air Act Amendments. As each agency tracks releases in a different fashion, we calculated each agency's top ten and determined an average; the result is this list.

placed into one or more of these categories by examining some chemical and physical properties. The major determining factor in a hazard that a chemical presents is vapor pressure. If the chemical does not have a vapor pressure, the risk of entering an environment with this spilled material is very low unless responders touch or fall into the material.

Another fact HAZMAT teams can use to their advantage is that the majority of the incidents to which they respond involve flammable and combustible liquids or gases. Although not designed as such FFPC offers ample protection for a rescue situation. HAZMAT teams and many first responders have great detection devices for these types of materials and can determine the true risk during an operation. **Table 5-3** lists the top ten chemicals spilled in this country every year. This is a HAZMAT team's bread and butter, and the team should be comfortable handling these materials. First responders should know the locations in which these materials are stored and used. The response to an incident involving these materials should be no different than a response to a bedroom fire.

SITE MANAGEMENT

Effective management at the site of a hazardous materials incident can alleviate the threat with a minimum of injury and property damage. Key management decisions are establishment of isolation **zones** and the need to evacuate or shelter citizens in place.

Figure 5-3 The truck to the right rear-ended the front truck trapping the driver in the truck on the right. In the trailer to the left are 55-gallon drums, some of which are leaking. Using PPE, air monitoring devices, and protection lines, crews continued with the rescue. HAZMAT crews evaluated the scene and handled and isolated the problem drums. *(Photo courtesy of Baltimore County Fire Department)*

Figure 5-4 Rescue crews are extricating the driver of a second vehicle involved in a rear-end collision. The driver remained conscious throughout the extrication and was severely entangled. Complicating the rescue was the presence of a truck full of hazardous materials containers. *(Photo courtesy of the Baltimore County Fire Department)*

CASE STUDY

Risk-Based Response: A risk-based response philosophy is based on the use of air monitors to guide responders safely. The monitors guide the responders as to which situations are safe to proceed into and when the use of PPE is appropriate. Unfortunately, many first responders do not have all of the required air monitors for a high level of protection. When confronted with rescue situations in potentially hazardous environments, first responders should use full protective equipment, including SCBA, as well as air monitors. They should also consult with the HAZMAT team, who can help guide them to the appropriate level of action. All emergency response involves some risk; steps need to be taken to manage (and thereby minimize) that risk. A truck carrying food rear-ended another tractor trailer stopped in the middle lane of the highway. The truck was traveling at an estimated 65 mph when it hit the stopped truck. The truck it hit was carrying a mixed load consisting of mostly 55-gallon drums. The driver of the stopped truck was not hurt and was able to remove his shipping papers.

The driver of the food truck was still alive but was extremely entangled in the wreckage, as can be seen in **Figure 5-3** and **Figure 5-4**. He was conscious and alert, and initial vital signs were stable.

The trucks were tangled together, and the door to the truck carrying the drums was torn away from the truck. Drums had shifted onto the food truck and were laid over the front of that truck. The first responders saw the placards on the first truck and requested a HAZMAT assignment. When they approached the truck to evaluate the driver, the responders saw the drums in a precarious position in the back of the truck, as shown in **Figure 5-5**. They already had FFPC on but also donned SCBA. They obtained the shipping papers from the driver and consulted with the HAZMAT company, which was about 20 minutes away. The contents of the drums contained mostly flammable liquids and some combustible liquids. The first responders were advised to use a combustible gas indicator, continue with full PPE, establish foam lines, and begin the rescue.

Upon arrival the HAZMAT company members met with the IC to evaluate the scene. They confirmed the monitoring being done by the first responders and began to evaluate the other parts of the load. Whenever the rescue companies moved a part of the dash of the truck, one of the leaking drums would increase its flow. The HAZMAT companies secured the leak, moved the drum, and secured the remainder of the drums. They examined the rest of the load to make sure there were no other problems and continued to monitor the atmosphere. Once the victim was removed, the rescue companies moved away and the HAZMAT team **overpacked** and pumped the contents of the other drums into other containers. The risk category was fire, and the PPE chosen was appropriate for the risk category. At no time were any flammable readings indicated during the rescue, although some were encountered during the transfer operation. This was an example of the various disciplines working together to rescue a victim in a hazardous situation.

Figure 5-5 Leaking drums were found during this rescue operation. Rescue crews had a foam line standing by and were air monitoring for flammable levels. When HAZMAT crews arrived, they took over air monitoring, quickly removed the leaking drums, and secured the other drums. When the driver was extricated, HAZMAT crews took care of the remaining drums. *(Photo courtesy of the Baltimore County Fire Department)*

Establishment of Zones

The term "zone" is used to refer to areas that are established to identify the various isolation points, such as those shown in **Figure 5-6.** These zones are referred to as the **hot, warm,** and **cold zones** and in many cases are identified by the use of colored barriers or cones. A comparison of the EPA designations is provided in **Table 5-4.**

Hot is usually identified by the color red, warm by the color yellow, and cold by the color green, as shown in **Figure 5-7.**

SAFETY The most important zone that needs to be established is the isolation area, and this needs to be done by the first arriving responder.

Figure 5-6 Proper isolation and work zones are needed to ensure the safety of responders.

TABLE 5-4

Hazard Area Designations	
Standard designation	EPA designation
Hot zone	Exclusion zone
Warm zone	Contamination reduction
Cold zone	Support zone

This area normally becomes the hot zone, after the arrival of the HAZMAT team. This isolation **hazard area,** or hot zone, is the area immediately around the release, which requires the use of PPE. Entering the hazard area exposes personnel to the substance, and in some cases if contact is made, they would become contaminated. The minimum that should be established is this isolation area, and the distances for this area can be determined by the use of the *Emergency Response Guidebook* (ERG). The minimum average distance established by this book is 330 feet (101 meters), which should be the absolute minimum for an unknown material. The distance for this area is obviously determined by the local situation, and it is difficult to make blanket statements about recommended distances. If it is convenient and easy to isolate a block, then do so; if isolating the block may create havoc in the community, then 330 feet (101 meters) may be recommended. Keep in mind that if there are no indicators of an actual release, this is for an unknown material. Any indications of a flowing spill or a vapor cloud will dramatically increase this distance.

If time permits and resources are available, then responders should attempt to set up the warm and cold zones. These zones are usually established by the HAZMAT team after its arrival and setup. It is important that the first responders position themselves in an upwind and uphill position, as shown in **Figure 5-8.** At a water-based spill, responders should operate upwind and upstream. The first arriving apparatus should set up in that position, but if it is not, responders should move it to that position as quickly as possible. Although it cannot always be accomplished, being in an upwind position is the first priority and uphill is the second. If responders can only be upwind but are forced to set up in the downhill position, they should make sure to extend the distance. A common mistake is that the first arriving apparatus does not communicate the on-scene wind directions and the best route of travel to allow for the other apparatus to arrive and stage at the upwind and uphill position. All apparatus should be positioned so that they are pointed away from the incident so in the event of a catastrophic release or other emergency, the apparatus can be

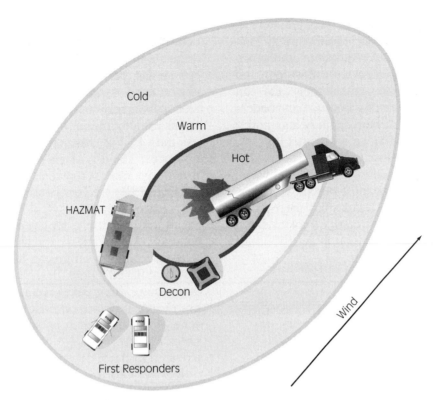

Figure 5-7 The establishment of zones is usually based on the types of hazards that may be presented. For general chemical spills, zones are hot, warm, and cold.

moved without turning around or backing. Other considerations for the hot zone are topography, accessibility for responding units and other resources, weather conditions, and bodies of water, especially those for drinking water. Setup areas for the HAZMAT team should accommodate decontamination areas and dress-out locations. Other exposures must be examined, and part of the isolation process is limiting possible ignition sources. Public exposure potential must be considered, and this can be broken down into specific

Figure 5-8 The best position for responders is uphill and upwind from the release.

time frames such as short term (minutes and hours), medium term (days, weeks, and months), and long term (years and generations). Although emergency responders will usually only deal with short-term and medium-term incidents, the actions they take can result in long-term exposures for both the responders and the public. Sewer lines, both sanitary and storm, must be taken into account along with other utilities, such as cable and phone, both above ground and below ground. Transportation corridors, such as highways, rail lines, ports, and airports, must be considered as well, as the incident may involve or affect these areas. The internal areas of buildings can be broken down into zones, but floor drains, ventilation ducts, and air returns must be taken into account.

SAFETY The Importance of Isolating the Scene: The first arriving responders can make a hazardous materials release easier to manage if they begin the isolation process. When chemicals are released there is great potential for liability and lawsuits are probable. Persons interested in joining these lawsuits may try and gain access to the perceived hazard area. In many cases, the insurance companies will try and settle with any potential victims prior to any lawsuits and will write checks to people who were involved, even if they were only in the proximity. One of the most difficult areas to attempt to control is a mall or a shopping center where hundreds of people may be involved. A simple discharge of pepper spray will result in a number of real victims and usually generate a few more suspect victims. Everyone must be treated and possibly transported, which can create a strain on the EMS system. Early isolation and evacuation can help reduce the number of potential victims.

The warm zone is set up after the arrival of the HAZMAT team and is usually where the decontamination area is established. In some cases the warm zone is also extended to allow the HAZMAT team to set up. In unusual situations the warm zone is an area where some type of PPE may be required. Depending on the circumstances, it may be affected by a wind shift or catastrophic failure. The establishment of these zones can be arbitrary and is based on the best judgment of the IC or HAZMAT team. There is no magic line of hot and warm zones, nor could one be established. The cold zone is the area where the IC post is established and where the first responding companies will be positioned. All support operations such as medical and rehab are set up in this area, and

movement between the zones is controlled at access control points. No PPE is required for the cold zone as there should be no chance for chemical exposure in this area. A person should be assigned to act as security for each of these zones to ensure that only authorized personnel enter these areas. Some HAZMAT teams employ monitoring to establish the zones and use the LEL, or exposure values, to determine these distances. The use of air monitors to establish these isolation points is crucial, as the distances provided in the DOT ERG are very conservative and are based on worst-case scenarios. Real-time, on-the-scene air monitoring cannot be replaced by plume projections, estimations, or models depicted in a text. The distance that the zones are set up can be based upon the theories established by risk-based response. The materials that require the greatest isolation are those that have high vapor pressure or exist as gases in their natural state. A material such as sodium hydroxide (sometimes referred to as lye or caustic) only requires an isolation distance of a few feet.

It is very corrosive and skin contact would cause severe irritation and burns if not washed off immediately, but it has a very low vapor pressure. The vapor pressure of sodium hydroxide is 1 mmHg at 1390°F (754°C), which means that the spill would have to be heated to 1390°F (754°C) to produce even a small amount of vapor; that is considerably less than the vapor pressure of water (17–25 mmHg). With this material, it is unlikely that a large isolation distance would need to be established, nor would any evacuations probably be necessary. A chlorine release from a railcar, such as the one discussed in the Graniteville Case Study, presents an extreme risk to the community: It may exist as a liquid in a railcar, but once in the atmosphere rapidly changes to a gas. Chlorine is not only poisonous but is an oxidizer and a corrosive material, and it has a very high vapor pressure of 5,168 mmHg. A substantial evacuation or shelter area in place is required for chlorine. One advantage is that chlorine is easily detected, and a true hazard area can be established using air monitoring.

Evacuations and Sheltering in Place

NOTE Making a decision to evacuate or to shelter in place can be a difficult decision for an emergency responder. No matter the decision, there are usually political ramifications—right, wrong, or indifferent.

The IC will be bombarded by the public, media, and coworkers about the evacuation decision. To a first responder, unfortunately, little assistance is readily

GRANITEVILLE, SOUTH CAROLINA CHLORINE CASE STUDY

On January 6, 2005 at 2:39 a.m. a train was traveling northbound and due to a misaligned switch ran into a parked train. The northbound train had 42 freight cars, 17 of which were empty (which can have residue) and 25 were loaded. Of the 42 cars 14 had hazardous materials or hazardous materials residue. When the trains collided 16 cars derailed. Of the cars that derailed three were tank cars of chlorine, one of which was leaking. The leaking chlorine car was carrying 13,830 gallons (52,352 liters) of chlorine, which is a liquefied gas. It has an IDLH value of 10 ppm, and is very corrosive. Compounding the resulting chlorine release is the fact that Graniteville lies in a shallow valley, and a stream is parallel to the railroad tracks. Chlorine is heavier than air and remains low to the ground. The leaking chlorine was accumulating and lying in the stream bed. Within the first few minutes the chlorine cloud extended 2,500 feet (762 meters) to the north, 1,000 feet (305 meters) to the east, 900 feet (274 meters) to the south, and 1,000 feet (305 meters) to the west. From the time of the first 911 call it was apparent that there was a chemical release, and the fire chief requested additional resources including a HAZMAT team. When the fire chief arrived he developed difficulty breathing and had to abandon his position. Several of the first arriving law enforcement officers (Deputy Sheriffs) drove into the chlorine cloud and required medical treatment. Eleven minutes after the collision the fire chief requested that the emergency notification system be activated, and for citizens to **shelter in place.** Adjacent to the area of the derailment were several industrial facilities in which there were numerous reports of persons trapped or missing. Several decontamination areas were established and the local hospital was notified to expect patients. In addition to the chlorine cars there was a car of sodium hydroxide that had been derailed. To effect rescue of persons in the hazard area firefighters in FFPC rode in pickup trucks that were used to transport victims. Later in the morning the Sheriff made the decision to evacuate those in the 1-mile radius around the crash which resulted in 5,400 persons being evacuated from their homes. About 21 hours after the crash HAZMAT crews were able to place a polymer patch on the leaking chlorine car. Two days later the sodium hydroxide was pumped from one of the derailed cars. During this operation the patch on the chlorine car failed releasing vapors. By reducing the vapor pressure in the leaking car, the leak was reduced and plans were developed for a more permanent patch. This new patch was applied the next day, on January 12th. By midnight on January 18th the offloading of the leaking rail car was complete. Nine persons died as a result of the escaping chlorine gas, one of whom was the engineer. At least 554 persons were taken to local hospitals. Of the 554 who were treated 75 were admitted for further treatment. Of the fatalities resulting from chlorine exposure, one was the engineer, six were from the adjacent milling facility to the west and north of the accident, and one was in a residence south of the site. Only the engineer lasted more than a few moments; all of the other fatalities occurred in the first few minutes of exposure. Six firefighters were treated and released, one was held for treatment for several days. Two sheriff's deputies were treated and released. As a result of chlorine exposure two engines, one ambulance, and a service truck were damaged beyond repair, costing $630,000.

Think about the impact of this incident on this small community; persons were killed, many were injured, and the economic impact was significant. Since the accident businesses and factories have closed. Although in this case there was a violent rupture of the chlorine car, think about an accidental release from a valve. The response to a chlorine (or comparable chemical) must be quick, efficient, and safe as the impact can be very dramatic. This incident occurred in a relatively low population area; think about the impact in a major metropolitan community. Using recognition and identification skills, proper protective clothing, and making appropriate isolation and evacuation decisions are keys to survival for both the responders and the community.

available. The only resource is the conservative DOT ERG, which may not apply in your specific situation. The best way to determine evacuation or sheltering in place is real-time air monitoring that can establish the exact hazard area. Consultation with the HAZMAT team and the local conditions guided by the DOT ERG can establish a starting point. If the incident is at a SARA Title III facility, the jurisdiction's emergency plan should have some recommendations regarding evacuation and that facility. Some plans provide a checklist to assist with the evacuation decision-making process. Close coordination with your local emergency management agency is essential to make the outcome successful. If you make the decision to evacuate, a suitable location needs to be found; transportation may be required and accommodations need to be established for evacuees. This type of assistance is usually provided by the emergency management coordinator, who is usually a good point of contact, and by the coordinating agency during an evacuation. The HAZMAT team can also run a plume projection using ALOHA through the CAMEO computer program, as shown in **Figure 5-9,** but that is also conservative and may cause a greater evacuation than is necessary.

Chemical vapor plumes have characteristic shapes as determined by the computer models, such as ALHOA model. These standard plumes are to be used for worst-case scenario and may not apply to your locality. Although the plumes are computed for a variety of topography and weather, each local area differs, and only local weather conditions and air monitoring results can provide truly accurate results. The standard shapes for plumes are provided in **Figure 5-10.**

Some studies have determined that in most cases sheltering in place is safer than evacuation. Just imagine evacuating the most populated area of your community. How would you notify all citizens? In most

Figure 5-9 Most HAZMAT teams use computer software to project the travel of a vapor cloud.

communities the police department may have this responsibility, but do they have the resources to accomplish this task? Is there an emergency alerting system for TV or radio in your community? In urban

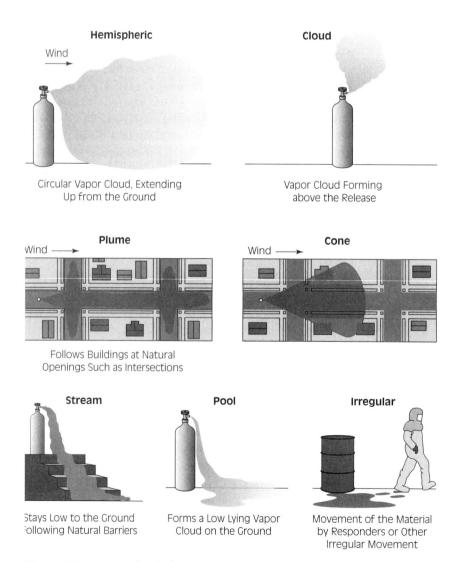

Hemispheric

Wind →

Circular Vapor Cloud, Extending
Up from the Ground

Cloud

Vapor Cloud Forming
above the Release

Plume

Wind →

Follows Buildings at Natural
Openings Such as Intersections

Cone

Wind →

Stream

Stays Low to the Ground
Following Natural Barriers

Pool

Forms a Low Lying Vapor
Cloud on the Ground

Irregular

Movement of the Material
by Responders or Other
Irregular Movement

Figure 5-10 Standard plumes or vapor clouds may form after a gas is released. The exact type varies with topography and buildings in the area.

or metropolitan areas a few blocks can require the evacuation of thousands of people. The worst-case scenario using the DOT ERG is an evacuation distance of 4,000 feet (1,219 meters) by 7 miles (11 km), which in a city could encompass 25,000 to 100,000 people in an evacuation. It would be nearly impossible to evacuate that many people, without widespread panic and chaos.

NOTE Typical evacuation situations can cause injuries and fatalities, even when the evacuation is announced days ahead of time.

Certainly, evacuation is required when an explosion is probable, explosives are involved, or the container may BLEVE or rupture with violent consequences. Evacuation may be the best option when the release will continue for a period of time, more than a few hours. However, an evacuation may not be recommended for a hospital, nursing home, jail, or other facility in which rapid removal is not practical. In these types of occupancies, air monitoring and control of HVAC are advisable; in addition, having a liaison remain at the facility who is in radio contact with the IC is important. The chemical properties will have an effect on the decision, as some materials will rise into the atmosphere and dissipate quickly, while others may stay low to the ground, causing evacuation problems. The type of leak must be assessed. Can it be quickly and easily controlled, or is there a probability that it cannot be controlled by the HAZMAT team? Will evacuating the citizens subject them to a higher level of the chemical than keeping them in place?

The ACGIH has provided some planning levels for the chemicals that require planning under the Clean Air Act Amendments (CAAA). These planning levels are on three tiers and provide actual levels that can be used for emergency planning. As the permissible exposure limits and threshold limit values are established for an 8-hour exposure during normal conditions, they do not have much applicability during an emergency release. These **emergency response planning (ERP)** levels are designed to assist with the emergency planners' preparation of the community emergency plan and would be useful in determining the evacuation zone. When making the decision to evacuate or shelter in place, the flow of information to the public and the media is essential. A lack of information can bring disastrous results. When sheltering in place, the citizens should shut all windows and doors, shut off air handling systems, and stay tuned to a TV or radio station. Without a continual information flow from the incident, citizens become frustrated and make attempts to find out the information for themselves. When dealing with larger facilities such as high rises, hospitals, nursing homes, schools, or jails, it may be best to put an emergency responder at that location to be a direct link to the incident, so that any questions can be immediately answered and fears alleviated. The emergency management office can also establish a rumor control hotline, which can be used to answer questions about the incident from concerned citizens and family members.

COMMON INCIDENTS

This section provides an overview of common incidents and the types of releases in each DOT hazard class. This listing is far from complete and focuses on the most common incidents or situations that have a great potential impact to the community. The recommendations provided here are only suggestions, and local policies and procedures should be followed.

Types of Releases

Chemical incidents can be classified based on how the chemicals are released from their container. In many cases the manner in which the release occurred provides clues to the successful mitigation of the incident. The type of release can be classified as a breach in a container or a release within a containment system. Although confusing, the NFPA provides two distinct categories for both breaches and releases.

In assessing chemical events there are several ways at looking at the potential release of a chemical from a container. Either the chemical itself is stressed or the container is stressed. In an incident involving a gasoline tanker that is rolled over at 55 mph (88.5 kph) the container is stressed, and the contents are likely to come out of the container. In an incident in which a paint waste is reacting within a drum, the material is stressed, but if the drum is sealed tight, the container will be stressed due to the chemical action creating pressure.

There are three general types of stress: thermal stress, mechanical stress, and chemical stress. Thermal stress is the addition of heat or cold to a container; cold stress on a container can be as damaging as heat stress. Placing a metal drum into a pool of a liquefied gas would result in the drum becoming brittle and easily cracked. Dropping a drum is an example of mechanical stress. Putting a corrosive in a metal drum is an example of chemical stress. Both pressure and non-pressure containers can breach under a variety of conditions. **Figure 5-11** provides examples of each type of container breach. The most common container breaches are punctures and closures opening up. It is uncommon to find a container that is disintegrated or has runaway cracking, splits, or tears.

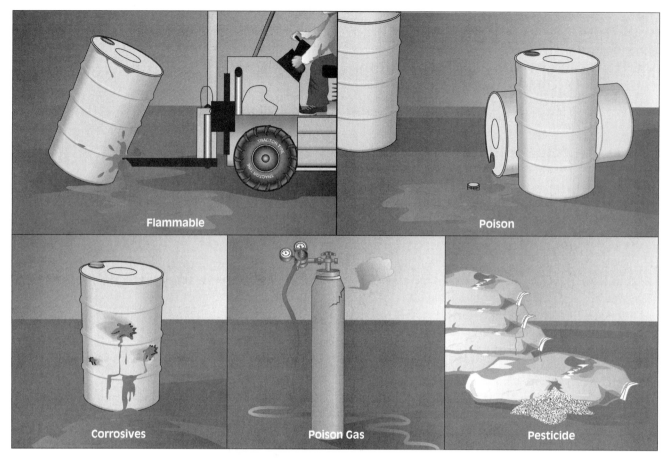

Figure 5-11 Examples of container breaches.

Many factors are involved with container breaches, but the most common breaches are nail punctures in drums, forklift punctures, dropped drums, and closures not used or not in place. Many incidents have been quickly mitigated by the use of a drum lid. On tank trucks a common breach point is the frangible disk, which may rupture due to overfilling or a quick stop causing the liquid to slosh and rupture the disk, releasing some of the contents. The incident is quickly handled by the replacement of the disk.

The methods in which chemicals can be released through a **containment system** are detonation, violent rupture, rapid relief, and a spill or leak. When some of these methods release a chemical, the action can be violent and have catastrophic consequences. When a container such as a propane tank, as the one shown in **Figure 5-12,** detonates, it can travel upward of a mile. If a container has a violent rupture, it means that the container was under pressure, causing the violent release. Much like a BLEVE, a violent tank rupture can send the tank or portions thereof a considerable distance. A container that has a rapid relief hopefully releases enough pressure to bring a margin of safety to the incident. This is only true if the cause of the increased pressure is removed and the pressure does not continue to climb within the container. If a relief valve is operating, this constitutes a true emergency condition

and failure to bring the pressure under control may mean catastrophic consequences. A relief valve operating is allowing the product to escape; if the material is flammable, it may find an ignition source and ignite. If the release ignites, this may cause the internal pressure to increase even more, causing more product to ignite. A relief valve should never be stopped, nor should a fire coming from the relief valve be extinguished. The only time to extinguish a fire coming from any valve (relief or otherwise) is if the flow of the material can be stopped with full assurance that it will be successful.

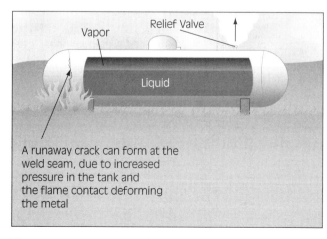

Figure 5-12 Propane tank detonation.

Extinguishing any other fire, including those that are impinging on a tank or container, should be a high priority as should hose lines directed to cool the container to prevent the relief valve from operating.

Spills and leaks are two events that result in loss of the product. A spill is defined as a loss of product from a naturally occurring opening, such as a **bung** on a drum or a leaking valve. A leak is defined as a release from an unintended opening, such as a puncture in the side of a drum. The result of a spill or a leak is that product is released and can create problems for responders.

Explosives

SAFETY The general rule with explosives is if the fire is near or is affecting the explosives, then emergency responders should isolate the area and back away. Life safety is the top priority.

All persons should be removed from the area, and a defensive operation should be established. A lot of other considerations come into play if the fire is not directly affecting the explosives. A brake fire on a truck carrying explosives is one example. As long as the fire has not reached the cargo area, the fire may be extinguished without incident. When making the decision to fight a fire in this situation, water must be applied quickly and in large quantities. Crews should be limited, and adjoining areas evacuated. Water should be applied to cool the cargo and then extinguish the fire. The recommendations in the DOT ERG, as shown in **Figure 5-13,** are a good starting point for isolation and evacuation when involved in fire situations. Some recommendations provided by the federal Bureau of Alcohol, Tobacco, and Firearms (ATF) are provided in **Table 5-5.** Much like the ERG, these recommendations are conservative and actual distances may vary. It is recommended that responders

Name of Material	Guide No.	ID No.	Name of Material	Guide No.	ID No.
Compressed gas, poisonous, flammable, n.o.s. (Inhalation Hazard Zone A)	119	1953	Compressed gas, poisonous, oxidizing, n.o.s.	124	3303
Compressed gas, poisonous, flammable, n.o.s. (Inhalation Hazard Zone B)	119	1953	Compressed gas, poisonous, oxidizing, n.o.s. (Inhalation Hazard Zone A)	124	3303
Compressed gas, poisonous, flammable, n.o.s. (Inhalation Hazard Zone C)	119	1953	Compressed gas, poisonous, oxidizing, n.o.s. (Inhalation Hazard Zone B)	124	3303
Compressed gas, poisonous, flammable, n.o.s. (Inhalation Hazard Zone D)	119	1953	Compressed gas, poisonous, oxidizing, n.o.s. (Inhalation Hazard Zone C)	124	3303
Compressed gas, poisonous, n.o.s.	123	1955	Compressed gas, poisonous, oxidizing, n.o.s. (Inhalation Hazard Zone D)	124	3303
Compressed gas, poisonous, n.o.s. (Inhalation Hazard Zone A)	123	1955	Compressed gas, toxic, corrosive, n.o.s.	123	3304
Compressed gas, poisonous, n.o.s. (Inhalation Hazard Zone B)	123	1955	Compressed gas, toxic, corrosive, n.o.s. (Inhalation Hazard Zone A)	123	3304
Compressed gas, poisonous, n.o.s. (Inhalation Hazard Zone C)	123	1955	Compressed gas, toxic, corrosive, n.o.s. (Inhalation Hazard Zone B)	123	3304
Compressed gas, poisonous, n.o.s. (Inhalation Hazard Zone D)	123	1955	Compressed gas, toxic, corrosive, n.o.s. (Inhalation Hazard Zone C)	123	3304
Compressed gas, poisonous, oxidizing, corrosive, n.o.s.	124	3306	Compressed gas, toxic, corrosive, n.o.s. (Inhalation Hazard Zone D)	123	3304
Compressed gas, poisonous, oxidizing, corrosive, n.o.s. (Inhalation Hazard Zone A)	124	3306	Compressed gas, toxic, flammable, corrosive, n.o.s.	119	3305
Compressed gas, poisonous, oxidizing, corrosive, n.o.s. (Inhalation Hazard Zone B)	124	3306	Compressed gas, toxic, flammable, corrosive, n.o.s. (Inhalation Hazard Zone A)	119	3305
Compressed gas, poisonous, oxidizing, corrosive, n.o.s. (Inhalation Hazard Zone C)	124	3306	Compressed gas, toxic, flammable, corrosive, n.o.s. (Inhalation Hazard Zone B)	119	3305
Compressed gas, poisonous, oxidizing, corrosive, n.o.s. (Inhalation Hazard Zone D)	124	3306	Compressed gas, toxic, flammable, corrosive, n.o.s. (Inhalation Hazard Zone C)	119	3305

Figure 5-13 ERG includes some basic explosives information that provides a good margin of safety when dealing with these incidents.

TABLE 5-5

Bomb Threat Standoff Distances

Threat description	Explosive capacity	Lethal air blast range	Mandatory evacuation distance	Desired evacuation distance
Pipe bomb	5 lbs.	25 ft.	70 ft.	850 ft.
Briefcase or suitcase	50 lbs.	40 ft.	150 ft.	1,850 ft.
Compact sedan	220 lbs.	60 ft.	240 ft.	915 ft.
Sedan	500 lbs.	100 ft.	320 ft.	1,050 ft.
Van	1,000 lbs.	125 ft.	400 ft.	1,200 ft.
Moving van or delivery truck	4,000 lbs.	200 ft.	640 ft.	1,750 ft.
Semi-trailer	40,000 lbs.	450 ft.	1,400 ft.	3,500 ft.

Explosive capacity—based on the maximum volume or weight of explosives (TNT equivalent) that could reasonably be hidden in the package or vehicle.

Lethal air blast range—the minimum distance personnel in the open are expected to survive from blast effects. It is based on severe lung damage or fatal impact injury from body translation.

Mandatory evacuation distance—the range to which all buildings must be evacuated. From this range to the desired evacuation distance, personnel may remain in the building (with some risk) but should move to a safe area in the interior of the building away from windows and exterior walls. Evacuated personnel must move to the desired evacuation distance.

Desired evacuation distance—the range to which personnel in the open must be evacuated and the preferred range for building evacuation. This is the maximum range of the threat from flying shrapnel/debris or flying glass from window breakage.

Source: Developed by the ATF, with technical assistance from the U.S. Corps of Engineers. Supported by the Technical Support Working Group (TSWG), a research and development arm of the National Security Council Interagency working group.

always consult with their local bomb technician for advice.

Other incidents that involve explosives include assisting the bomb squad with a suspected explosive device. Close coordination with the bomb squad is required to make sure that the operation is conducted safely. Although the distance varies with some jurisdictions, radios should not be used within 500 to 1,000 feet of a suspected device. Many bomb squads are detonating suspected devices in place, which may start a fire or cause a structural collapse. Other techniques are to "disrupt" the device or render it safe by the use of a water cannon. These techniques could cause the device to detonate. The time to discuss these scenarios and some plans of action is prior to the incident, not during the incident. At these types of incidents involving a suspected device, be aware of the potential for a secondary device, designed to injure the responders. When operating at the scene of an explosion, it is important to have the bomb squad search the scene for secondary devices, as shown in **Figure 5-14.** Chapter 7 covers secondary devices in more depth.

Other incidents may involve a shipment of explosives that are involved in an accident; in most cases these incidents present little risk to responders or the community. In this case the bomb squad is a great

Figure 5-14   Handling explosives requires extensive training and equipment.

Figure 5-15 This truck has separate tanks containing ammonium nitrate and fuel oil. In the event of a crash, they may mix, but without an initiating charge the mix is unlikely to detonate. *(Courtesy of the Maryland Department of the Environment)*

technical resource and should be used to evaluate the scene prior to moving any explosives. In some cases an explosives truck, such as the one shown in **Figure 5-15** carrying ammonium nitrate, may be involved in an accident. The fuel oil on board the vehicle could mix with the ammonium nitrate, creating ANFO. Although without an initiation charge or other substantial energy source the material is unlikely to explode, the ammonium nitrate does present a toxicity hazard and care should be taken around spills of this material. In incidents of this type the bomb squad should also be consulted. Chapter 7 discusses some possible terrorism scenarios, and explosives are used in almost all of the incidents. HAZMAT and bomb squad duties are becoming more and more interlaced.

SAFETY It is not uncommon to have explosives brought to the responders by well-meaning citizens. If the explosives are outside or in the citizen's vehicle, leave them there and call for the bomb squad.

These explosives may include hand grenades, pipe bombs, dynamite, blasting caps, ammunition, or distress flares. Responders should not handle these materials but should request that the bomb squad handle the situation. Depending on the type and size of device, isolation and evacuation may be necessary.

SAFETY It is a good general practice not to bring unknown materials into the responders' work area.

Citizens may bring in old chemicals, some of which may be explosive and shock sensitive. With picric acid, for example, the simple removal of the cap or vibration of the container is enough force to initiate the explosion of the container, which may produce enough force to cause fatal injuries to the person who attempts to open or set down the container. Another common chemical that may be brought in are containers of old ether (ethyl ether), which is outdated after a year in storage. Opening this container may also bring fatal consequences due to the formation of peroxide crystals under the lid. These crystals can be shock sensitive and can detonate by the **shearing effect** when opening the container. The job of handling these containers usually falls to the bomb squad with a HAZMAT team interface. In some cases toxic materials may be brought in, which can contaminate the person who accepts the package and cause severe health problems for firefighters in the station. Depending on your state regulations, if you accept any package from a citizen, you may end up owning the package and may be required to pay for the proper disposal of the item. This can amount to a costly good deed.

Gases

Incidents involving gases include both flammable and nonflammable gases, with the most commonly released gases being flammable. Also in this category are poisonous gases; see the poisons section later in this chapter for more information on poisonous gases. Luckily, many responders are carrying gas detection devices that will warn of potentially explosive atmospheres involving flammable gases. Depending on the setup, the detector may warn of oxygen-deficient atmospheres, which may be caused by the release of a nonflammable gas.

The two most commonly released flammable gases are natural gas and propane. Common propane bottles are shown in **Figure 5-16.** Although used for the same purposes, natural gas and propane have differing characteristics that can affect a response to a gas leak. Both gases are odorless when they exist in their natural state; when moved through a distribution system, an odorant is added to them. In major interstate pipelines it is possible that the gas may be traveling without an odorant added.

Figure 5-16 These 20-pound cylinders, which many homeowners have, can create large fireballs and explode with considerable force if involved in fire.

NOTE The largest difference between natural gas and propane is the vapor density. Propane is heavier than air and will stay low to the ground trying to find an ignition source. Natural gas, on the other hand, will rise in air and should dissipate quickly.

Dissipation of gases is weather dependent, and responders may encounter weather conditions that affect these characteristics. Under some conditions the gas may travel the length of a pipe following the path of least resistance, usually coming up inside a building. The leak may not be detected for a few days or weeks, until it surfaces. This is most common in areas of the country where the ground freezes and limits the upward movement of the gas. In many states specific training is offered on natural gas and propane emergencies, in most cases provided by gas companies.

Incidents involving these gases are commonplace. For natural gas the most common incident involves a ruptured gas main. Much like a municipal water system that is set up in a grid system, the gas system is set up on a grid, with larger distribution pipes and smaller delivery pipes. The pipes leading into a home are typically 1/2 inch to 1 inch (1.27 to 2.5 cm) in diameter. Distribution pipes may be 2 inches up to 36 inches (5 to 91 cm) depending on the region. First responder actions at these types of incidents generally involve isolation and protection, followed

by standby until the line can be shut off. Gas detection devices should be employed to determine the true hazard area, and adjacent buildings should be checked for gas. When checking buildings, air monitoring should be done at a variety of points, including the highest point, which should be checked prior to declaring a building safe.

Another significant issue with the release of flammable gases which also applies to vapors coming from flammable liquids is the presence of ignition sources. Pilot lights, electrical switches, electrical contacts, and some radio and cell phone transmissions are also possible sources of ignition. Static electricity is a concern when there is movement of persons or the chemical. Attempts should be made to eliminate these sources of ignition. The electricity supplied to a building can be cut off at the pole which is a safer alternative than shutting off the electricity inside the hazard building. At vehicle accidents hot engines, vehicle batteries, and static electricity are potential ignition sources.

CAUTION At no point should first responders jump into a hole to attempt to shut off a leak, nor should underground valves be shut in an attempt to stop the leak.

Gas line valves may actually be keeping the amount of gas reduced; moving the valve may increase the amount of gas escaping. If the leak is on the outtake site of the meter, it is acceptable to shut off the valve on the meter. The meter should be locked out and tagged out, and only the gas company should turn the flow of gas back on. Some departments carry special tags that mark the system as being out of service, and some actually lock the system in the off position. When gas company employees arrive, they repair the leak and place the system back in service. When dealing with pipes and electrical systems in an industrial setting, an OSHA regulation prescribes the procedure for shutting off a system and marking it so that the system is not accidentally turned back on prematurely. This regulation is known as the *lock out, tag out* regulation and applies to many emergency service situations.

Propane releases generally involve cylinders, although it is possible to have a pipeline propane leak. A number of pipelines running across many states carry both natural gas and propane. Other than flammability, the fact that propane sinks and stays low to the ground increases the hazard. If the cylinder is releasing liquid propane, then the liquid is also becoming very cold, below 0°F (−18°C). A common incident involving propane cylinders, especially those 17-pound (7.71-kg) cylinders designed for home barbecue grills, is overfilling the cylinder. Cylinders are designed to be filled to 80 percent of capacity to allow for expansion of the gas. If the cylinder is filled more than 80 percent and the temperature increases, the gas may escape the relief valve.

For fires involving propane tanks, there is the potential for a BLEVE, an event that can be catastrophic to responders. A BLEVE can occur with any flammable gas storage tank but is usually associated with propane tanks. If fire impinges on the tank and increases the temperature within the tank, the pressure also increases. As the temperature increases toward the boiling point, even more vapors are being produced and the pressure is increasing. At a certain point the pressure increases enough to cause the relief valve to operate. When this relief valve operates, the pressure in the tank will force gas from the tank and in most cases cause an ignition of the releasing gases. Under these circumstances this is good for responders as gases that have not ignited can cause more problems. If the pressure in the tank is not reduced, the potential is that the pressure inside the tank will increase and overwhelm the relief valve and the tank will fail. When pressurized tanks fail, they can travel for a considerable distance. Although no one can predict exactly when a tank will fail, some clues may indicate that a BLEVE is forthcoming: increased flame height or appearance that the flames coming from the relief valve are under a higher pressure. The sound of a relief valve is deafening, but when the pressure increases, the pitch of the sound will get higher and may become louder.

SAFETY If the tank becomes discolored or distended or loses shape, then an immediate withdrawal is indicated.

When making the decision to fight a propane tank fire, a large quantity of water needs to be applied quickly and continuously. Responders should concentrate water flow on the vapor space of the tank, and the fire should not be extinguished unless responders are certain the flow of gas can be stopped. When preplanning, first responders should establish a flow of water on the tank in excess of 500 gpm (1,893 lpm) within a few minutes of arrival. BLEVEs are responsible for a number of responder deaths each year, and in some cases whole alarm assignments have been killed and seriously injured during BLEVEs. A container does not need to have a flammable liquid in it to BLEVE. Any heated liquid produces vapors that will cause the pressure inside a tank to increase and may cause the container to fail. This would be called a violent tank rupture (VTR) and would not have the accompanied fireball associated with a BLEVE. A drum of water that is heated can create a VTR and can be very dangerous to people nearby.

Some cars now are powered by natural gas or propane; the most common are cars that are part of a fleet of vehicles, including government and utility vehicles. To determine if a vehicle is powered by one of these gases, look for a sticker that reads CNG, LP, propane powered, or natural gas powered. Sample stickers are shown in **Figure 5-17.**

(A)

(B)

Figure 5-17 Vehicles that use alternative fuels such as natural gas, propane, or electric are marked to indicate the fuel source. Photo A shows a CNG sticker on the rear bumper, indicating that the car is fueled by compressed natural gas. The sticker in photo B is beside the front driver's quarter panel and indicates this vehicle is environmentally friendly alternatively fueled. Cars with alternative fuels typically still have a gasoline tank, and there may be gasoline present in emergencies.

Figure 5-18 A hydrogen tube trailer was involved in a collision, which caused a rupture that was ignited and impinged on the high pressure tanks. *(Courtesy of the Maryland Department of the Environment)*

Figure 5-20 This propane tank truck had a slight rollover in a toll booth. The tank was damaged and the propane was flared off. It is not possible to know the extent of the damage to the tank, and righting the tank could expose the tank to stresses which could cause a catastrophic release. *(Courtesy of the Maryland Department of the Environment)*

Although not common now, there are cars powered by hydrogen. This gas is extremely flammable and has an unusual risk. Hydrogen is also transported by tank truck, rail, and IMO. Hydrogen presents an unusual risk when burning: it burns with an invisible flame and does not produce any smoke, which can be a major concern as shown in **Figure 5-18** and **Figure 5-19.** Occasionally the heat waves can be seen above a hydrogen flame, but in many cases a responder would identify a hydrogen fire when they are too close.

When propane tank trucks overturn as shown in **Figure 5-20,** the truck will usually not leak, but the situation is still quite dangerous. It is impossible to determine the exact damage to the tank, and it is not known how much increased pressure could cause a violent release of the tank contents. If the ambient

temperature is increasing, so is the pressure in the tank. It is possible that the relief valve has been damaged during a rollover and may not function as designed. It would be dangerous to right any liquefied gas tank without the product being transferred or, in the case of propane, the contents flared, as shown in **Figure 5-21.**

Other common gas releases involve carbon dioxide, chlorine, and ammonia. Although carbon dioxide (CO_2) is a common fire extinguishing agent, it is also

Figure 5-19 The tanks suffered severe damage, but they did not explode. The pressure in tube trailers can be very high and if the tanks explode the damage could be extensive. *(Courtesy of the Maryland Department of the Environment)*

Figure 5-21 A propane tank truck was involved in a collision and rolled over. Shown are two flares that are burning off the propane, which has to be done prior to the tank being righted. Many HAZMAT response teams carry flaring equipment to flare small grill–sized propane tanks up to rail cars. *(Courtesy of the Maryland Department of the Environment)*

used for the distribution of beverages and is found commonly in restaurants, bars, and convenience stores. An increase of CO_2 in a building can make people sick and is becoming a common **sick building chemical.** Although carbon dioxide is not commonly monitored, when dealing with emergency response to a **sick building,** responders should determine CO_2 levels. Both chlorine and ammonia have good warning properties and have distinct odors and should be easily identified by people in a building. In most cases the odor is so irritating that people self-evacuate the building prior to hazardous levels being built up. Both chlorine and ammonia are extremely corrosive and present a great risk to those who are not wearing protective equipment. Both gases will attack metals, sometimes very quickly and destructively. Both are also very reactive, and if other chemicals are also released may react violently. Ammonia also has the unique characteristic of being flammable. Early in the hazardous materials response business, two Shreveport, LA firefighters died in an ammonia leak. They were attempting to stop the leak when a flash fire erupted and ignited the ammonia. With the exception of NFPA 1991 certified suits, chemical protective clothing does not offer flash fire protection. Certified suits only offer very limited protection in these circumstances.

Flammable and Combustible Liquids

By far, this is the leading category for the most common type of releases, as it includes gasoline and diesel fuel, as shown in **Figure 5-22.** Responders across the country respond to these types of incidents thousands of times a day.

Figure 5-22 A common incident during the winter are traffic accidents involving home heating oil trucks. *(Courtesy of the Maryland Department of the Environment)*

SAFETY Gasoline is the leading chemical when it comes to chemical accident fatalities, and transportation leads in those fatalities. Unfortunately, due to gasoline's familiarity many emergency responders do not adequately protect themselves when responding to incidents of this type.

In some cases, responders themselves may improperly use the chemicals in the station, unaware of the hazards. Gasoline has between 1 percent and 5 percent benzene, which is a confirmed carcinogen to which responders are commonly exposed. When you smell gasoline, you are receiving an exposure to benzene, an exposure that is repeated on a regular basis. When dealing with small spills, the tendency is to use little or no protective clothing, but the body is continually being exposed to this toxic material. In addition to its toxicity, gasoline is also very flammable and can be ignited by the static electricity generated by responders' clothing as they approach an incident. Although a lesser concern with diesel fuel, fuel oil, or kerosene, toxicity remains a hazard. In many cases, a lower level of concern has resulted in responder injuries. Although in normal circumstances it is difficult to ignite these materials, if the temperature is increased, they can be easily ignited. On a day with a temperature below 70°F (21°C), the potential for ignition is low, but a spill onto blacktop when it is 100°F (38°C) outside adds considerable risk for a fire as a larger quantity of vapor will be produced on the pavement, which is likely in excess of 110°F (43°C). In years past it was commonplace to just flush spills of flammable or combustible liquids down the storm drain. This practice for the most part has been discontinued, as it causes severe environment damage and only moves the problem to another location. Responders in many areas of the country absorb all the material and dispose of it in an environmentally sound manner. To provide the best protection for the environment the material must be picked up. Discharging oil (includes fuels) is a violation of the federal Clean Water Act, and emergency responders could face serious fines if the materials were flushed into a waterway.[2]

Flammable Solids/Water Reactives/Spontaneously Combustible

When dealing with materials in these categories, specific identity and emergency response information are crucial. Consultation with the HAZMAT team is important, as a wrong tactic can be devastating to the

TO FIGHT OR NOT TO FIGHT?

A common situation involves overturned or burning tanker trucks as shown in **Figure 5-23** and **Figure 5-24.** The most common product in these trucks is gasoline, and when one of these trucks is involved in an accident, a fire is usually the outcome. Two examples of gasoline tanker fires are depicted in **Figure 5-25** and **Figure 5-26.** When first-responding companies arrive, they have a tendency to try to extinguish the fire. Without larger quantities of firefighting foam, it is very difficult to extinguish a tanker fire. If, upon arrival, the responders find that the truck is well involved and is not near any exposures or impacting an adjacent community, the best course is usually to let the fire continue to burn, as shown in **Figure 5-27.**

When attempting to extinguish this type of fire, the considerable runoff usually contains both gasoline and firefighting foam. Both of these chemicals are not good substances to allow to enter a waterway. If the fire is extinguished, a large amount of gasoline at an elevated temperature may remain. Once extinguished, the truck must have foam reapplied every few minutes to ensure that the fire does not reignite, as shown in **Figure 5-28.** It is then necessary to remove the hot gasoline from the truck and pump it into another truck, a dangerous proposition. If, on the other hand, the fire is allowed to burn, nothing would remain that could harm responders or the environment. Although it looks horrible, the resulting black smoke is in reality less damaging to the environment than a liquid spill. The smoke is predominately carbon, which provides the thick black smoke.

If the fire is threatening a community or adjacent structure, then all attempts should be taken to protect the structure from the fire. In cases where the truck is wrecked under a bridge, a critical structure in the community, the structure needs to be protected. The cost of a bridge usually is in the millions, and responders should assess the financial impact in terms of hard dollars and inconvenience if it were closed for a year for reconstruction. If units arrive and are confronted with a large fire, the initial water should be directed toward the bridge. When sufficient foam and water arrive, firefighters should then extinguish the truck and continually apply foam. The application of foam at a nonfire incident usually causes friction between firefighters and environmental agencies. Foam can damage the environment, and even if "environmentally safe" foam is used, the breakdown products of the fuel are not environmentally safe. Environmental agencies would prefer the minimal use of foam. Another concern is that when a truck is overturned, it must be drilled and pumped out prior to righting the truck; the application of foam makes the situation very slippery and creates additional hazards. Foam should only be applied on the gasoline, and coating the top of the truck is not necessary when the gasoline is on the ground. The use of air monitors to determine when the foam blanket is breaking down is recommended, as is the application of the least amount of foam to provide a foam blanket. An example of other hazards that may exist at a tanker rollover is described in **Figure 5-29.**

Firefighting Foam

The use of a foam product is a useful tactic in flammable liquid firefighting, and has benefits in non-fire situations as well. Water and fuels typically are not a good mix, and fighting a flammable liquids fire with water is typically not successful. The primary issue is that water and fuel do not mix—the fuel floats on top of the water. So when flowing large amounts of water onto a flammable liquids fire the water increases the volume and the flaming fuel gets moved around by the water streams. If the water streams are flowed into the fuel causing it to splash up into the air, this increases the flammable vapors in the air, which creates a larger fire. Fuels or flammable liquids are in a chemical category known as hydrocarbons, comprised of hydrogen and carbon. In the hydrocarbon family there are several smaller groups, such as aromatic hydrocarbons, ketones, alcohols, and others. Most hydrocarbons are not water soluble or do not mix with water well. These are also known as being non-polar, which is a chemical term that means that the material does not mix with water. Substances that are polar or known as a polar material are ones that mix with water. Alcohols are polar substances as they mix easily with water. When a hydrocarbon catches fire there are two potential issues when using water: If it is a polar substance the water will mix with the hydrocarbon increasing the volume of the flammable material. If the material is non-polar substance then the flaming liquid will float on top, and the agitation will increase the size of the fire. A responder is not able to determine if a liquid is polar or non-polar by sight. One has to identify the material and then through research determine if it is polar or not. Any material that is an alcohol, such as ethyl alcohol, is polar, as is methyl and isopropyl alcohol. One would think that the dilution of the alcohol would extinguish the fire but this is not the case. It takes considerable water to reduce the flammability of a polar substance. To aid in extinguishing a flammable or combustible liquid fire we use firefighting foams.

There are a number of foams available for fighting such fire, and each has advantages and disadvantages.

(continued)

TO FIGHT OR NOT TO FIGHT? *(continued)*

Foams are classified as low expansion, mid expansion, and high expansion. Low expansion foams have expansion ratios up to 20:1 and are the most common foam used in flammable liquid firefighting. Mid-expansion foams have expansion ratios from 20:1 to 200:1 and are used for chemical vapor suppression. High expansion foams have expansion ratios of greater than 200:1 and require a high expansion foam generator. This type of foam was used in the past for basement fires and has applicability in ship fires. Depending on the type of material or type of situation will depend on which type of foam is best. The use of foam is a skill that requires practice and training should be implemented prior to any use of foam in a fire situation. Special equipment is used to apply foam, and it requires several additional steps above using a water hose. One of the most common places where foam is used is for aircraft rescue firefighting (ARFF), as planes carry substantial amounts of fuel. Airport firefighters use vehicles known as crash trucks, which are large fire engines designed to flow large amounts of foam and water very quickly. Crash trucks are designed for airport firefighting operations, and can be used for other purposes, such as a gasoline tank truck fire. Many fire departments have foam engines that are engine companies that carry foam and water. They typically do not have the capacity of a crash truck, but could handle a tank truck or small tank fire. In other cases engine companies may have smaller foam tanks and the piping necessary to flow foam. For other situations the equipment can be carried separately and can be attached to any fire engine pump and it can then flow foam. When foam is desired there are several items that are required. In addition to water, one needs a foam nozzle, a foam proportioner (eductor), and foam concentrate. For some foams a specific foam nozzle is not always required, but the use of one improves the efficiency of the foam. One of the common fallacies with using foam is that a single engine company with 15 gallons of foam can successfully extinguish a 9,000-gallon gasoline tanker fire. In order for a foam attack to be successful large quantities of foam should be on hand before initiating an attack. The same 15 gallons of foam can be useful in covering a gasoline spill to prevent ignition of the gasoline vapors. The foam manufacturers provided foam calculators that can determine the amount of foam required for a fire or a spill.

Foam is comprised of water, foam concentrate, and air bubbles and in simple terms can be equated to liquid soap. The foam concentrate is introduced into the water stream and mixed with water. The device that introduces the foam to the water stream is known as foam eductor, which can be separate or part of the foam proportioner. Foam is introduced at a certain percentage into the water stream, and to produce an acceptable foam stream the foam and water percentage must be according to the manufacturers recommendations. The percentage may vary but most foams are used at a 3 percent or 6 percent concentration. Foam can be used to extinguish fires, or can be used to prevent fires by suppressing flammable vapors. For liquids that are on fire the foam and water combination cools the burning liquid, and places a barrier on top of the liquid excluding oxygen. Foam is applied in a number of ways: indirectly in a sweeping method raining the foam down on the seat of the fire, or on top of the spilled material. Damming and diking are key product control tactics that need to be implemented simultaneous to the firefighting operations. Flowing foam onto a street-based flammable liquid fire will push the flammable liquid farther down the street and potentially into storm drains, culverts, or other diversions. Foam is not that effective on moving liquid fires, and efforts should be made to reduce the spread of the liquid. Foam can be banked off of objects, such as a wall, or rolled into the spill or fire. The object of banking is to slow the introduction of the foam into the liquid spill or fire. Coating a side of a gasoline tank truck does not offer any protection, nor offers any other benefit. The coating of personnel is not effective either and does not offer any additional protection. Foam is very similar to soap and is very slippery and presents a tremendous slip hazard. When a foam blanket is applied one cannot see the ground and someone may trip or fall into a hole. When HAZMAT crews are operating on top of a tank truck, foam that has been applied to the tank truck adds additional risk to their operations and does not offer any protection. Another method of applying foam is to apply the foam at the edge of the spill or fire and roll it into the flammable liquid. It is then pushed farther by the application of more foam.

There are two major groups of foam, natural and synthetic, with the most common being synthetic. Within these groups there are two major classes, class A and class B foam for class A and B fires. Class A foams are used in wild land firefighting, in addition to structural firefighting. Class A foams are surfactants that reduce the surface tension of the water, making it easier for the water to soak into the class A material. Class A foams are also known as wetting agents, or light water. Class B foams are used for flammable and combustible liquid firefighting.

(continued)

TO FIGHT OR NOT TO FIGHT? *(continued)*

Combustible liquids are more difficult to ignite, but once on fire they are difficult to extinguish and if heated will act much like a flammable liquid. There are several different types of class B foams, and the majority include aqueous film-forming foam (AFFF), fluoroprotein foam (FPF), and protein foam. In addition to those foams there are foams that can be used for polar substances such as alcohols. The foams incorporate "alcohol resistant" into their name. There are two main types, alcohol-resistant AFFF (AR-AFFF) and alcohol-resistant fluoroprotein (AR-FP). There are several other types of foam but the ones mentioned are in the majority. The manner in which AR- AFFF is used depends on the percentage of foam used. When used at a 3 percent concentration it is very similar to standard AFFF, but when used at a 6 percent solution it has alcohol resistance. This alcohol resistance comes from an additive in the foam that seals the top of the polar flammable liquid. One

of the reasons to use foam is to add a layer on top of the liquid which prevents vapor production. One caution though is that foam layers break down after a few minutes. Foam must be reapplied to ensure that the foam blanket remains intact. One way to determine if the foam blanket is intact is to use a detection device that can detect the vapors coming from the liquid. As flammable gas sensors may be damaged by foam it is best to use a photoionization detector (PID). The PID is more sensitive and can detect when the vapors are escaping the blanket. Making sure the foam blanket remains intact is an important consideration as well, as a responder sloshing through the foam is breaking the blanket allowing the vapors to escape. If the liquid was on fire the temperature of the liquid is high and re-ignition is very easy. Plunging the foam stream or water stream into the foam blanket also disturbs it and can result in a fire. Some foam characteristics are noted in **Table 5-6.**

Figure 5-23 A diesel tank truck cab caught fire impinging on the cargo tank. A quick and aggressive response by the Washington, DC and Prince Georges County Fire Departments was able to knock down the fire before the contents were ignited. If the tank had become compromised firefighting would have been very challenging as the tank truck was on a significant incline and burning fuel would have traveled down the highway and most likely into storm drains. *(Courtesy of Mark E. Brady, PGFD)*

Figure 5-24 Hazardous materials incidents require significant planning. As the fire was being extinguished response crews must be thinking about the resources required if the entire tank became involved, and a plan if there were a release of burning fuel. Plans to block storm drains and contain any runoff fuel on this significant downhill area were necessary. Another truck and the ability to transfer the fuel was also necessary as the tank was damaged in the accident and the fire. (Courtesy of Mark E. Brady, PGFD)

Figure 5-25 This photo is a graphic representation of the remnants of a gasoline tanker fire. All of the gasoline was consumed in the fire and the environmental damage was very slight. The truck wrecked in an area that was not impacted by the fire and it could be allowed to burn. *(Courtesy of the Maryland Department of the Environment)*

Figure 5-26 This tanker had to be extinguished as it had slammed into and remained under a major bridge. The majority of the tank was intact, and in most cases only parts of the tank melted down, exposing the gasoline inside. Out of the 9,000 gallons it was carrying, between 3,500 and 4,000 gallons of hot, boiling material was left on the truck. This created a very dangerous situation, especially during the fuel transfer operation. The reapplication of foam on a regular basis in this case was essential; the truck reignited after it had been extinguished. *(Courtesy of the Maryland Department of the Environment)*

Figure 5-27 Upon arrival this gasoline tank truck was well involved and heavily damaged. As there were no exposures the environmentally correct decision was made to let the fire burn. Considerable resources would have been required to extinguish this fire. *(Courtesy of Mark E. Brady, PGFD)*

Figure 5-28 Although a necessary evil when dealing with flammable liquids, responders should minimize the amount of foam applied at a spill. It can cause further environmental damage and slip, trip, and fall hazards.

Figure 5-29 Not all hazards come directly from vehicles or chemicals when responding to highway accidents. This diesel fuel truck was offloaded safely, but the next day all of the response crews came down with poison ivy; some cases were very severe. *(Courtesy of the Maryland Department of the Environment)*

TABLE 5-6

Firefighting Foam Comparison					
Foam Type	Quick Knockdown	Vapor Suppression	Heat Resistance	Use on Polar Solvents (Alcohols)	Notes
Protein	No	Yes	Yes	No	Easily blown away, short shelf life
Fluoroprotein	Fair	Yes	Yes	No	Can be injected subsurface in storage tank situations
AFFF	Yes	Fair	No	No	Can use regular nozzle, needs regular re-application
AFFF-ATC or AFFF-AR	Yes	Fair	No	Yes	Can use regular nozzle, needs regular re-application
Fluoroprotein Film Forming Foam (FFFP)	Yes	Yes	Yes	Fair	Combination-type foam

community. Most emergency responders have some experience with flammable solids, as road flares are classified as flammable solids. In most cases flammable solids are difficult to ignite, but once ignited they burn vigorously and are difficult to extinguish.

The water-reactive group can be defined in two ways. When some chemicals get wet, a violent reaction such a fire or an explosion may result. This is commonly what most people think of as water reactive. The second method is that once a reactive metal is ignited, the application of water may create a violent reaction. Magnesium is one example of this type of reaction. Once ignited, it burns vigorously, and when water is applied, it may explode, as shown in **Figure 5-30, Figure 5-31, and Figure 5-32.** Some materials such as calcium carbide are shipped as water-reactive materials. When water is applied to calcium carbide, the reaction may amount to a little bubbling, something that would not be considered violent.

However, the gas that is being released by the bubbles is acetylene gas. When pure acetylene gas is produced in this manner, it is unstable and reactive. Any nearby ignition source will produce a fire or explosion. Thus, the actual application of water in itself does not produce a significant problem, but the action of the water may create additional concerns.

Materials that are spontaneously combustible are usually transported in a manner to keep them stable. To remain stable they may require specific temperatures or have a material added to them to keep them stable. White phosphorus, which is usually covered with water, is an example of a spontaneously combustible material, as it ignites in the presence of air. If your community has a facility that uses flammable metals, it is a good idea to have a stockpile of metal extinguishing agent such as metal-x or lith-x, commonly referred to as class-D extinguishing powder. Most HAZMAT teams have a limited supply of this type of extinguishing agent and usually rely on a local industrial resource for more. Sand that is known to be dry (it has been stored inside in a closed container) can also be used, but most sand contains moisture.

SAFETY The use of water on flammable metals is not recommended and can cause severe injuries to the hose crew.

The metal has a tendency to explode when water is applied, sending hot metal and other fragments flying

(A)

(B)

Figure 5-30 Photo A shows a fire in which a measuring cup of magnesium shavings was added to burning straw and diesel fuel. The fire looks normal and does not indicate any unusual materials. Photo B shows what happens when a slight water spray is added over the fire. There is a dramatic increase in the size of the fire, and the resulting detonation spreads white hot particles of metal in the pan.

and producing other hazardous chemicals. The water will break down and release hydrogen gas, which will increase the amount of heat and flame produced. When sodium is put into water it will detonate violently, and it also makes sodium hydroxide, which is a very corrosive material.

Figure 5-31 A truck of magnesium was involved in an accident, and a tow truck driver using a welding torch ignited the magnesium. The resulting fire was very intense and if water would have been applied the results would have been explosive. *(Courtesy of Cambria County, PA Emergency Services)*

Oxidizers and Organic Peroxides

This is another class of materials for which responders should request help early in the incident. Both of these types of materials can have explosive characteristics and can lead to fatal mistakes. Organic peroxides can react explosively even if the responders have taken no action. A well-known oxidizer is ammonium nitrate.

Figure 5-32 Sand that has been stored inside and is known to be dry is one of the choices for attempting to extinguish a magnesium fire. Any moisture in the sand increases the rate of burning. *(Courtesy of Cambria County, PA Emergency Services)*

By itself it is relatively harmless, although it does have some toxicity. It cannot explode unless it is mixed with a fuel in the presence of an initiating charge. Although all three items should not be transported together, there may be occasions that they might be on the same truck. In other situations the oxidizer does not require any oxygen to produce a fire, as it will provide its own. Oxygen is labeled with a specific oxygen placard or can be shipped with an oxidizer placard. Oxygen itself cannot be ignited, but it will greatly intensify a fire when it is involved in amounts greater than 20.9 percent, the typical atmospheric level. Liquefied oxygen (LOX), which is a cryogenic, presents even more hazards in addition to supporting combustion. On asphalt or another hydrocarbon, LOX makes the surface shock sensitive so that it can detonate if compressed, such as by a firefighter walking on it. Another concern is the absorption of oxygen in FFPC from a LOX release, which can present a flammability problem. Caution should be used and the gear should be given time to air out, but this is a very rare occurrence. Most hospitals have large upright cryogenic LOX storage tanks, as do many nursing homes. An example is shown in **Figure 5-33.** Smaller in-home

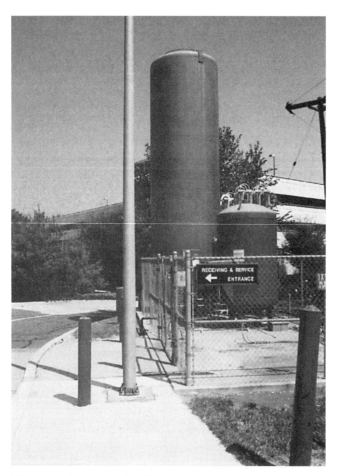

Figure 5-33 A leak of liquid oxygen on asphalt can present a shock-sensitivity problem and increase the risk of a fire.

versions are available as well, presenting a large fire risk in a residential home, and a freezing and contact hazard if knocked over.

Another commonplace situation involving oxidizers is an incident caused by pool chemicals. Most of these chemicals are oxidizers or may react with oxidizers and can be involved in a chemical reaction or fire. If the materials do not ignite, a dangerous situation can develop; a large chlorine vapor cloud can be created from just a handful of pool chemicals, which can produce disastrous effects on a neighborhood. Specific advice from the HAZMAT team is required prior to handling or disposing of any of these types of chemicals. There is a lessened level of concern with these types of chemicals because they are considered household products, but they do present a large hazard to the community.

Poisons

Class 6 refers to poison liquids, and we will include poisonous gases from class 2 (gases). When dealing with these types of materials, including "stow away from food stuffs" and "marine pollutant" materials, keep in mind that they are toxic to both humans and the environment. Materials labeled as poisonous should be considered very toxic, as the DOT reserves this class for highly hazardous materials. The predominate hazard a chemical poses places it into one or another category. There has been some improvement with the subsidiary placarding and labeling category, but there is still room for more.

Poisonous gases may also be accompanied with a label on a bulk container "toxic by inhalation," which would be listed on the shipped papers as TIH. By the definition these materials present a risk to humans and the community and require extra precautions. Some incidents with these types of materials are from pesticides and agricultural chemicals. Although most household materials are of lesser concentrations, there are rare occasions in which higher strength materials have been used. In several incidents, insect spraying companies (or individuals) have used full strength agricultural products in residential homes, requiring evacuations lasting a few weeks and removal of some homes due to the contamination.

SAFETY Technical grade pesticides are very dangerous and should be handled with care.

Incidents involving home pesticides and home fertilizing trucks are common, but in most cases this material is diluted to 1 percent or less to be ready for home application. The fertilizers and herbicides carried on these vehicles present little or no risk to responders or the environment unless in concentrated quantities. The insecticide components are most likely pyrethroid based such as permethrin (Pounce™, Dragnet™, Astro™); cypermethrin (Ammo™ and Cygon™); cyfluthrin (Baythroid™); bifenthrin (Tallstar™ and Biflex™); or natural pyrethrum (Pyrenone™). These all have a very low toxicity for mammals such as responders, but are highly toxic to aquatic life. Any spills from home fertilizing trucks and/or termite treatment tank spills must be prevented from reaching storm sewers, drainage ditches, and other waterways. Diking of spills is important.

On rare occasions, chloropyriphos (Dursban™) or a diazinon may be the insecticide component; both of these are no longer registered for residential use, but existing inventories may continue to be consumed. Both of these are hazardous to mammals (responders) and pets. The initial responders should consult with the HAZMAT team if the area is still wet before taking action.

Many home pest control vehicles may carry a mixed load of premixed and undiluted concentrates. When these vehicles are involved in accidents and their contents are released, first responders should use FFPC and SCBA to effect rescues. Anyone who comes in contact with any leaked materials should be decontaminated until the specific pesticide is identified.

Radioactive Materials

Although considerable emphasis was placed on radiation in the early days of HAZMAT, due to a low number of incidents involving these materials, some of the emphasis has been redirected to other areas. It is true that incidents involving these types of materials are rare, but they do have a significant impact on the well-being of the community, and with the potential now for terrorism, radiation is being emphasized again. Radioactive materials are commonly used in the community, in smoke detectors, in ground imaging equipment, and in the medical community. Although many people express concern when around radioactive materials, most radioactive materials are not harmful even for a lengthy exposure.

SAFETY Keeping radioactive material in the container intended for such use is paramount to reducing the exposure levels of the responders.

Radiation is divided into two categories: ionizing and nonionizing radiation. Radioactive materials that emit alpha, beta, gamma, and neutron forms of radiation are ionizing types. The ionizing forms of radiation are able to make changes in atoms, which can lead to health problems with humans. Nonionizing radiation takes the form of sunlight (visible light), microwaves, and radio waves. These forms of energy are not as strong as their ionizing counterparts, but high amounts or repeated exposure (e.g., to sunlight) can cause health problems.

Some radioactive materials, such as those coming to and from a nuclear power–generating facility, can present some risk to the community. Radioactive materials that present a risk to the community are shipped in high-strength containers designed to withstand a substantial crash and to be involved in a fire situation without any release of radioactive materials. These shipments are usually well tracked and, when involved in an incident, specialized help is usually already on the way before local responders call for it. Knowing who is the local contact for radiation emergencies is key to an effective response when dealing with radioactive materials.

SAFETY Keeping the material in the container that it is intended to be in is paramount to reducing the exposure levels of the responders.

When dealing with potential radiation hazards, the adage "time, distance, and shielding" should always be followed. This basic principle of radiation exposure should be applied to all chemical exposures. Firefighters must be careful to limit their time around a radiation source, keep their distance, and wear some type of shielding; then they can be protected against most forms of radiation. Most radiation exposure guidelines are based on a 1-hour exposure time. By reducing their time, firefighters reduce their exposure. By doubling their distance, they can reduce their radiation exposure to one-fourth of the original exposure. Protective clothing and SCBA can provide some shielding for some forms of radiation, while more substantial forms of shielding such as lead or concrete may be needed for more dangerous types of radiation. Persons who have been exposed to a radiation source are not radioactive. They may suffer some significant health problems, but they do not present a risk to others. Persons who have radioactive material on them, such as a radioactive powder, are contaminated and may present a risk to others; this would be considered external contamination. Alpha and beta radiation sources can contaminate someone, and are inhalation and ingestion hazards, which is considered internal contamination. Neutron and gamma radiation do not create contamination issues with regard to humans. Neutron radiation sources do have the ability irradiate some materials, but the chances of

WEIRD INCIDENTS

Not all chemical investigations involve toxic chemicals, but until the HAZMAT team can determine the identity of the product, first responders should isolate the area and deny entry.

One HAZMAT team was called to assist an engine company at a marine terminal. The engine company reported that a sea box placarded flammable was leaking an unknown colorless liquid. Upon arrival the engine company isolated the area, established the hot zone, obtained the shipping papers from the company, and began the research process. Nothing on the shipping papers seemed to fit the leak. The HAZMAT team chemist, who was also on scene, was equally confused. The HAZMAT team suited up and proceeded to investigate the truck with air monitors. At the back of the truck was a clear colorless liquid running from the corner of the truck onto the pavement. The liquid seemed to be evaporating and leaving a crystal-like material on the ground, as shown in **Figure 5-34.** None of the air monitors indicated any problems. The HAZMAT responders opened the back of the truck and revealed several 55-gallon drums. One drum was tipped on its side and was found to be leaking **(Figure 5-35).** The HAZMAT team entered the truck to check the label and spelled out, "H, O, G, C, A, S, I, N, G, S," which produced a confused look on everyone's face. This hog casings (intestines) incident took a few minutes to wrap up after realizing that the crystal-like material was salt from the brine in which the hog casings were stored.

About a month later, again at the marine terminal, the same HAZMAT was sent to investigate a ship that was having a "problem" with some containers. When responders arrived, they found that a ship had lost twenty-three sea containers at sea and now had seventeen containers leaning at an extreme angle. All but one of the containers contained food products; the one of concern was a container of phosphorus pentachloride, a very reactive material. From information obtained through the Coast Guard, Maryland Department of Environment, and the ship's crew, the HAZMAT team determined that the container of chemicals was involved but not damaged. The team boarded the ship to get a better view of the containers, which are shown in **Figure 5-36.** On the deck was an unusual odor, one that was faintly familiar. The responders all looked at each other with a quizzical look, trying to determine what the odor was; they were concerned that they might have been misled as to the safety of the ship. One of the bottom containers was crushed and the door was open. Responders looked inside to find the source of the odor—more hog casings. In fact, the sixteen other containers held more hog casings, with one HAZMAT container in the middle. Dealing with the HAZMAT container became a serious issue after it was determined how to offload the other containers, but it was interesting that hog casings, which are used to make sausages, were involved in another marine terminal event.

Figure 5-34 This photo shows the resulting leak from a sea container. In the middle is solid material that was left as the liquid evaporated. *(Courtesy of the Maryland Department of the Environment)*

encountering a neutron source are small, except in war or a terrorist attack. Simple decontamination with soap and water is needed to reduce the injury to the person. When dealing with radioactive materials it is best to try to contain the runoff so as not to further contaminate the incident scene. The best protection against radiation is to keep the radiation source in its protective container.

Corrosives

Behind the flammable and combustible categories, the next most common incident will probably involve a corrosive. Common corrosives are sulfuric acid, hydrochloric (muriatic) acid, nitric acid, and sodium hydroxide. Both of the acids are transported as liquids, but the sodium hydroxide may be transported wet or dry. Both sulfuric acid and sodium hydroxide have little or no vapor pressure and present little risk outside of the immediate spill area.

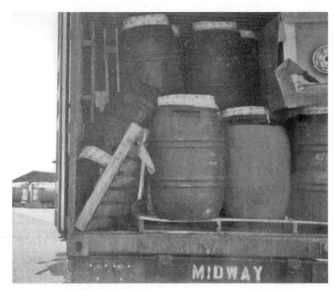

Figure 5-35 This photo shows the back of a sea container, with a leaking drum on the left. One never knows what conditions will be found when opening the back of a sea container. In this case there was an upside-down drum and a knocked-over washing machine. More "junk" was found farther up in the front of the container mixed in with the hazardous materials and hog casings. In some instances wooden braces at the back of the sea box require the use of tools to remove. The easiest removal method may be a chain saw, as long as flammability is not a concern. *(Courtesy of the Maryland Department of the Environment)*

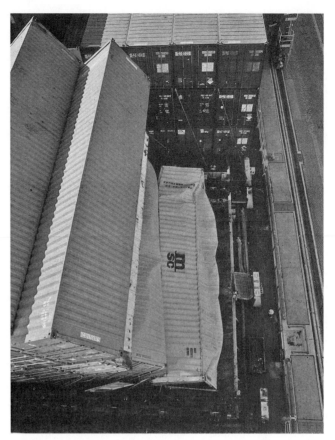

Figure 5-36 On this container ship these 17 containers were listing severely; only one container in the middle contained hazardous materials. It was a difficult operation to remove the containers without toppling the entire pile and damaging the HAZMAT container. *(Courtesy of the Baltimore County Fire Department)*

Hydrochloric acid, on the other hand, has high vapor pressures and may require some additional isolation and evacuation. Nitric acid is less common, but is easy to detect as it has a characteristic vapor cloud, as shown in **Figure 5-37.** When released it forms a brown vapor cloud that is comparable to when bromine is released. If it is exposed to a little water, even high humidity, it can react violently to release chlorine, hydrochloric acid, and other hazardous gases. If flooded with water, it is relatively safe. In the event of the discovery of pool chemicals reacting, dump them into a pool. In effect, you will have created a "shock treatment" to the pool, a situation the owner can deal with, but at the same time you have averted a potentially dangerous gas release. To handle any of these materials, chemical protective clothing is required; after some time they will eat a hole in FFPC. A common misconception is that spilling sulfuric acid on your hand will make your fingers fall off. Although this is an exaggeration, any splash of these materials should be washed off as soon as possible; actual burns will take a couple of minutes to take place and will progress in severity if the chemical is not quickly washed off. This applies for these materials only. Some uncommon acids, such as oleum or sulfur trioxide, will cause immediate burns upon contact and disfiguring injuries over a period of time. Chemical neutralization may be the best choice to handle a corrosive spill. This operation needs to be performed by a HAZMAT technician who has donned appropriate levels of PPE. When neutralizing, heat and violent reactions may be possible, so this action should be taken only after consultation with a chemist. Although one would think that simple dilution would take care of an acid spill, this is not true. A 100-gallon acid spill may require thousands, if not hundreds of thousands, of gallons of water to bring the spill back to a neutral pH level.

Other Incidents

It is impossible to outline each specific action first responders should take at a chemical release. The first rule if responders suspect that hazardous materials

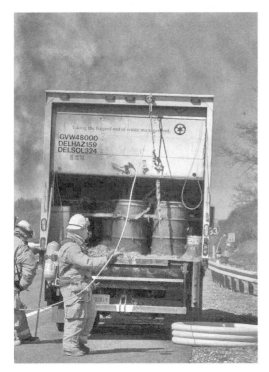

Figure 5-37 The shipping papers did not indicate the presence of nitric acid. The vapor cloud (which is brown) is a result of a chemical reaction. When the HAZMAT team opened the back of the truck they were greeted with these vapors. Bromine and red fuming nitric acid are the two chemicals that produce this type of vapor cloud. *(Courtesy of the Maryland Department of the Environment)*

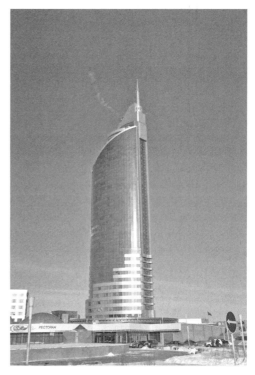

Figure 5-38 It is difficult to determine the origin of unusual odors in a high rise building. There are many occupants, offices, and potential sources.

are involved is to isolate the area and evacuate people in the immediate area until further information is available. Request the assistance of the closest HAZMAT team as its members will be needed for their technical expertise and equipment. First responders are not trained or equipped to provide detection or characterization of unknown materials. Depending on senses such as smell is playing Russian roulette.

SAFETY The absence of a vapor cloud or odor does not mean a potentially deadly material is not present.

Many toxic materials are odorless and colorless and provide little warning prior to causing lethal effects. It is not the job of first responders to seek out the cause of a chemical release nor to provide an identification. If the information is presented to them or can be obtained without risk, then it is of benefit, but otherwise it is the HAZMAT team's job to perform those tasks.

Common incidents that first responders may get involved with are odor complaints that may be classified as sick buildings, such as the one shown in **Figure 5-38.** The use of the HAZMAT air monitors is key to survival with these types of incidents. Always remember that your nose is not sufficient to determine immediately dangerous levels. When involved in sick buildings or odor complaints, the best course of action for first responders is to determine the validity of the complaint. First responders should interview the people involved and assess the likelihood that they are in fact experiencing the problems they state. If a number of people who work in a variety of locations throughout the building are complaining of similar signs and symptoms and report an identical odor, then the complaint is probably valid. Once it is determined that the call is valid and not related to a sunny Friday afternoon, first responders should don SCBA and request the services of the HAZMAT team. The occupants of the building should be removed, and the windows and doors should be shut. The HVAC system should be shut down so as not to remove the unknown material. These steps are crucial in the success of the HAZMAT team finding the source of the problem.

First responders may also get called to gas leaks inside a building, and their ultimate safety hinges on air monitoring, which is further explained in Chapter 6. It is impossible to determine a level of gas in a building based on smell. Once in an environment that has a gas odor, the body desensitizes itself to the odor, and responders will not be able to smell the gas anymore, even if the amount has been increasing and moving closer to a potential explosion.

Gas grills are prone to propane leaks, and the leak can reach an ignition source, causing severe damage to the adjacent buildings. Many HAZMAT teams carry pipe flares to burn off the gas and prevent an explosion. First responder actions should be limited to isolation and evacuation and preparing standby hose lines. If the vapors are traveling toward other homes, then the use of hose lines with fog nozzles to move the vapors is appropriate.

DECONTAMINATION

The task of **decontamination** in some cases may fall to a person trained at the first responder operations level. If a first responder is expected to perform decontamination of a HAZMAT team, specific training in that area is required along with training in the use of chemical PPE. Decontamination is the physical removal of contaminants from people, equipment, and the environment. It is important to note that the concepts provided here may be for one or all three of these areas of concern. If one type of decontaminating solution is effective for a tool, it does not mean that it can be used on humans. Prior to performing any decontamination procedure on humans, responders must consult with the HAZMAT team or a chemist. Some decon solutions use very dangerous materials, especially upon contact with skin. For equipment it may be acceptable to use large quantities of sodium hydroxide mixed with detergent, but this same solution would cause severe burns if placed on the skin. It is commonly recommended that no more than soap and water be used on victims. When using decon solutions on humans, caution and consultation are required.

There are two sources of contamination: direct contact or secondary contamination. Going into a spill area and putting your hand into a drum of blue paint is an example of direct contamination. If you leave the area and shake hands with your partner, this is an example of secondarily contaminating your partner with the blue paint.

Types of Decontamination

With the revision to the NFPA 472 standard there are three general types of decontamination levels: **emergency decon, technical decontamination,** and **mass decontamination**. One other type, fine decontamination,

Figure 5-39 For HAZMAT incidents a plan for decontamination is required. For some incidents a large-scale setup is required, but for most incidents the setup shown is generally sufficient. *(Courtesy of the Maryland Department of the Environment)*

is added here as it is an important type of decontamination for victims. In a full-scale incident, all four may be used, but in reality the majority of the incidents only require a minimal decon setup such as the one shown in **Figure 5-39.** Within each of the two large categories of decontamination, wet and dry decontamination, are four levels with some modifications. The two categories should be self-explanatory: one requires the use of water or other liquid solutions, while the other does not involve a wet process. The process of decontamination is chemical specific, and there are no absolute rules to the process. First responders should as a minimum have a good understanding of emergency decon procedures and develop a local procedure to handle this type of problem. Being confronted with a contaminated patient who is in need of decon is not the time to develop the procedure. With potential terrorism in mind, not only should responders think about the emergency decon of one person but of decontaminating thousands of people.

Emergency Decontamination

When a victim or a responder is contaminated with a hazardous material it is vitally important to remove the contamination as rapidly as possible. To delay the removal of the hazard could result in fatal consequences. No matter what type of emergency activity is occurring, responders and the IC should have some thought of emergency decontamination. Even responding to a medical call can result in the need for emergency decon; having a plan is essential for survival. Emergency decon can be as simple as washing someone off with a hoseline, though it can be further complicated by involving an unconscious patient on a backboard or stokes basket. For most chemicals,

Figure 5-40 One of the simplest forms of emergency decon is the use of hoseline.

plain water washing is sufficient to decontaminate the person once clothes are removed. By removing the patient's clothes, you can eliminate 60 percent to 90 percent of the contamination. Military research involving warfare agents, even toxic nerve agents, shows that if the victim has been exposed to an aerosol, just having the person stay in fresh air for 15 minutes can be an effective decon method. It would always be recommended that a person in this situation be decontaminated with a soap and water solution.

The common methods of emergency decon involve the use of a water line and some method of runoff control, such as shown in **Figure 5-40.** Once through emergency decontamination, the victim should receive technical or fine decontamination, as emergency decontamination is not likely to remove all of the contaminant.

NOTE When people's lives are on the line, runoff is not a primary concern, but when not under those conditions, attempts should be made to recover any runoff. When dealing with radioactive decontamination efforts should be taken to contain runoff.

There are a number of manufacturers who can provide decontamination containment. Alternatives for containing runoff is the use of an inflatable children's pool that can easily be carried on a fire truck. Another method involves the use of a tarp or plastic that is spread between two ladders. If none of these methods is possible, responders should use a tarp or plastic to catch the runoff as best as possible. If all other methods are not available, responders can seek natural areas that would collect the runoff. This type of decon should not be performed over storm drains or near areas of environmental concern. An industrial plant

may have safety showers, and it is acceptable to use these for decon. Prior to using the showers, ask the people from the facility where the runoff goes. In most cases the runoff will go to the facility's own water treatment area and can be processed by that system. Caution should be used when the safety shower drains straight to the sanitary system; prior to using that shower, contact the local sewer authority.

Another general rule with regards to emergency decon applies to the removal of corrosives from a victim. Large quantities of water should be used. When the water mixes with the corrosive, there is the potential for a heat increase, and massive amounts of water will eliminate any potential for further injury. Besides using large quantities of water on corrosives, it is necessary to continue the flushing for a minimum of 20 minutes. No matter how bad the temptation is for EMS providers to remove the patient, the best treatment for the burns is the water flush, especially for burns caused by a base, such as sodium hydroxide.

Technical Decontamination

The first step in technical decontamination is the removal of the majority of the contaminants. It is usually the first rinsing station within a full decon setup, as shown in **Figure 5-41.** This step is known

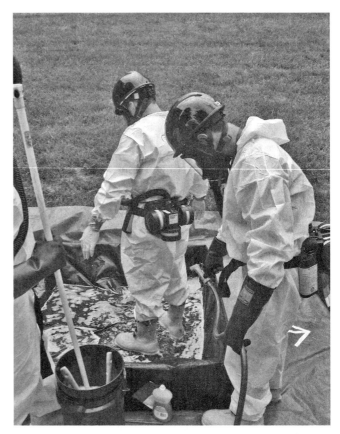

Figure 5-41 Gross decontamination is used to remove the majority of contamination.

as gross decon and in many cases it is just part of the whole process, but it can be the only step in a minor setup. Some HAZMAT teams use shower setups for gross decon, while others may use hoselines or pump tanks. As this is the most likely location for contaminants, attempts should be made to recover any runoff. When part of a multistep process entry team members usually perform this step on themselves without any assistance. When dry decon is used, there may be occasions that overgarments such as booties or extra gloves are used. When leaving the hot zone, there is a container or location to put dirty boots and gloves. For many materials dry decon is an acceptable use of PPE and resources. When using any type of decon, the bottom line is to make it safe for the responder to get out of the PPE and to create a safe situation for the responders assisting with PPE removal. When using reusable PPE, decon needs to be more formal, but even extensive cleaning can be done to the PPE after the incident and removal of the responder.

Further steps in the technical decon process are scrubbing and cleaning to remove any residual contaminants, as shown in **Figure 5-42.** This is usually done after the gross decon, typically in a couple steps. Attempts should be made to recover any runoff, but this is not as critical as with gross decon. It is during this scrubbing process that decon solutions are typically added to the process. These solutions are chemical specific. This process may involve showers, hoselines, or pump sprayers, and usually one to two people dressed in a lower level of PPE perform this activity. When performing decontamination it is important to pay particular attention to the areas most likely to be contaminated: hands and feet.

Figure 5-42 Formal decontamination is used to remove any further contamination that may remain after gross decon.

Fine Decontamination

This form of decon is not performed by first responders, but by hospital-based personnel. This cleaning removes all contaminants from the body. It would be nearly impossible to clean eyes, ears, fingernails, and other areas of the body in the street, but this process can be accomplished in the better conditions of the hospital. To perform fine decon according to OSHA, the staff must have proper training to perform these activities and must be trained to the operations level, although some training to the technician level is preferred. Some hospitals have designated decon areas with a separate entrance, separate ventilation system, and a holding tank for runoff control. Through the LEPC plan, hospitals are required to be identified to accomplish decontamination of patients. Every hospital should have a policy to cover a walk-in contaminated patient, and emergency responders should test these plans on a regular basis. Regular training and interface are required to keep this type of system functioning well within a community. Responders should ensure this capability is well run and trained, as they are likely candidates to become patients one day.

Mass Decontamination

Most HAZMAT teams' decontamination setup is designed primarily to decontaminate team members. It is designed to handle two to four people dressed in chemical protective clothing. Although most of these systems can handle civilians, they are intended to handle the needs of cleaning responders' PPE. When decontaminating civilians, several issues arise: clothing, privacy, weather impact, and inconvenience. Mass decontamination requires that plans must be developed to handle thousands of people. Communication will be difficult, and bullhorns or loudspeakers should be employed to address large crowds. Keeping contaminated victims informed helps to keep them calm. Some mass decon setups are provided in **Figure 5-43.** Most of the mass decontamination issues arise when terrorism events are discussed. More information on possible terrorism scenarios is provided in Chapter 7. If victims are contaminated with military nerve or blister agents, speed is of the essence. If decontamination is not performed on symptomatic contaminated patients, their chances for survival are slim. The hardest part of this process is identifying those who need decontamination and those who need **psychological decontamination.** If victims are symptomatic and a credible threat is presented, decontamination should be accomplished expeditiously. First arriving units

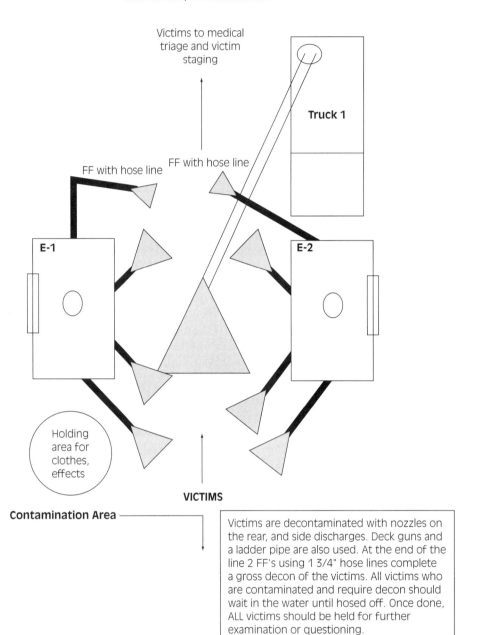

BASIC PLAN
Baltimore County Fire Department
Hazardous Materials Response Team
Mass Casualty Decontamination

Victims to medical triage and victim staging

Truck 1

FF with hose line

FF with hose line

FF with hose line

E-1

E-2

Holding area for clothes, effects

VICTIMS

Contamination Area

Victims are decontaminated with nozzles on the rear, and side discharges. Deck guns and a ladder pipe are also used. At the end of the line 2 FF's using 1 3/4" hose lines complete a gross decon of the victims. All victims who are contaminated and require decon should wait in the water until hosed off. Once done, ALL victims should be held for further examination or questioning.

Figure 5-43 Mass decon setups.

ADVANCED PLAN
Baltimore County Fire Department
Hazardous Materials Response Team
Mass Casualty Decontamination

Victims to medical
triage and victim
staging

Truck 1

FF's with 2 hose lines

E-1

E-2

VICTIMS

Contamination Area

Victims move through a staggered line,
like at an amusement park or bank
waiting line. Gross decon is completed at
the end of the line with two 1 3/4" hose
line. While waiting in line, nozzles
from the rear, side, and deck pipe are
washing the victims as well. Rope or
barrier tape can be used for the line.

Figure 5-43 *(continued)*

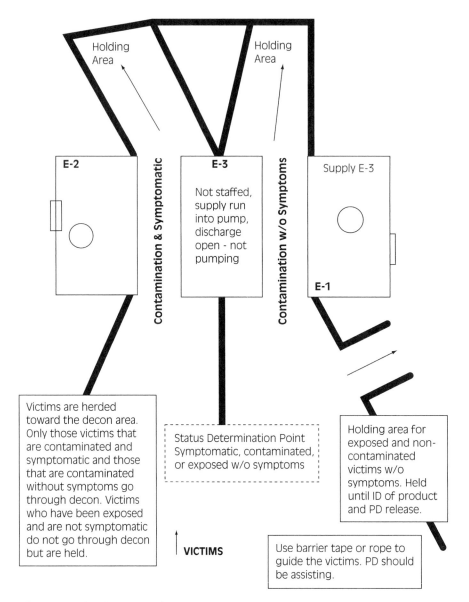

Herding Guide
Baltimore County Fire Department
Hazardous Materials Response Team

Holding Area

Holding Area

E-2

Contamination & Symptomatic

E-3

Not staffed, supply run into pump, discharge open - not pumping

Contamination w/o Symptoms

Supply E-3

E-1

Victims are herded toward the decon area. Only those victims that are contaminated and symptomatic and those that are contaminated without symptoms go through decon. Victims who have been exposed and are not symptomatic do not go through decon but are held.

Status Determination Point Symptomatic, contaminated, or exposed w/o symptoms

Holding area for exposed and non-contaminated victims w/o symptoms. Held until ID of product and PD release.

↑ VICTIMS

Use barrier tape or rope to guide the victims. PD should be assisting.

Figure 5-43 (continued)

Water Flow for Herding Guide
Baltimore County Fire Department
Hazardous Materials Response Team

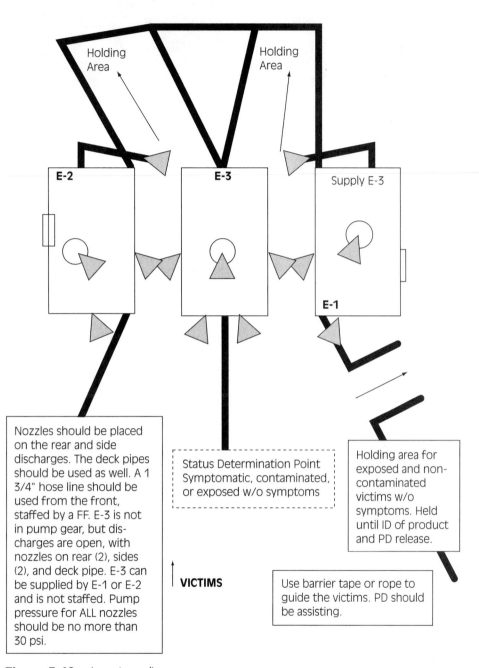

Holding
Area

Holding
Area

E-2

E-3

Supply E-3

E-1

Nozzles should be placed on the rear and side discharges. The deck pipes should be used as well. A 1 3/4" hose line should be used from the front, staffed by a FF. E-3 is not in pump gear, but discharges are open, with nozzles on rear (2), sides (2), and deck pipe. E-3 can be supplied by E-1 or E-2 and is not staffed. Pump pressure for ALL nozzles should be no more than 30 psi.

Status Determination Point Symptomatic, contaminated, or exposed w/o symptoms

Holding area for exposed and non-contaminated victims w/o symptoms. Held until ID of product and PD release.

VICTIMS

Use barrier tape or rope to guide the victims. PD should be assisting.

Figure 5-43 *(continued)*

need to establish this decontamination process. The best decontamination solution is water, followed by soap. One method that is built into the engine company is shown in **Figure 5-44.** This engine company even carries robes and slippers to be distributed after decontamination. Another method of mass decontamination is the use of mobile showers shown in **Figure 5-45** and **Figure 5-46,** which

can enable a number of victims to decontaminate themselves.

SAFETY The use of a bleach and water solution in terrorism decontamination is no longer considered an acceptable practice.

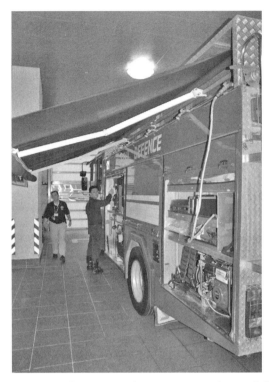

Figure 5-44 This engine has a canopy that unfolds to provide a decontamination corridor. The side of the engine is pre-piped with shower nozzles and garden hose extensions for additional water sprays. Sides can be added for privacy.

Studies from the military have shown that the use of bleach can actually cause more deaths and injuries among contaminated victims. Decontamination is chemical specific and unless responders can identify the specific chemical compound, they should use a simple soap and water solution. A simple hoseline

Figure 5-45 An example of a decontamination vehicle that has showers and can contain any runoff. *(Courtesy of the Maryland Department of the Environment)*

Figure 5-46 An example of the decontamination vehicle completely set up. The use of the tent adds an additional layer of privacy. *(Courtesy of the Maryland Department of the Environment)*

can be used to accomplish emergency victim decontamination; as more assistance arrives, the system can become more elaborate. When setting up the decontamination process it is important that patient backup be avoided. If there are delays, make sure that patients wait in water. After the patients exit the decontamination area, they should be conducted to holding areas. It is important to keep symptomatic patients away from nonsymptomatic patients. If these patients are mixed, then the nonsymptomatic victims will tend to pick up the symptoms due to psychological reasons.

Decontamination Process

Decontamination procedures may vary slightly, but most organizations have developed similar procedures. In **Figure 5-47** the photos show the basic process and outline the following steps: tool drop, gross decon, scrubbing, rinse, PPE removal, SCBA removal, clothing removal, body wash (emergency only), dry off, and medical evaluation including hydration. One should consider a final step—recordkeeping. While this may sound out of the ordinary, all exposure information including the suit worn and pertinent information should be logged for future use if needed. The variations include the use of shower systems, tents, number and location of steps, and personnel. The method of runoff collection also varies from team to team. **Figure 5-48** depicts the use of a swimming pool for containing runoff, while some teams use folding containment to provide a rectangular setup; it generally includes a pump sprayer and brushes. For most situations that involve solids and liquids a minimum type of decon should be set up. The HAZWOPER regulation requires that the IC have a plan for decontamination, which can vary from no

JOB PERFORMANCE REQUIREMENTS

Figure 5-47 Steps in the Decontamination Process

A The tool drop is usually the first stage in the decon setup. Tools are placed for the use of another entry team or may be collected for later decontamination.

B In most systems gross decon is the next step. Some response teams use showers or hoselines to accomplish this task.

C The next step is formal decon, which may involve two stations in which the responder is rinsed, scrubbed, and rinsed again. The solution depends on the chemical. This is the first step in which other responders may assist.

D After the wet portion the entry team removes PPE with the assistance of other responders.

E After the removal of the PPE, the entry team removes SCBA. It is usually placed into some type of containment system for later cleaning.

(continued)

JOB PERFORMANCE REQUIREMENTS
Figure 5-47 (continued)

F In some systems a fourth washing area is established for body washing. Some teams may set up a tent for this purpose.

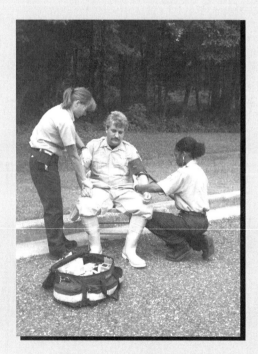

G The last step is medical evaluation and rehydration.

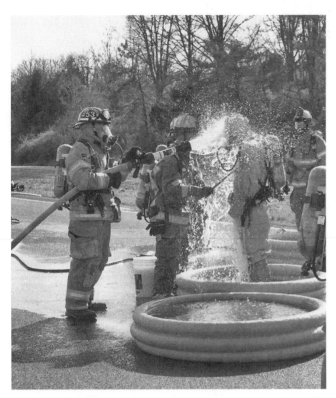

Figure 5-48 Some form of containment is required when performing decontamination. *(Courtesy of the Maryland Department of the Environment)*

setup through a multistep process as depicted in the decon photo series. For a material such as a technical grade pesticide, a full setup would be recommended with all stations established. For gases, in reality, no decon is required, but a minimum setup

should be established prior to entry in the event of an emergency and for psychological purposes. No matter the type or size of the setup that will be used, it should be organized and ready prior to any entry into a hazard area.

METHODS OF DECONTAMINATION

Table 5-7 lists general categories of decontamination that apply to humans, equipment, and the environment, although they may not apply directly to all three. Be sure to consult with the HAZMAT team or a chemist prior to using any of these methods on a human, as they may be more dangerous than the contaminant.

ABSORPTION

The spilled material is picked up by the absorbent material, much like a sponge. Common absorbent materials include ground-up newspaper, clay, kitty litter, sawdust, charcoal, and poly fiber. The absorbed material does not change, and the volume of material may increase. The whole mixture should be treated as waste and will not change any characteristics, such as flammability. Compatibility needs to be researched prior to using these materials, as some materials could cause a reaction.

TABLE 5-7

Methods of Decontamination			
Type	**Used on humans**	**Used on equipment**	**Used on environment**
Absorption	Yes	Yes	Yes
Adsorption	Yes	Yes	Yes
Chemical degradation	No	Yes	Yes
Dilution	Yes	Yes	Yes
Disinfection	Yes	Yes	Yes
Evaporization	Yes	Yes	Yes
Isolation and disposal	No	Yes	Yes
Neutralization	Under specific conditions	Yes	Yes
Sterilization	No	Yes	Yes
Solidification	No	Yes	Yes
Vacuuming	Yes	Yes	Yes
Washing	Yes	Yes	Yes

ADSORPTION

With this method the material bonds to the outside of the adsorption medium. Activated carbon is the most common adsorbent and should be available locally at a chemical facility. Many chemical facilities have drums, bags, or totes of activated carbon set up to receive liquids within a process, and those materials may be available in the event of an emergency.

CHEMICAL DEGRADATION

The ability to degrade a chemical varies from compound to compound, and is much like neutralization. To degrade a chemical another chemical is added, or the chemical is exposed to the elements, which breaks down the chemical into less hazardous components.

DILUTION

The ability to dilute a contaminant depends on the chemical structure of the spilled material. Dilution is not effective with many chemicals. Imagine being coated with used motor oil: Flushing with water will not remove all of the contamination. When dealing with corrosives, large quantities of water are required, sometimes hundreds of thousands of gallons. Some recent military studies have stated that plain water can be "effective" and that some bleach and water solutions can cause greater problems for patients. When time and resources permit, soap can be added, as this option is shown to be of additional benefit in the decon process.

When dealing with humans a 0.5 percent bleach and water solution can be used for some etiological contaminants, although contact time is needed for success. It was initially thought that bleach solutions would also be effective for biological contaminants, but it has been determined that plain water is just as effective for humans. This is due to the contact time required for effectiveness; the exposure of skin to bleach solutions causes burns if left on for more than a short time. For equipment or the environment, disinfection can be accomplished using a higher percentage of bleach or other chemical.

EVAPORIZATION

Is allowing a chemical to evaporate changing its state of matter? When a solid or a liquid is allowed to remain in the open, depending on its vapor pressure it may change to a vapor. A chemical that approaches its boiling point will also evaporate, moving the material to the vapor state. The material does not disappear—it merely is changing its state of matter.

ISOLATION AND DISPOSAL

One of easier forms is to isolate the contaminant, collect it using appropriate protective clothing, and then dispose of the contaminant. The disposal should follow appropriate federal, state, and local regulations.

NEUTRALIZATION

This method is usually reserved for corrosive materials but also is used here to describe the procedure to reduce the toxicity of a poisonous material. It is highly recommended that responders consult with a chemist prior to performing this activity. Many chemicals used to reduce this risk are in themselves hazardous, though usually in a much lesser fashion than the contaminant. An emergency responder may avoid mixing a neutralization solution using sodium hydroxide as it is very corrosive, but with some chemicals it is extremely useful in small amounts in reducing the hazard of some of the contaminant. In general, neutralization is exothermic and can produce violent reactions, although these are extremely rare.

STERILIZATION

There are two primary methods of sterilization: One is through a combination of steam and high heat, and the other is chemical sterilization. The use of steam and high heat is effective for items that are contaminated with etiological materials. Chlorine dioxide has been used to decontaminate buildings for anthrax contamination. Chlorine dioxide is very corrosive and is considered to be a very concentrated form of bleach.

SOLIDIFICATION

A solidification agent is used to make a liquid material into a solid and may alter the spilled agent. In some cases, however, the solidification agent will not have any effect on the agent and will reduce the hazard and enable later sampling. Chemical compatibility needs to be confirmed prior to the use of a solidification material.

VACUUMING

For dusts, fibers, or other solid materials, a vacuum would certainly reduce the hazard. An ordinary shop vac would not be recommended for this task, however, as it is not filtered well enough to keep the exhaust from blowing agent back into the air. Use a vacuum that is equipped with a high efficiency particulate accumulator (HEPA) filter.

Specialty vacuums are made exclusively to pick up mercury without the mercury vapors being released into the air.

WASHING

Typically washing is done with soap and water. This is one of the more effective decontamination solutions, but in reality the soap and water merely remove the contaminant. The contaminant may still present a risk, but is removed from the person or item. Washing can be low pressure or high pressure, and can also include elevated temperature liquids. Under the direction of a chemist, other solutions may be added to the washing solution. To test the effectiveness of decontamination, responders can use air monitors, particularly a PID, to determine if the contaminant has been removed. Other detection devices can be used for other materials. For the ultimate assurance, responders can take **swipe samples** and send the swipes off to the laboratory for analysis. As part of the termination procedures the use of detection devices or other tests for the effectiveness of decontamination should be noted in the report. The type of decontamination and the specific types of decontamination solutions should also be recorded. The report should include any potential exposures to hazardous materials as well—an activity log for the responders documenting their actions. These records should be kept for at least 30 years past the last date of employment for the responders. In addition to being NFPA requirements, they are also part of the OSHA HAZWOPER regulation.

SUMMARY

The chlorine rail car accident in Graniteville is an important lesson. Many of the topics in this chapter directly relate to that incident. Although that incident caused a significant impact on that community and many lives were lost, the chlorine release was not as pervasive as it could have been. Protective actions are used for a variety of purposes to protect the public and, in many cases, the responders. The management of a chemical release is not an easy task, as multiple agencies are often involved in accomplishing a variety of tasks at the same time. Unlike most fires or accidents, political pressure may be associated with a chemical release, especially if there is the potential that thousands of people may require evacuation and relocation. The determination for decontamination can also be a difficult decision, as most departments do not have the ability to decontaminate large numbers of victims. Choosing the method of

decontamination is often complex, as there are many possibilities. The best options for protection are to provide prompt isolation and prevent further escalation of the exposed population. For those persons who are contaminated, a water-based decontamination is usually recommended.

KEY TERMS

8-Step Process© Management system used to organize the response to a chemical incident. The elements are: detect, estimate the harm, choose an objective, identify the action, do the best you can, and evaluate your progress.

Bung Cap of a drum. The cap varies in size but the most common is 3/4 and 2 1/2 inches.

Containment system Provided as a backup system in the event of a tank failure. If the tank fails, the containment system is designed to hold the released product.

DECIDE Management system used to organize the response to a chemical incident. The factors of DECIDE are: detect, estimate, choose, identify, do the best, and evaluate.

Decontamination Physical removal of contaminants (chemicals) from people, equipment, or the environment; most often used to describe the process of cleaning to remove chemicals from a person.

Emergency decon Rapid removal of material from a person who requires immediate cleaning. Most emergency decon setups use a single hoseline to perform a quick, gross decon.

Emergency response planning (ERP) Levels that are used for planning purposes and are usually associated with preplanning for sheltering in place or evacuation zones.

Evacuation Movement of people from an area, to another area that is considered to be safer. Persons are evacuated when they are no longer safe in their current area.

Fine decon More detailed form of decontamination, usually performed at a hospital by staff who are trained and equipped to perform decon procedures.

Formal decon Washing and scrubbing portion of the decontamination process. The process is usually repeated and is performed by a decon crew.

GEDAPER© Management system used to organize the response to a chemical incident. The factors are: gather information, estimate potential, determine goals, assess tactical options, plan, evaluate, and review.

Gross decon Step in the decontamination process that removes the majority of the chemicals through a flushing of the person. The gross washing employs large amounts of water.

Hazard area   Term used to describe the dangerous area around a release; also called the isolation area or hot zone.

Hot, warm, and cold zones   All zones of isolation used at chemical releases, from the most dangerous to the safest areas.

Isolation area   Area in which citizens and responders are not allowed; may later become the hot zone/sector as the incident evolves. This is the minimum area that should be established at any chemical spill.

Overpacked   Response action that involves placing a leaking drum (or container) into another drum. Some drums are made specifically to be used as overpack drums and are oversized to handle a normal-sized drum.

Psychological decontamination   Used to make patients who are not contaminated feel better, producing psychological cleaning.

Responsible party   Person or company responsible for the spill or release, who is ultimately responsible for the cleanup of the incident.

Sector   See *Zone.*

Shearing effect   Tearing of the molecules, which can be equated to being ripped apart.

Shelter in place   Form of isolation that provides a level of protection while leaving people in place, usually in their homes. People are usually sheltered in place when they may be placed in further danger by evacuation.

Sick building   Term associated with indoor air quality; occupants in a sick building become ill as a result of chemicals in and around the structure.

Sick building chemical   Substances that cause health problems for occupants of a structure.

Span of control   Number of workers that one commander can effectively manage. When the number of workers exceeds the span of control, the supervision must be divided among more commanders.

Technical decontamination   The process used to perform decontamination of persons or equipment.

Zone   Area established and identified for a specific reason; typically a hazard exists within the zone. Zones are usually referred to as a hot, warm, and cold to provide an indication of the expected hazard in each zone. Some agencies use the term *sector.*

REVIEW QUESTIONS

1. What is the first priority of any incident management system?
2. What zone or sector should be established first?
3. When dealing with a flammable liquid spill, what criteria could be used to establish the hot zone?
4. What level incident would be used to describe a chemical release in which 2,000 people were evacuated?
5. What is the minimum decontamination that should be set up for every incident?
6. What are the four types of decontamination?
7. In what step of decon are hospitals involved?
8. To decide to make a rescue depends on what information?
9. What causes the most common breach in a 55-gallon drum?
10. If fire has reached the cargo portion of a truck carrying DOT Class 1.1 materials, what tactics are used to fight the fire?
11. How can air monitors be used in the determination of hazard zones?

ENDNOTES

[1]In most states the fire department IC is in charge of the incident until the IC relinquishes command to another agency. Check your local and state regulations for more information. Although some responders leave when the cleanup contractor begins work, a safer policy may be to maintain a presence as long as the scene is considered hazardous. The emergency responders are solely responsible for public safety, a responsibility that cannot be delegated. The fact that a group has been "hired" to work in a hazardous environment does not relieve the responders of their responsibility. In many cases the contractor will use appropriate measures to protect employees and the adjacent community; in those cases the responders may choose not to remain on site. In cases where the contractor does not have the ability to protect workers through the means of air monitoring, as an example, someone must provide oversight. In some states this can be handled by the state or local environmental agency. But ultimately the responsibility for public safety is an emergency response function, and a contractor is part of that "public" we are sworn to protect.

[2]Any location such as a storm drain, ditch, or culvert that eventually leads to a waterway is defined as a waterway.

Suggested Readings

Bevelacqua, Armando S. S., *Hazardous Materials Chemistry,* 2nd ed., Delmar Cengage Learning, Clifton Park, NY, 2006.

Bevelacqua, Armando S., and Richard Stilp, *Hazardous Materials Field Guide*, Delmar Cengage Learning, Clifton Park, NY, 1998.

Hawley, Christopher, *Hazardous Materials Air Monitoring/ Detection Devices,* 2nd ed., Delmar Cengage Learning, Clifton Park, NY, 2007.

Hawley, Christopher, *Hazardous Materials Incidents*, 2nd ed., Delmar Cengage Learning, Clifton Park, NY, 2004.

Henry, Timothy, *Emergency Decontamination for Hazardous Materials Responders*, Delmar Cengage Learning, Clifton Park, NY, 1999.

Koonce, Ric, *Firefighters Handbook,* 2nd ed., Delmar Cengage Learning, Clifton Park, NY, 2004

Lesak, David, *Hazardous Materials Strategies and Tactics*, Prentice Hall, Upper Saddle River, NJ, 1998.

National Foam, Kidde Firefighting, A Firefighters Guide to Foam, 2002, Exton, PA.

Noll, Gregory, Michael Hildebrand, and James Yvorra, *Hazardous Materials Managing the Incident,* Red Hat Publishing, Chester, MD, 2005.

Schnepp, Rob, and Paul Gantt, *Hazardous Meterials: Regulations, Response, and Site Operations*, Delmar Cengage Learning, Clifton Park, NY, 1999.

Stilp, Richard, and Armando Bevelacqua, *Emergency Medical Response to Hazardous Materials Incidents*, Delmar Cengage Learning, Clifton Park, NY, 1997.

Stilp, Richard, and Armando S. Bevelacqua, *Emergency Medical Response to Hazardous Materials Incidents*, Delmar Cengage Learning, Clifton Park, NY, 1997.

CHAPTER 6

PRODUCT CONTROL AND AIR MONITORING

OUTLINE

▮ Objectives
▮ Introduction
▮ Defensive Operations
▮ Air Monitoring at the First Responder Operations Level

▮ Meter Terminology
▮ Carbon Monoxide Incidents
▮ Summary

▮ Key Terms
▮ Review Questions

(Courtesy of the Maryland Department of the Environment)

STREET STORY

In January 1983 I was faced with the realization of how little I knew about chemical emergencies. It was a hard lesson to learn. I was called at home around 4 A.M. to respond to the downtown area of my city for a fire that involved chemicals. It was reported that several firefighters had been injured. En route I was in radio contact with several officers on the scene and was informed that there had been an explosion and that four firefighters had been burned.

Once on the scene, I learned that a firefighter who had driven a car in my wedding party was being sent to a burn center in New York City. Not only was I dealing with a difficult incident, but it also now had an emotional aspect tied to it. Even with my limited knowledge, I was given the job of informing and advising the incident commander of the potential hazards involved with several of the chemicals that had been identified. The fire was caused by an overpressurization of a reactor vessel. Because of the explosion, no product control was initiated, and to compound the problem, the property was adjacent to the Long Island Sound. This incident was causing environmental damage to the land, water, and air, and four firefighters had already been injured. My cadre of tools included two reference books and sixteen hours of hazardous materials training. After several difficult hours of trying to identify, evaluate, and provide useful information to both the IC and the hospital, I came to the conscious decision that I simply did not have the right tools nor the required knowledge to be very helpful.

This one incident changed the way I would respond to chemical emergencies for the rest of my life. This incident taught me the importance of good identification skills, the requirement to understand every term as it relates to the hazards of the product, and the absolute need for air monitoring devices. To keep our troops safe, being armed with knowledge is the only way to face any type of emergency.

—Street Story by Frank Docimo, Special Operations Officer, Turn of River
    Fire Department, Turn of River, Connecticut

OBJECTIVES

After completing this chapter the reader should be able to identify or explain:

■ Equipment that can determine hazardous environments and isolation areas (O)

■ Available defensive operations for a release (O)

■ The various methods of damming, diking, diverting, and other defensive operations (O)

■ Detection equipment available for the first responder and equipment that a HAZMAT team may carry that could be used at a chemical release (O)

INTRODUCTION

The use of product control techniques can provide a quick reduction in damage done to the community and the environment. The reduction of surface area over which the product has spread provides a direct reduction in the danger to responders and the community. By implementing these types of measures, first responders can provide an extreme advantage to the overall mitigation of the incident. The use of air monitoring devices is becoming more commonplace, and many first responders have some type of air monitoring device available to them. This section provides a brief overview of this complex subject and concentrates on the knowledge first responders should have. To ensure responder safety, basic air monitoring must be accomplished, even for natural gas leaks, gasoline spills, or any other incident in which responders may be exposed to hazardous materials.

DEFENSIVE OPERATIONS

According to the NFPA three large tasks fit into operations level tactical response: basic air monitoring, containment, and confinement. All of these can be combined into product control, which is one task that persons trained to the operations level can perform. The task of containment and confinement should be performed away from the hazard area and in the smallest safest area. To perform defensive operations, responders do not need to get close to the actual spill. The activities that fit into product control operations are **absorption, diking, damming, diverting, retention, dilution, vapor dispersion, vapor suppression,** and the use of **remote shutoffs.** The type and level of PPE required for each of these tasks vary with the chemical involved, but a minimum would be firefighting PPE and SCBA. If making a

dam a considerable distance away from the chemical release and if no contact with the material is possible, a lesser level could be used. The tasks described in this chapter are among common tasks assigned to responders trained to the operations level. These critical tasks can protect the community and the environment. It is not necessary for first responders to carry all of the equipment cited in this chapter, but they must know its location and how to improvise, adapt, and overcome.

Absorption

First responders are often asked to handle absorption of a spilled material, and on a daily basis they probably absorb a large quantity of fuel and oil products using a product similar to the one depicted in **Figure 6-1.** Knowing the type of absorbent materials that a department carries is important. Some absorbent materials will not pick up water, while others absorb any liquid with which they come in contact. When trying to remove fuels from water, using an absorbent that does not pick up water is essential. Absorbents vary from clay kitty litter to ground-up newspaper, corn husks, and synthetic fibers. Their weight and absorbent capabilities vary, and not all can be used interchangeably. They may be available in loose bulk form or as pads, such as

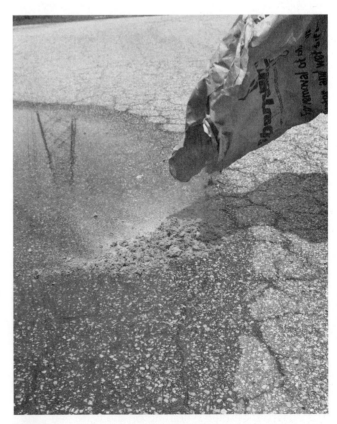

Figure 6-1 A common defensive action is the application of absorbent to a spilled material.

those shown in **Figure 6-2,** or as large rolls or in boom style. The recommended method to hold the absorbent material is to construct a filter fence such as the one shown in **Figure 6-3.** The filter fence can be made of "rat wire" or other small-weave fencing material. Wooden stakes can be used to hold the fence in place; in rocky soil metal stakes are recommended. Some examples of how to use common materials are shown in **Figure 6-4** and **Figure 6-5.** Always construct multiple systems so that in the event of a failure the spilled material can be caught by a backup system. Also available on the market

Figure 6-4 When time permits, more advanced methods of damming a waterway can be used. This system is more advanced than most but takes time and resources to accomplish. *(Courtesy of the Maryland Department of the Environment)*

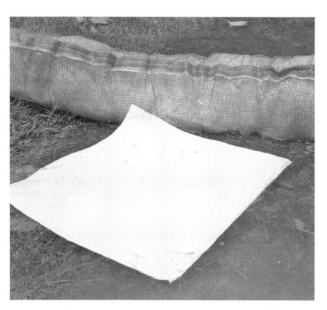

Figure 6-2 Absorbent pads will pick up oil from the water but will not absorb the water, allowing them to float.

Figure 6-5 The use of natural surroundings and materials assists in the construction of a filter fence used to skim material from the top of the water. In this case a snow fence is used to filter the water and hold the absorbent material. A fallen tree adds support to the operation. In some larger spills HAZMAT crews have felled trees to assist in this process. *(Courtesy of the Maryland Department of the Environment)*

are solidification agents that encapsulate the spill and microbiological oil-eating bugs. Both of these agents are expensive, and absorption materials are not generally used by emergency services due to their cost and the quantity required. It is best to carry a variety of absorbent products in several different styles.

Diking and Damming

The quantity spilled will determine to what extent these operations are performed. The actions required at a 50-gallon spill differ from those of a

Figure 6-3 A screen fence can be used to collect the absorbent material floating on the water, and the absorbent will collect the spilled material. The water flows through the fence below the absorbent.

500,000-gallon release. The two possible locations for diking, damming, and diverting operations are a spill on the ground or a spill on a waterway. A spill on the ground is much easier to control than one on the water, but the principles are the same. When creating a dike or a dam, you are either stopping the flow of the material or keeping it from a specific area. Except for small streams it is nearly impossible to stop a body of water from flowing. If the spilled material is soluble with water and the waterway has a good flow, it is impossible to collect the spilled material unless responders have heavy equipment and a large area available to create a very large dam.

Commonly, first responders use earth, sand, or rocks for dikes or dams, depending on what is available. Having local contacts available around the clock for these materials is important, and some jurisdictions have emergency contacts for dump truck loads of sand or dirt. Local or state highway departments are good places to start when these items are required. Heavy equipment such as front-end loaders or track

hoes can also be used to create dams or dikes. When constructing a dam or a dike, it is best to follow the rule of three and construct three setups. Start at the farthest point away from the spill and work back toward the spill. It may be necessary to establish the first containment miles away from the spill. If you start near the spill, you are likely to be overtaken by the spill and be forced to play catch-up in additional PPE.

Two basic types of dams can be created: overflow and underflow dams, as shown in **Figure 6-6.** The specific gravity of the spilled material, or whether the material is floating on top of the water or traveling on the bottom, will determine the type of dam that must be constructed. An overflow dam allows the water to overflow the dam and contain the spilled material at the base of the dam. An underflow dam allows the water to flow under the dam and collect the material on top of the water. Most spilled materials will be floating on the top of the water and require the use of an underflow dam. The materials required for either style dam are

Figure 6-6 The overflow dam allows clean water to flow over the dam, and collects the spilled material at the base of the dam. The underflow dam collects the spilled material on top and allows the water to flow through the bottom of the dam.

(A)

(B)

Figure 6-7 (A) Overflow dam. (B) Underflow dam.

Figure 6-8 First responders have equipment such as hard sleeves to establish over- and underflow dams.

shovels, dirt or sand, and pipes. Pipes should be 4 inches in diameter at minimum, and larger is better. The higher the flow or the more water, the bigger the pipe and the more pipes that are required. A good rule of thumb is to have enough pipes to cover two-thirds of the width of the waterway, which would be the minimum amount required. Standard PVC pipe is adequate and for emergency situations, the hard sleeves (suction) on a fire engine company are a good substitute. The photos in **Figure 6-7** and **Figure 6-8** show the process of setting up both styles of dams.

For an overflow dam, one tendency is to establish a dam, place the pipes on top, and consider the dam finished. It is best to set the pipes on top of the dam and then continue with a layer of dirt/sand on top of the pipes to avoid the possibility that the water will erode the dam and the containment will collapse without this extra layer of dirt/sand. **Figure 6-9** is another example of how to construct an underflow dam.

Diverting

To divert a running spill, almost anything can be used, such as dirt, sand, absorbent material, tarps, and

Figure 6-9 Another method used in the construction of an underflow dam. Note the number of pipes that have been used to move the water. *(Courtesy of the Maryland Department of the Environment)*

hoselines (see **Figure 6-10**). The most common use of diversion is to keep a running spill from entering a storm drain. A ring of dirt around the storm drain will keep the spilled material out but does not stop the flow of the product. It may be impossible to stop the flow of a large quantity of the spilled material, but it would be possible to keep it from the storm drain, as the protective measures in **Figure 6-11** demonstrate.

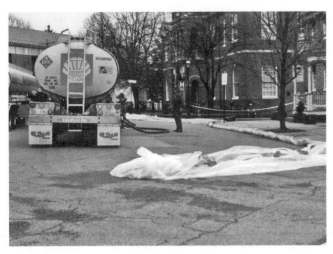

Figure 6-10 Plastic filled with sand is used to create a berm to hold back a potential spill from a catastrophic release from the tank truck. The larger the potential, the more substantial the device should be. This tank truck could hold between 9,000 and 10,000 gallons, which if dumped at one time would probably overtake this system. *(Courtesy of the Maryland Department of the Environment)*

Figure 6-11 If it cannot be controlled, a spill should be diverted around areas such as sewers.

Figure 6-12 This harbor boom has a 12- to 18-inch weighted skirt hanging below the waterline to hold back floating product. It is called a harbor boom as it is commonly used to boom around a ship while it is docked in the harbor.

Figure 6-13 Another use of harbor boom and a floating skimmer shown in the upper left of the boomed area. The skimmer pulls floating liquids, such as fuel oil, off the water for recovery. This type of equipment is usually provided by a cleanup contractor. *(Courtesy of the Maryland Department of the Environment)*

Figure 6-14 This floating boom collects the spilled material that is floating on top of the water. Like pads, the boom does not absorb water and should stay afloat even when saturated with the spilled material.

Diversion can even be used for spills on the water. By using a solid boom, such as a harbor boom, material can be diverted into another area or even contained. This type of boom is shown in **Figure 6-12** and **Figure 6-13.**

Another method of diversion is to dig a trench alongside the waterway to collect the spilled material off the top of the water. For large spills this is effective if there is a collection vehicle such as vacuum truck standing by to collect the spilled material as it spills into the diversion. These types of vehicles are available from hazardous waste cleanup contractors. The most common method is to use a floating absorbent boom, which will absorb the material off the water, as shown in **Figure 6-14.**

Figure 6-15 Small retention areas can be dug by hand, but larger spills require larger equipment. Responders should have established contacts for such equipment. *(Courtesy of the Maryland Department of the Environment)*

Retention

The most common method of retention is digging a hole, either by hand or by machine, like the track hoe shown in **Figure 6-15.** To catch a running spill, the best method is to dig a hole large enough to collect the spill and then divert the spill into that containment area. If time permits, place plastic sheeting in the bottom, overlapping the containment area. This will assist in limiting the spread of contamination.

NOTE The ability to create a large enough containment area is paramount to success.

If unable to build a large enough containment, responders should concentrate on building several dikes and dams in a row to collect the material. As with dikes and dams it is best to start the farthest from the spill and create several other areas closer to the spill.

SAFETY When digging retention areas, be aware of underground utilities such as water, electric, sewer, cable TV, phone, or natural gas.

Dilution

Much like dilution in decontamination, this is not always the solution to pollution. If a water-soluble material is in a waterway, then for all purposes it is being diluted. If the waterway is small, there may be occasions when the fire department may be needed to add some water to the spilled materials. Simple flushing of a material into a waterway is no longer an acceptable tactic, and it should not be considered. Items such as fuels, oils, or other hydrocarbons are not usually water soluble and cannot be diluted by water. Dilution may be an effective response for some specific chemicals when combined with other containment tactics.

Vapor Dispersion

This topic can create some confusion for responders. By their nature, firefighters like to use water, usually in large quantities. In many occasions the application of water will make the situation safer than without a water spray. In incidents involving a severe threat to life, such as an adjacent nursing home that cannot be evacuated, using a water spray to disperse vapors would be a good idea. If, on the other hand, the release is occurring in a remote area away from any population, a water spray to disperse vapors, such as the situation shown in **Figure 6-16,** is not necessary. To be effective the material being released must be water soluble or the vapor cloud must be able to be moved by the water streams. Although the DOT ERG suggests for many spills to "use water spray to reduce vapors or divert vapor cloud," this course is not recommended except in life-threatening scenarios. Some departments use water sprays at natural gas leaks, which is not necessary as it further complicates the repair of the leak. Many leaks occur in a hole that was created, and the use of water spray fills the hole with water or ice and makes repair dangerous, complicated, or impossible. In addition, the use of water may actually knock down vapors that normally rise up and

Figure 6-16 The use of a water spray is not necessary unless the vapor cloud is threatening the public. In some cases it creates more hazards than those it may eliminate.

dissipate quickly. When using a water spray, you may knock down the vapors, but you also may create another substance. When anhydrous ammonia is released into the air and a water spray is applied, the resulting runoff is ammonium hydroxide, an extremely corrosive liquid that itself would require containment and cleanup. Natural gas, propane, chlorine, and sulfur dioxide are not water soluble, and the use of water vapor simply relocates the vapors to another area.

NOTE It is important to know what material is leaking prior to the application of water, as the water spray may cause more problems. It is very important to consult with the HAZMAT team prior to utilizing this tactic.

Vapor Suppression

With the use of firefighting foams, vapor suppression is another tactic that first responders can use. The type of material spilled will depend on the type of foam that should be used, as not all foams will work on all products. See Chapter 5 and *The Firefighter's Handbook* for more information regarding foam and its application. The use of foam to suppress vapors is generally limited to flammable liquid spills, but some chemical foams are available for other types of materials. Before applying foam, responders should ensure that the material is contained, that the application of foam will not cause any further problems, and that the foam is compatible with the spilled material. As with foam to extinguish a fire, responders should make sure they have enough foam stockpiled to keep a layer of foam on the spilled material.

Remote Shutoffs

Shutting off valves is not usually a tactic used by first responders, but there are some exceptions to this rule. On most tank trucks and at some fixed facilities, remote shutoffs can be operated by the first responders. Each type of truck is different, but in general an emergency shutoff is located behind the driver's side of the cab, and another is near the control valves. The shutoffs are mechanically operated and may be self-operating in the event of a fire. They are usually well marked as "emergency shutoff" and in an easy-to-find location. At some facilities, typically those with loading racks, there may be a remote shutoff near the entrance so that in the event of an emergency first responders can shut off the flow of product.

AIR MONITORING AT THE FIRST RESPONDER OPERATIONS LEVEL

The use of air monitoring is one of the most important tasks to accomplish when responding to chemical incidents. Air monitoring can keep responders alive. The NFPA 472 Committee recognized this and incorporated a number of air monitoring competencies at the operations level. The new objectives are added as mission-specific competencies and are optional. If firefighters are going to use air monitors at the operations level, the mission specific competencies should be used for training. Many responders purchasing detectors to assist with carbon monoxide alarms are realizing associated benefits. Depending on the type purchased, it may also be used to detect flammable gases, oxygen levels, and one or two toxic gases. An example is shown in **Figure 6-17**. Many responders use these instruments for flammable gas leaks and for confined space entries. The use of air monitoring is one of the most crucial components in the response to chemical incidents. Air monitoring can keep responders alive, but first responders must have the proper training and equipment to ensure their safety. When responding to situations that involve unknown materials, HAZMAT responders need pH detection for corrosives, a combustible gas detector for the fire risks, a photoionization detector (PID) for the toxic risks, and a radiation monitor for radiation risks. Unfortunately many responders only rely on one or two of these detectors. The most common detector is a three-, four-, or five-gas instrument, and the other instruments are not available or not used. By not using a method of detecting toxic materials, responders (and the public) can be lulled into a false

Figure 6-17 Air monitors, especially those called three- and four-gas instruments, are being used more commonly by first responders.

sense of security. Most combustible gas detectors are not made or designed to measure toxic gases in air, and they only read those types of gases when they become fire hazards. Many flammable gases are very toxic at considerably lower levels than their lower explosive limit.

Some new combustible gas detectors measure in both ppm and percent of the LEL, although the parts per million (ppm) range usually starts at 50 or higher. Many first responders have purchased air monitors, but may not have a full understanding of how they work.

NOTE The use of air monitoring and sampling equipment can be complicated and requires practice.

This text will provide information related to air monitoring and its importance in emergency response. Most responders have a basic understanding of combustible gas indicators, but even that knowledge can be limited. Our society is becoming much more sophisticated and the citizens (our customers) expect more. This section presents some concepts on air monitoring, monitoring strategies, information on how the monitors work, and their uses. The field of air monitoring technology is ever changing and new technology emerges each year, although the basic principles of safe decision making remain the same. When purchasing air monitors, it is important to understand the basic features of the instruments so that the purchase benefits the organization.

NOTE The general rule when purchasing an air monitor is to figure one-fourth of the purchase price into the annual budget for upkeep and repairs. It is necessary to keep the instrument calibrated and repaired on a regular basis.

Regulations and Standards

Within OSHA's HAZWOPER Regulation (29 CFR 1910.120) are few specific requirements for air monitoring, even for the HAZMAT technician, but air monitoring is the principle safety element throughout the document. OSHA specifies that the IC must identify and classify the hazards that are present on a site. The use of air monitoring is the primary key to fulfilling this obligation, as is the staffing of a HAZMAT technician trained to use detection devices. The NFPA 472 document also has some requirements for air monitoring, but like the OSHA regulation, these are fairly generic. The 2008 edition of NFPA 472 has

included some new mission-specific competencies that provide training objectives for the use of air monitors. The technician level was also revised with some additional weapons of mass destruction (WMD) monitoring competencies. There have always been air monitoring competencies at the technician level, but as technology has improved the NFPA 472 needed to address these new technologies. The WMD-specific monitoring competencies were added as mission specific, much like the air monitoring competencies at the operations level.

It is important for first responders to understand how monitors work and what their deficiencies are, as they may place their lives on the line depending on the readings of an air monitor. First responders are more affected by the confined space regulation as it applies to air monitors than they are by the HAZMAT regulation. Unfortunately, it is likely that a lack of understanding and/or maintenance will play a factor in future responder fatalities and injuries based on the large number of departments using these instruments.

Air Monitor Configurations

Most responders purchase an instrument that is known as a three-, four-, or five-gas instrument, commonly referred to as a **multi-gas detector,** such as the one shown in **Figure 6-18.** In other words, it samples for three to five gases. A typical detector can monitor oxygen levels, LEL, and the presence of two or three toxic gases. Typical toxics are carbon monoxide (CO) and hydrogen sulfide (H_2S), and if a fifth gas can be sampled, most departments choose chlorine or sulfur dioxide. For most responders, the CO and H_2S are more than sufficient. Although these instruments are considered expensive, in the grand scheme of things they are inexpensive compared to other detection devices used by HAZMAT teams.

Figure 6-18 These instruments are four-gas instruments and measure LEL, oxygen, carbon monoxide, and hydrogen sulfide.

Many responders may purchase a device with little or no training, which can be detrimental to the successful use of the instrument. In addition to obtaining proper training, departments buying a detector must purchase several additional items to keep the instrument functioning. Typically these items are covered in a special "kit" price and include a battery-powered sample pump, calibration gas, and hardware. The calibration gas usually has an expiration date and will require replacement in 6 months to a year. The sensors in the detector need regular calibration, and the instrument should be turned on and used on a regular basis, for a period specified by the manufacturer. Rechargeable batteries, depending on the charging system, may need replacement on occasion. The sensors themselves will need replacement, although most manufacturers are providing 2-year warranties for the sensors. In general the LEL sensor will last 4 to 5 years, the oxygen 12 to 18 months, and the CO and H_2S detectors 18 to 24 months. Other toxic sensors may last 6 to 18 months depending on the type. Most of the perceived problems with instruments are created by a lack of maintenance and adequate training.

NOTE No matter what a salesperson says, no device can conclusively identify an unknown material.

METER TERMINOLOGY

To comprehend fully the use of air monitors, the responder must understand the basic terminology that is generic to all monitors and applies across the board. More detailed information related to specific monitors is provided in later sections.

Bump Test

Two terms that need further explanation are bump test and calibration as they are both related. A **bump test,** also known as a field test, entails exposing a monitor to known gases and allowing the monitor to go into alarm and then removing the gas, as shown in **Figure 6-19.** Exposing the monitor to a known gas allows responders to verify the monitor's response to the gas.

Most manufacturers provide bump gas cylinders, which contain the gases required to check their instruments. When bump testing, follow the manufacturer's recommendations. In most cases the bump test is used to ensure that the alarms function as intended and the instrument is reading. By regularly calibrating and bump testing, users can determine

Figure 6-19 A bump test assures that the instrument responds to known gases. The detector should reach alarm levels to verify that it is functioning.

how accurate the monitor will be in the field. Some instruments react very well to bump testing and display the levels as provided on the bump gas cylinder, while others will be off slightly. When responding to confined spaces, responders are required by OSHA to bump test the instrument prior to entry into the confined space.

Calibration

Calibration is used to determine if a monitor responds accurately to a known quantity of gas. When new sensors are installed, they will usually read higher than they are intended; calibration electronically changes the sensor to read the intended value. As the sensor gets older, it becomes less sensitive and calibration electronically raises the value that the sensor displays. When a sensor fails, it cannot be electronically brought up to the correct value. Regular calibration gives a picture of the expected life of a sensor, as it will deteriorate over time.

The regularity of calibration is subject to great debate, as some departments calibrate daily, others every 6 months. The only guideline found in the regulations (for anything that requires air monitoring) is in the confined space regulation, which requires calibration according to the manufacturers' recommendations. Most of the written instruction guides from the manufacturers require calibration before each use. The International Safety Equipment Association (ISEA), an industrial association, has lobbied OSHA to add regulations governing calibration. The ISEA recommends bump testing prior to use and depending on the conditions where the device may be used recommends calibration every 30 days. The definition of calibration at this point is also subject to debate, as responders can verify the monitor's accuracy by exposing it to a known quantity of gas, but not perform a full calibration. The ISEA calls this type of test a functional test. Most response teams establish a regular schedule of calibration (weekly/monthly) and then perform bump or field checks during an emergency response. Check with the manufacturer as to what calibration/bump test policy it recommends.

Reaction Time

All monitors have a lag time or, as it is better known, reaction time. The reaction time varies depending on whether responders are sampling with a pump. Monitors without a pump operating are in what is called the *diffusion mode* and will generally have a 15- to 30-second time lag. Monitors with a pump have a typical reaction time of 7 to 30 seconds depending on the sensor type. Some instruments, primarily WMD detection devices, may take up to 90 seconds to react. Hand-aspirated pumps usually require ten to fifteen pumps to draw in an appropriate sample. Hand-aspirated pumps are not recommended because readings vary widely depending on how the user operates the hand pump. The goal is to provide a given amount of volume across the sensors.

NOTE When using sampling tubing, add 1 to 2 seconds of lag time for each foot of hose.

Be sure to follow the manufacturer's recommended lengths of hose to ensure that the pump operates correctly.

Recovery Time

In addition to reaction time, responders should also be aware of the monitor's recovery time. Recovery time is the amount of time that it takes the monitor to clear itself of the air sample. This time is affected by the chemical and physical properties of the sample, the amount of sampling hose, and the amount absorbed by the monitor. Some monitors take an extended period of time to clear if they are exposed to a large quantity of a gas. In some cases the instrument must be taken out of the environment and shut off, then started again. The reaction time will affect the overall recovery time.

Relative Response

When a gas monitor is purchased, it is set to read a specific type of gas, such as methane. If any other gas is sampled, the instrument will usually read the gas but at different values. Flammable gases have varied lower explosive limits (LELs) and therefore react differently with the flammable gas detector. The term *relative response* or *correction factor* (CF) is used to describe the way the monitor reacts to a gas other than the one it was calibrated for. Both of these terms are not commonly used by emergency responders. For this reason the LEL instrument is commonly used in a go/no-go situation. It cannot identify the gas that is present, and until the gas is identified, the exact amount of gas in the air cannot be determined. To maintain a high level of safety, each person operating an air monitor must have a basic understanding about relative response, as the monitor may lead you into a dangerous situation. Each detector has a correction factor that compensates for different types of gases. The monitor's manufacturer has tested the monitor against other gases and has provided the factors, which responders can use to determine the amount of gas present when sampling. The displayed reading is multiplied by the factor to provide the reading for the gas that is actually present. Examples of correction factors are shown in **Table 6-1.**

At a xylene spill responders using an ISC TMX-412 calibrated for pentane may take a reading 50 percent of the LEL. The correction factor for xylene is 1.3, which is multiplied by the LEL reading.

$$\text{Detector reading} \times \text{correction factor}$$
$$= \text{actual LEL reading}$$
$$50 \times 1.3 = 65$$

Thus the actual percent of LEL is 65 percent, a situation that is worse than the instrument's reading. On the other hand, if responders use the same instrument for a spill of propane, the factor is 0.8. Using the same scenario, the actual level is 40 percent of the LEL ($50 \times 0.8 = 40$), a safer situation than reported by the detector.

Oxygen Monitors

Oxygen is one of the most important gases for which to sample. Responders need it to survive, and the other instruments need it to function correctly. Normal air contains 20.9 percent oxygen; below 19.5 percent is considered oxygen deficient and a health risk, and above 23.5 percent is considered a fire risk.

If an oxygen drop is noted on the monitor, responders should assume that one or possibly more than one contaminant is present, causing the reduced oxygen levels. In oxygen-deficient atmospheres, any combustible gas readings will also be deficient and unreliable. In an oxygen-enriched atmosphere, the combustible gas readings will be increased and inaccurate. Most oxygen-enriched atmospheres may be due to a chemical reaction involving oxidizers and

TABLE 6-1

Correction Factors (Calibrated to Pentane)		
Gas being sampled	ISC TMX-412 factors	MSA 261 factors
Hydrogen	0.5	0.6
Methane	0.5	0.6
Acetylene	0.7	0.8
Ethylene	0.7	0.8
Ethane	0.7	0.7
Methanol	0.6	0.7
Propane	0.8	0.9
Ethanol	0.8	N/A
Acetone	0.9	1
Butane	0.9	1
Isopropanol	1	1.1
Pentane	1	1
Benzene	1	1.1
Hexane	1.2	1.3
Toluene	1.1	1.2
Styrene	1.1	N/A
Xylene	1.3	N/A

Always use response factors supplied by the manufacturer, and keep in mind these are laboratory estimates. ISC factors are for Industrial Scientific Corporation LEL sensor 1704A1856-200 calibrated with pentane. The MSA factors are for the Mine Safety Appliances MSA 360/361 calibrated with pentane.

typically signal a dangerous situation. Monitoring for oxygen is important, and the alarm values of 19.5 percent and 23.5 percent are set by OSHA regulation, but the reality is that responders may be in danger before these values are reached. As an example, an oxygen drop to 20.8 percent can be a very dangerous situation depending on the contaminant. A drop to 19.9 percent would not result in an alarm, but considerable contaminants may be present in the air. **Figure 6-20** provides additional information regarding this topic.

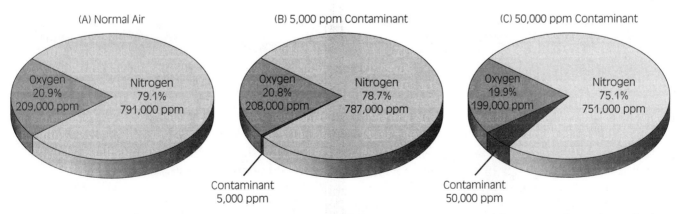

Figure 6-20 Normal air contains 20.9 percent oxygen, 78.1 percent nitrogen, and 1 percent of a variety of other gases. We will only be using oxygen and nitrogen for this example, as the 1 percent of the other gases would further complicate the explanation. Normal air is represented by (A) and when the atmosphere is contaminated both the oxygen and nitrogen will be reduced, as shown in (B) and (C). In (B) the contaminant would drop the oxygen to 20.8 percent and in (C) the level of oxygen would be 19.9 percent. In both cases the oxygen sensor would not alarm, but a contaminant at 5,000 ppm or 50,000 ppm has great potential to cause serious injury or death.

Oxygen Monitor Limitations

There are three types of oxygen sensors: two versions of lead wool technology and solid polymer electrolyte (SPE) sensors. In all of the sensors there is a chemical reaction, and the calibrated sensors are able to determine the level of oxygen in the air. Most oxygen sensors last only 12 to 18 months, because they are always working. An oxygen sensor is an electrochemical sensor that has two electrodes within a gel-like material. When oxygen passes through the sensor, it causes a chemical reaction within the sensor, creating an electrical charge and effecting a readout on the monitor. As long as the sensor is exposed to O_2, it will cause a reaction within the electrolyte solution sealed within the sensor. This is the most commonly replaced sensor, and departments should plan on annual replacement, although some may last longer. Exposures that can damage oxygen sensors are chemicals with lots of oxygen in their molecular structure, including carbon dioxide, and strong oxidizing materials, such as chlorine and ozone. The problem with CO_2 is that it is always present in the air, and the higher the percentage, the faster the sensor will deteriorate. The optimal temperature for operation is between 32°F (0°C) and 120°F (49°C). Between 0°F (–18°C) and 32°F (0°C) the sensor slows down (the electrolyte is like a slushy, gelatin material); temperatures below 0°F can permanently damage the sensor. The operation depends on absolute atmospheric pressure and calibration to the atmospheric pressure responders will be sampling. Responders should calibrate the sensor especially at the temperature, pressure, and weather conditions of the area in which they will be sampling.

Flammable Gas Indicators

Flammable gas indicators (FGIs), also referred to as combustible gas indicators or sensors, have been used by the fire service and industry for many years.

However, the term *flammable gas indicator* or FGI is more appropriate, as these detectors only detect flammable gases and vapors. The term *combustible* comes from their use in coal mines, and is outdated. All types of sensors work, some better than others in different situations. It is important to understand how each of these sensors work, as some departments may only be able to afford to purchase one type. To detect a wide variety of chemicals effectively, it may be necessary to have more than one type of sensor. Most of the new FGIs are used to measure the LEL of the gas for which they are calibrated. The greatest majority of FGIs are calibrated for methane (natural gas) using pentane gas. When calibrated for methane, the FGI sensor will read up to the LEL; some new units will shut down the sensor when the atmosphere exceeds the LEL. This is an important consideration, as the longer the sensor is exposed to an atmosphere above the LEL, the faster it will deteriorate. The FGI reads up to 100 percent of the LEL; thus, if a FGI calibrated for methane reads 100 percent, the actual concentration in that area is 5 percent for methane (the LEL for methane is 5 percent). If the FGI reads 50 percent, then the concentration for methane is 2.5 percent. Any flammable gas sample that passes over the sensor will cause a reaction. How much of a reaction depends on the gas. The EPA has established action levels that provide a safe layer expressly because of the relative response curve problem. It has factored in the response curve to provide guidelines, outlined in **Table 6-2.**

TABLE 6-2

EPA Air Monitoring Guidelines		
Atmosphere	**Level**	**Action**
Flammable gas	<10% LEL	Continue to monitor with caution.
	>10%	Evacuate space (OSHA confined space).
	10%–25% LEL	Continue to monitor, but use extreme caution, especially as higher levels are found (EPA).
	>25%	Explosion hazard, withdraw immediately.
Oxygen	<19.5%	Monitor with SCBA; FGI values are not valid (EPA); OSHA requires SCBA.
	19.5%–25% (OSHA >23.5%)	Continue monitoring with caution; SCBA not needed based on O_2 content only. OSHA requires evacuation of the confined space.
	>25% (EPA)	Explosion hazard, withdraw immediately.

Figure 6-21 From left to right are examples of flammable gas sensors: catalytic bead, metal oxide, and infrared.

Flammable Gas Sensor Types

The basic principle behind most FGIs is that a stream of sampled air passes through the sensor housing, causing a heat increase and conversely creating an electric charge and causing a reading on the instrument. Three types of gas sensors are shown in **Figure 6-21.** It is important that when purchasing or using a monitor that responders know the sensor type. Readings can and do vary among the four. Responders' safety depends on that instrument, so they must understand how it works.

CATALYTIC BEAD

The **catalytic bead** sensor is the most common sensor in use today. A catalytic bead sensor is wire between two poles with a solid bead of metal in the center. There are two wires and beads in the sensor: one burns the gas, and the other is used to compare the heat increase or decrease. The beads are coated with a catalytic material that aids the burning off of the gas. These sensors replaced the wheatstone bridge sensor which was a coiled piece of platinum wire between two poles.

METAL OXIDE SENSOR

The **metal oxide sensor** is commonly referred to as an MOS, or a tin dioxide sensor. It is a very sensitive detector, which causes several perceived problems. If used and interpreted correctly, this sensor can provide many clues to many chemical responses. The MOS is a semiconductor in a sealed unit with a Wheatstone bridge surrounded by a metal oxide coating. Heater coils provide a constant temperature. When the sample gas passes over the heated bridge, it combines with a pocket of oxygen created from the metal oxide. This reaction

causes an electrical change, which produces the reading.

This sensor is very sensitive and will pick up almost anything that crosses it, which can be confusing to some users. The MOS requires regular maintenance and calibration, more so than other types of sensors. Most MOS FGIs do not provide a readout of the percentage but an audible warning or a number from within a range. The range extends from 0 to 50,000 units on a relative scale. If you spill a tablespoon of baby oil on a table and pass a MOS over it, the sensor will report a small reading in the range of 5 to 30. If you take a MOS into a room with 5 percent methane, the reading will be within a range for a flammable gas, probably near 45,000 to 50,000 units. Some monitors allow the MOS to read in percentages of the LEL in addition to the general sensing range of 0 to 50,000 units. The MOS reacts to tiny amounts, which is an outstanding feature of the monitor, as most detectors are not nearly this sensitive. The MOS is typically not the first choice of departments that can afford only one FGI, because it can produce somewhat erratic readings. However, the MOS sensor does have great applicability for a HAZMAT team or a department that already has another type of LEL sensor.

INFRARED SENSORS

The **infrared sensor** is common in industry. But not all that common in the emergency response field. It uses a hot wire to produce a broad range of light wavelengths and a filter to obtain the desired wavelength; a detection device on the other side of the sensor housing measures that light. The light that is emitted from the hot wire is split, one through the filter, the other to the detection device to be used as a reference source. When a gas is sent into the sample chamber, the gas molecules will absorb some of the infrared light and this light will not reach the detection device, creating a difference in the amount of light reaching the two detectors. The amount of light reaching the detection device is compared, and a comparable reading is provided. The big advantage of infrared is that it does not require oxygen to function and can read in oxygen-deficient atmospheres. The device also is not affected by temperature, nor is it easily poisoned by high exposures. Disadvantages include cost and many cross-sensitivities. This is another sensor that should not be purchased as a stand-alone unit, but should be considered after the purchase of a catalytic bead or Wheatstone bridge sensor.

Toxic Gas Monitors

This section will describe the sensors that are commonly used in three-, four-, or five-gas units (LEL, O_2, toxic/toxic/toxic) and are usually used to measure carbon monoxide (CO) and hydrogen sulfide (H_2S). Toxic sensors are available in a variety of gases, including chlorine, sulfur dioxide, hydrogen chloride, hydrogen cyanide, nitrogen dioxide, and many others. The most common unit sold today by far is a four-gas unit that measures LEL, O_2, CO, and H_2S; its popularity is a direct result of the confined space regulation issued by OSHA. Many people understand the first three sensors but don't understand why H_2S was chosen as the common fourth gas. OSHA does not specify H_2S, but it is a common gas found in confined spaces, and most confined space entries (and consequently most deaths and injuries) take place in sewers and manholes. H_2S is commonly found in sewers and any place else organic materials are rotting. Thus, a common recommendation is to stick with H_2S as the fourth or fifth sensor unless a department has a specialized need for some other gas detection. Response teams must consider carefully which toxic sensors they most require. They must weigh the cost of making this decision. There are other more cost-effective methods of detecting toxic gases such as chlorine or sulfur dioxide. The average cost for a sensor is $400 to $500, and the equipment is usually guaranteed for 1 year. The calibration gas is $300 to $400 and is only good for 6 months before it expires. The sensitivity of these detectors is another issue: If you take an instrument into an atmosphere with more than 20 ppm chlorine, you will ruin the unit.

Most toxic sensors are electrochemical sensors with electrodes (two or more) and a chemical mixture sealed in a sensor housing. The gases pass over the sensor, causing a chemical reaction within the sensor and an electrical charge, which causes a readout to be displayed. All toxic sensors display in parts per million. Some toxic sensors use metal oxide technology and react in the same fashion as an MOS.

Tactical Use of Multi-Gas Detection Devices

It is important to understand the types of sensors that are in the multi-gas detection device so that responders can make effective tactical decisions. There are differences among the three flammable gas sensors. The metal oxide sensor (MOS) will react very quickly and will detect low levels of flammable gases, whereas the other flammable sensors have a 7- to 10-second delay in reading the gas levels, and do not have as low of a detection threshold. The MOS does not report readings that would be considered accurate, and it will fluctuate considerably. In high humidity situations it detects water vapor and gives erroneous readings. MOSs have an advantage in some situations but disadvantages in others. They are useful when determining if there is a contaminant in the air, but are not accurate for determining LEL levels. The catalytic bead and infrared are both accurate for determining LEL levels.

The other consideration when using flammable gas detectors is identifying the gas or vapor that is present. Most flammable gas sensors are calibrated for methane and will only read methane accurately. They will detect all other flammable gases, but the readings will not be accurate. Using the correction factors make the device accurate. Any reading on the flammable gas sensor indicates that there is a flammable gas or vapor present. Most detection devices alarm at 10 percent of the LEL for methane. This reading is only accurate for methane, if the device was calibrated for methane. There are correction factors that exceed a factor of 3, which means your monitor readings are off by a factor of 3. This is not dangerous if the meter reads 10 percent as the actual reading would then be 30 percent of the LEL. If the meter reading is 40 percent, then multiplied by a factor of 3, the actual reading is 120 percent and well into the flammable range—in some cases above the UEL for some materials.

When an LEL sensor alarms at the 10 percent level for an unidentified material, first responders should re-evaluate their presence in the building or in this level of flammable vapor. They should retreat and take steps to reduce the level of flammable gas. They should immediately consult with their local HAZMAT team for additional tactical decisions. Citizens should be removed as rapidly as possible from the immediate area. Detection devices can help determine the size and boundary of the hot zone. If the gas is identified as methane and the meter is set to read methane, the reading of 10 percent is of concern and the source must be identified as rapidly as possible. If the reading at the ground floor door of a three-story apartment building is 10 percent of the LEL, no matter where the source is the levels on the top floor will be extremely high, and probably well into the flammable range. Any spark could ignite the vapors causing a massive explosion in the building, putting responders at great risk. The electrical company should be immediately contacted to shut off the electricity to the building as rapidly as possible. The electricity should be shut off away from the building, preferable at the closest pole or transformer.

As for the toxic sensors, they are calibrated to their specific gases, and are accurate for only those gases. Both CO and H_2S react to some other gases and will give readings for about 35 other materials. There is no way for a first responder to determine which gas may be present. Victims may be able to provide some clues; for example, CO is odorless, so if victims or bystanders reported an odor, then CO is not a likely culprit. The exception to this would be if the victims reported a vehicle exhaust odor. CO makes up approximately 7 percent of vehicle exhaust. If H_2S is present victims will report a rotten egg odor; if they do not report this odor, H_2S is not the culprit. Both CO and H_2S are toxic, and any exposure should be avoided. Citizens should be removed from the area if these gases are found. Carbon monoxide is produced only from incomplete combustion so this source must be located to ensure the occupants' safety.

Other Detectors

In most cases these other detectors will be used by a HAZMAT team, but the first responder should be aware of the devices and their capabilities. In areas without a HAZMAT team or where the team has unusually long response times, the first responder may want to consider the purchase of these items. As the other instruments are complicated, so are these instruments. First responders are capable of beginning the detection process, but in almost all cases further detection capability is required.

PHOTOIONIZATION DETECTORS (PID)

The **photoionization detector (PID)** is sometimes referred to as a *total vapor survey instrument;* an example is shown in **Figure 6-22.** Because of their ability to detect a wide variety of gases in small amounts, PIDs are becoming essential tools of HAZMAT response teams. The PID will not indicate what materials are present, much the same as the FGI will not identify the specific material present. Used as a general survey instrument, the PID can alert responders to potential areas of concern and possible leaks/contamination. These instruments look for toxic

Figure 6-22 The photoionization detector (PID) is used by HAZMAT teams to detect toxic gases in the air. It is a very sensitive instrument and measures small amounts of gases in the air.

materials in the air and are essential to determining exposures to many toxic materials. The flammable gas detector will start reading toxic materials in a range of 50 to 500 ppm, which for some gases is extremely high and could pose immediate risks for a responder. The PID starts to read at levels near 0.1 ppm, which is very sensitive. There are PIDs that are sensitive enough to measure in parts per billion. Because of their sensitive nature, they can detect small amounts of hydrocarbons in the soil. As sick building calls are on the increase, the PID is a valuable tool in identifying possible hot spots within the building.

SAFETY The dose makes the poison, and so responders should have some understanding of dose levels. Most toxic materials report dose levels in parts per million (ppm) or parts per billion (ppb) levels. These typically are indicated for airborne inhalation hazards. Most first responder toxic sensors report their readings in ppm. Some alarms are for instantaneous readings, some for a 15-minute exposure, and others for an 8-hour exposure. An alarm with a CO sensor in an industrial application is different than one found in a home. Workplace exposures may be legal at certain levels, and some sensors alarm lower than these legal levels. Responders should consult with their local HAZMAT team when confronted with exposure levels that cause their meter to alarm.

RADIATION DETECTION

Many first responders carry some form of radiation detection, as shown in **Figure 6-23.** There are two primary types: pagers/dosimeters and detection devices. Many agencies are using radiation pagers or comparable dosimeters. Radiation pagers provide an indication that a radiation level above background was detected. It alerts the user that a radiation source is nearby. The dosimeter shown in **Figure 6-24** also alerts when levels above background are indicated and provides a continual reading of radiation dose as well. The ability to know the dose level is crucial to responder safety. Detection devices can read radiation dose rates and may also provide radiation count rates. Some devices also indicate the type of radiation source present.

COLORIMETRIC SAMPLING

Colorimetric tubes are used for the detection of known and unknown vapors. As sick building calls become more frequent, the use of colorimetric tubes by HAZMAT teams has greatly increased. Colorimetric sampling consists of taking a glass tube filled with a reagent, usually a powder or crystal, that is placed

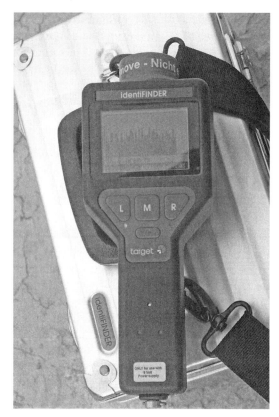

Figure 6-23 This radiation monitor has the ability to detect gamma radiation and can also identify the type of radioactive isotope.

Figure 6-24 This radiation pager alerts the user to the presence of gamma and x-ray radiation. Some detectors also alert to the presence of neutron radiation. This pager also determines the dose rate being received, while other pagers only alert to the presence of radiation and won't identify the dose being received.

Figure 6-25 A standard colorimetric tube is a glass tube with a crystalline material that changes color when a certain type of gas is present. These sensors can be used to identify unknown gases and can provide a level of a known gas.

into a pumping mechanism that causes air to pass over the reagent. A colorimetric sampling device is shown in **Figure 6-25.** If the gas reacts with the reagent, a color change should occur, indicating a response to the gas sample. Detection tubes are made for a wide variety of gases and generally follow the chemical family lines (i.e., hydrocarbons, halogenated hydrocarbons, acid gases, amines, etc.). Although the tubes may be marked for a specific gas, they usually have cross-sensitivities, or react to other materials, which at times is the most valuable aspect of colorimetric sampling.

One system that has been developed involves the use of bar-coded sampling chips. This system, called the **chip measurement system (CMS),** involves the insertion of a sampling chip into a pump. The pump recognizes the chip in use and provides the correct amount of sample through the reagent. Using optics and a light transfer system, the reflective measurement provides an accurate reading of the gas that may be present. Technology related to the detection of hazardous materials is growing at a rapid rate. There are many detection devices that are available for HAZMAT teams; some are simple indicators and others are laboratory-style devices. Some of the other instruments used by a hazardous materials team include pH detection, flame-ionization detectors, a variety of warfare agent detection devices, and characterization sample kits. The detection of unknown materials is complicated even for some HAZMAT teams. When first responders are confronted with a possible chemical release—no matter how insignificant they think it is—they should request or as a minimum consult with their closest HAZMAT team.

CARBON MONOXIDE INCIDENTS

The CDC estimates that more than 40,000 people are treated for carbon monoxide (CO) poisoning each year. In the U.S. about 500 people die from accidental CO poisoning, and it is responsible for 50 percent of the fatal poisonings. More than 2,000 people die each year from intentional CO poisoning. Many communities as part of the fire protection code require the installation of carbon monoxide (CO) detectors, as shown in **Figure 6-26.** Since CO detectors first became available to homeowners in the mid-1990s, the fire services are the responses to carbon monoxide detector alarms. In 1995 the city of Chicago ran several thousand carbon monoxide detector alarms in one day, due to an inversion, which kept smog, pollution, and carbon monoxide at a low elevation within the city. As carbon monoxide is colorless, odorless, and very toxic, it is important that first responders understand the characteristics of carbon monoxide and how home detectors work. As it is with other chemicals, carbon monoxide (CO) can be an acute or chronic toxicity hazard. It is only acutely toxic at high levels, typically at levels in excess of 100 ppm. At levels less than 100 ppm the hazard comes from a chronic exposure. The EPA has established an outdoor air quality level of 9 ppm for an 8-hour exposure, and 25 ppm for a

1-hour exposure. The Consumer Product Safety Commission recommends that a long-term (8 hours) exposure indoors be less than 15 ppm, and 25 ppm for an hour. Women who are pregnant and are exposed to CO are encouraged to seek medical attention. The fetal hemoglobin has a higher affinity for CO, and it is possible for the mother to be asymptomatic and the unborn child to suffer ill effects. CO is also a suspected teratogen and possible **abortifacient,** and chronic exposure can cause low birth weight. The fetus does not clear the resulting carboxyhemoglobin as well as the mother does. First responders' CO detectors typically alarm at 35 ppm. A chronic exposure to carbon monoxide can be hazardous since many people are exposed to CO in the home. The time people spend in their homes constitutes a considerable exposure. CO can only be detected by a specialized monitor. In extremely high concentrations it can be explosive. Exposure to CO causes flu-like symptoms, headache, nausea, dizziness, confusion, and irritability. Exposure to high levels can cause vomiting, chest pain, shortness of breath, loss of consciousness, brain damage, and death.

When exposed to levels of CO in a house, the residents may not be exhibiting any signs or symptoms. If CO is detected, they should seek medical attention because CO poisoning signs and symptoms may be delayed for 24 to 48 hours. Levels above 100 ppm are extremely dangerous, and residents should be medically evaluated. Women who are pregnant should be advised to be checked, as there have been fetal fatalities at relatively low levels of CO. Monitoring with a CO detector is essential to determine the possible exposure to CO. When people show only minor effects of CO poisoning and would normally be transported to the closest hospital, their home should be checked for CO levels. If high levels are found using a monitor, then alternative treatment, such as a **hyperbaric chamber,** may be the best treatment and should not be delayed. Many times a Pulse-Ox (oxygen saturation monitor) will be used incorrectly to determine a patient's O_2 level. Patients who have been exposed to CO will cause a Pulse-Ox to read 100 percent as the monitor reads the oxygen molecule in CO as being O_2. The elderly, children, or women who are pregnant are especially susceptible to CO and may have had a serious exposure without showing any effects.

If the first arriving units do not have a CO monitor and victims may be remaining in the residence, personnel are to don functioning SCBA when searching the residence. After determining no victims are present, crews are to ensure that the house is closed up and then wait outside for a CO monitor. Do not enter

Figure 6-26 This carbon monoxide detector is used in residential locations and uses a metal oxide sensor. This sensor also detects a wide variety of flammable gases that may be present such as methane or propane.

TYPES OF CO SENSING TECHNOLOGY

Biomimetic employs a gel-like material designed to operate in the same fashion as the human body when exposed to CO. This sensor is prone to false alarms, as it can never reset itself, unless it is placed in an environment free of CO, which in most homes is impossible. The sensor may need 24 to 48 hours to clear itself after an exposure to CO. The actual concentration of CO at the time the detector sounds may be low, but the exposure may have been enough to send the detector over the alarm threshold. If responding to an incident in which one of these detectors has activated, it should be placed in a CO-free environment until it clears. In the meantime, residents should rely on an alternate CO monitor.

The metal oxide detector is similar to sensors used in combustible gas detection but is designed to read carbon monoxide. How successful this design is in only reading CO is subject to debate. In some respects, the metal oxide monitor is superior to the biomimetic sensor, but it does have some cross-sensitivities and will react to other gases. For example, it is possible for this type of sensor to sound an alarm for natural gas, propane, and other flammable gases. It will react to the flammable vapors from nail polish remover. Responders using only a CO instrument may find themselves walking into a flammable atmosphere. Metal oxide detectors can usually be identified by the use of a power cord as the sensor requires a lot of energy. In most cases these sensors provide a digital readout. Once activated this sensor needs some time to clear itself, usually less than 24 hours.

An electrochemical sensor, which may also be referred to as instant detection and response (IDR), is the same type of electrochemical sensor that is in your three-, four-, or five-gas instrument. It has a sensor housing with two charged poles in a chemical slurry. When CO goes across the sensor, it causes a chemical reaction, which changes the resistance within the housing. If the amount is high enough, it will cause an alarm. This sensor provides an instant reading of CO and does not require a CO buildup to activate. It has an internal mechanism that checks the sensor to make sure it is functioning, which is a unique feature. Out of the three types of residential detectors, based on sensor technology, the electrochemical monitor would provide the best sensing capability.

---

an area with a CO detector that is activated without the use of SCBA. If crews find levels that exceed 35 ppm according to the monitor, they should use SCBA to continue the investigation.

Crews should be suspicious when responding to reports of an unconscious person or of "several persons down." They should not enter an area without SCBA if it is possible that CO (or other toxic gases) may be present. When several persons are reported to be ill in the same area, the cause is usually chemically related, and first responders should use extreme caution. An air monitor will ensure responder safety from the gases for which it samples. Not using an air monitor to check the atmosphere for contaminants is a dangerous practice.

Occasionally there are reports that people may be found unconscious due to a natural gas leak, but this is very unlikely. Natural gas, which has a distinctive odorant added to it, is nontoxic and will only asphyxiate a person by pushing oxygen out of an area. The only sign of this exposure is unconsciousness or death; any flu-like symptoms are due to CO poisoning. If the level of natural gas is high enough to cause unconsciousness, a very severe explosion hazard is present, and in reality an explosion would be imminent. It is possible that standard fire department air monitors will not pick up any CO. This is because the CO detectors purchased for the homes are made to detect small amounts of CO over a long period of time, and fire department detectors are instant readings that only pick up 1 ppm or more. The fact that responders do not pick up any readings does not mean the residents' detector is defective. Some factors that account for this discrepancy are low amounts of CO or a momentary high level that activated the alarm and dissipated prior to responders' arrival. The amount of time the residence is open dramatically affects readings. Crews are reminded to keep the residence as closed as possible so that the air monitor has a chance to establish the level of CO. As a reminder, any time you respond to unknown odors or sick building calls, it is important to remove people from the building and keep it closed up. As the amounts of toxic gases in sick building incidents are usually small, keeping them contained is very important. You cannot treat patients of toxic gas exposure unless you have a clue as to the source. For the patients' long-term health and your own well-being, you need to obtain quick, reliable gas samples.

In some parts of the country CO detectors are required just as smoke detectors are. Performance varies from brand to brand. The three basic sensing technologies are biomimetic, metal oxide, and electrochemical, each with advantages and disadvantages. Location, weather conditions, and the type of sensor determine what types of readings can be expected from a particular brand of detector.

Common sources of carbon monoxide include:

■ Furnaces (oil and gas)

■ Hot water heaters (oil and gas)

■ Fireplaces (wood, coal, and gas)

■ Kerosene heaters (or other fueled heaters)

■ Gasoline engines running inside basements or garages

■ Barbecue grills burning near the residence (garage or porch)

■ Faulty flues or exhaust pipes

SUMMARY

The use of defensive product control methods is a key component for the protection of the community and the environment. In most cases first responders have the equipment necessary to handle these tasks.

The limiting of spills will mitigate the incident sooner and prevent its spread. If first responders cannot stockpile the necessary equipment, then contacts should be made for those facilities that may have the materials. Under the requirements of the Oil Pollution Act (OPA) of 1990, certain facilities are required to maintain stockpiles of emergency equipment. First responders are also becoming more involved with air monitoring and are becoming more aware of the hazardous chemicals present. Flammable gases such as methane or propane and carbon monoxide incidents are the most common incidents that first responders are involved in. When using air monitors, first responders are reminded that they are not all encompassing and require training and experience. Understanding the action levels is an important consideration for safety—when is safe really safe? Detection devices require testing, maintenance, and regular repair and cannot be expected to function properly without this upkeep. To determine true safety first responders must use a range of instruments, something which is usually above the first responder level. When in doubt, responders should consult with their local HAZMAT team.

KEY TERMS

Abortifacient A chemical or material that can cause abortions.

Absorption Defensive method of controlling a spill by applying a material that absorbs the spilled chemical.

Biomimetic Form of a gas sensor that is used to determine levels of carbon monoxide. Typically used in residential CO detectors, it closely recreates the body's reaction to CO and activates an alarm.

Bump test Used to determine if an air monitor is working; will start alarm if a toxic gas is present. This quick check ensures the instrument responds to a sample of gas.

Calibration Setting the air monitor to read correctly. To calibrate a monitor, the user exposes it to a known quantity of gas and makes sure it reads the values correctly.

Catalytic bead The most common type of combustible gas sensor, which uses two heated beads of metal to detect the presence of flammable gases.

Chip measuring system (CMS) Form of colorimetric air sampling in which the gas sample passes through a tube. If the correct color change occurs, the monitor interprets the amount of change and indicates the gas level on a LCD screen.

Colorimetric tubes Used to detect both known and unknown types of gases. A form of detection for toxic gases, which changes color in the presence of a toxic gas.

Combustible gas indicators Detection devices used to measure flammable gases in air.

Damming Stopping a body of water, which at the same time stops the spread of the spilled material.

Diking Defensive method of stopping a spill; a common dike is constructed of dirt or sand and is used to hold a spilled product. In some cases, such as around a tank farm, in a facility a dike may be preconstructed.

Dilution Adding a material to a spill to make it less hazardous; in most cases water is used to dilute a spilled material, although other chemicals could be used.

Diverting Using materials that can move products around or away from an area; for example, several scoops of dirt can be used to divert a running spill around a storm drain.

Hyperbaric chamber Chamber used to treat scuba divers who ascend too quickly and need extra oxygen to survive. The chamber, which recreates the high pressure atmosphere of diving and forces oxygen into the body, is also used to treat carbon monoxide poisonings and smoke inhalation.

Infrared sensor Monitor that uses infrared light to determine the presence of flammable gases. Light is emitted in the sensor housing and the gas passes through the light; if it is flammable, the sensor indicates the presence of the gas.

Metal oxide sensor (MOS) Metal oxide coated LEL sensor that is heated to detect the presence of flammable and other gases.

Multi-gas detector Air monitor that measures oxygen levels, explosive (flammable) levels, and one or two toxic gases, such as carbon monoxide or hydrogen sulfide.

Photoionization detector (PID)   Air monitoring device used by HAZMAT teams to determine the amount of toxic materials in the air.

Remote shutoffs   Valves that can be used to shut off the flow of a chemical; the term *remote* is used to denote valves that are away from the spill.

Retention   Digging a hole to collect a spill; can be used to contain a running spill or collect a spill from the water.

Vapor dispersion   Intentional movement of vapors to another area, usually by the use of master streams or hoselines.

Vapor suppression   Defensive strategy to minimize the production of vapors. For example, flammable liquids produce hazardous vapors that could ignite; firefighting foams are used to suppress the vapors and not allow them to move by forming a barrier on top of the spill.

REVIEW QUESTIONS

1. What type of dam would be required for a fuel spill in which the fuel has a specific gravity of less than 1?

2. What type of dam is required for a chemical spill in which the material has a specific gravity of greater than 1?

3. What are two key items needed to construct an underflow or overflow dam?

4. Who should be consulted prior to using vapor dispersion techniques?

5. What is the normal configuration for a multi-gas detector that is used by first responders?

6. If you have a detector that is calibrated for methane, and you respond to a propane release, will the instrument detect propane? If yes, what do you need to know about propane and the detection device?

7. How often should detectors be calibrated?

8. In regard to safety and the ability to detect various toxic gases, how would you rate the use of a multi-gas detector (LEL, O_2, CO, H_2S)?

   a. Provides a high margin of safety

   b. Provides a low margin of safety

Suggested Readings

Bevelacqua, Armando S., *Hazardous Materials Chemistry,* Delmar Cengage Learning, Clifton Park, NY, 2001.

Bevelacqua, Armando S., and Richard Stilp, *Hazardous Materials Field Guide,* Delmar Cengage Learning, Clifton Park, NY, 1998.

Bevelacqua, Armando S., and Richard H. Stilp, *Terrorism Handbook for Operational Responders,* 2nd ed., Delmar Cengage Learning, Clifton Park, NY, 2004.

Greingor, J. L., J. M. Tosi, S. Ruhlman, and M. Aussedat, Acute carbon monoxide intoxication during pregnancy, *Emergency Medicine Journal,* 2001, 18:399–401.

Hawley, Christopher, *Hazardous Material Air Monitoring and Detection Devices,* 2nd ed., Delmar Cengage Learning, Clifton Park, NY, 2006.

Hawley, Christopher, *Hazardous Materials Incidents,* Delmar Cengage Learning, Clifton Park, NY, 2004.

Henry, Timothy, *Emergency Decontamination for Hazardous Materials Responders,* Delmar Cengage Learning, Clifton Park, NY, 1999.

Lesak, David, *Hazardous Materials Strategies and Tactics,* Prentice Hall, Upper Saddle River, NJ, 1998.

Maslansky, Carol J., and Steven P. Maslansky, *Air Monitoring Instrumentation,* Van Nostrand, Reinhold, 1993.

Noll, Gregory, Michael Hildebrand, and James Yvorra, *Hazardous Materials: Managing the Incident,* 3rd ed., Fire Protection Publications, Oklahoma University, 2006.

Schnepp, Rob, and Paul W. Gantt, *Hazardous Materials: Regulations, Response and Site Operations,* Delmar Cengage Learning, Clifton Park, NY, 1999.

Stilp, Richard, and Armando S. Bevelacqua, *Emergency Medical Response to Hazardous Materials Incidents,* Delmar Cengage Learning, Clifton Park, NY, 1997.

Tomaszewski, Christian, *Carbon Monoxide Poisoning,* Post Graduate Medicine Online, 1999.

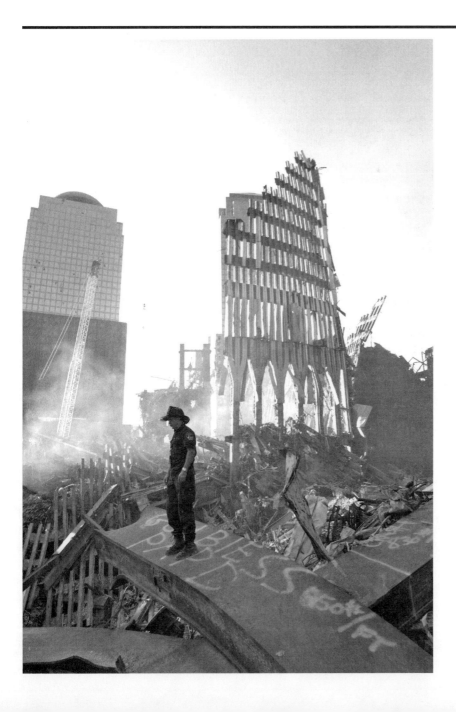

TERRORISM AWARENESS

OUTLINE

▌ Objectives
▌ Introduction
▌ Types of Terrorism
▌ Potential Targets
▌ Indicators of Terrorism
▌ Incident Actions
▌ General Groupings
 of Warfare Agents
▌ Detection of Terrorism
 Agents
▌ Federal Assistance
▌ Basic Incident Priorities
▌ Summary
▌ Key Terms
▌ Review Questions

STREET STORY

The emergency services is made up of a large, tight-knit family, and members of the emergency services hold a special bond.

On September 11, 2001, our family and our nation were attacked in a manner that was unprecedented. After the attacks in New York and Arlington, Virginia, and the plane crash in Shanksville, Pennsylvania, America was at war. Terrorists had struck, and in a few precious minutes we suffered a grievous loss. America changed that day, and the emergency services took a severe hit, but we persevered. During the first few minutes of the fire in the World Trade Center my thoughts were that it was a large fire, and that it would be a tough battle, but one the FDNY could win. When the towers collapsed, my heart sank. I knew that I had just witnessed hundreds of my family members being struck down. My thoughts were immediately for my good friends who work in FDNY and for all of the other emergency responders at that scene.

The events of that day caused the deaths of many emergency responders and an enormous amount of emotional damage to the survivors, family members, and population of the United States. But on that day, when citizens across the country dialed 9-1-1, emergency responders showed up . . . we proved that we were not defeated. We are emergency responders and the keystones of our communities. We are the first line of defense against terrorism in this country. We must remain focused on our mission, which is to protect the citizens in our community. With that said, we as emergency responders must look at the real threat that terrorism presents to our communities.

There are many events that occur in this country that present substantial risks to you as a responder and to the community. The events discussed in this chapter are events that can have catastrophic consequences for your community. You must prepare for these events and become tuned in to the potential for terrorism or other criminal behavior in your community. The tactics and motivation of terrorists have changed even since September 11th. Suicide/homicide bombers (SHB) are attacking commuter systems throughout the world. SHB attack locations where large numbers of people congregate. In many cases the profiles of these bombers indicate they are merely sympathetic to al Qaeda and are not directly associated with them. In Iraq terrorists are now using explosive devices combined with chlorine and other chemicals. The idea of suicide/homicide bombers is troubling, and the use of toxic industrial chemicals is a dangerous twist on terrorism response. Domestic terrorism is rearing its ugly head again, and although there have only been a few arrests, these individuals present a risk as they typically act alone. Every community has the potential to be attacked or to be a base of terrorist operations. Read the listing of the case histories, and you will learn that all communities are at risk. As you respond in your community, remember those who served on September 11, 2001, and dedicate your career to being the most knowledgeable and prepared firefighter that you can be. . . You never know what the future may bring.

—Street Story by Christopher Hawley, Baltimore County Fire Department (Ret.)

OBJECTIVES

After completing this chapter the reader should be able to identify or explain:

▪ Potential target locations (A)

▪ Indicators of potential terrorist activity (A)

▪ Incident actions to be taken at a terrorist attack (A)

▪ Additional hazards at a terrorist attack (O)

▪ Other specialized resources to assist with a terrorist attack (O)

▪ Methods of requesting federal assistance (O)

▪ Common agents that may be used in a terrorist attack (A)

INTRODUCTION

It is unfortunate that a chapter on terrorism needs to be included in this text. Until recently this would not have been necessary. Although we all have seen terrorism on the evening news, it was typically in places such as Northern Ireland, Beirut, or Israel. The United States had for the most part remained immune to the reign of the terrorist. This changed in February 1993 when the World Trade Center was bombed in New York City (see **Figure 7-1**). Even when this bombing occurred, the responders did not pay much attention, as the bombing was considered as a big city event or an incident that could only happen in New York City. Many HAZMAT

instructors tried to inform their students as to the ease of making such a bomb, but for the most part the warnings fell on deaf ears. The people who were found to be responsible for the bombing were controlled by an ideology from outside the United States, and so the thought was that it was an isolated international terrorism attack. When the Alfred P. Murrah Federal Building in Oklahoma City was devastated by a bomb on April 19, 1995, the United States emergency responders took notice. The attack on the Murrah Building, which is shown in **Figure 7-2,** was perpetrated by Timothy McVeigh and Terry Nichols, who did not have ties to another country and were natural-born citizens. It was perceived as an attack on our country from one of our own, not some unknown foreign nation. It brought terrorism home for the public and emergency responders alike. On September 11, 2001, the World Trade Center and the Pentagon were hit by three airplanes,

Figure 7-2   A truck bomb caused the devastation in the Oklahoma City bombing, killing 167 people and injuring 759. The damage extended several blocks in each direction, and 300 buildings were damaged. Fatalities occurred in fourteen separate buildings.

Figure 7-1   In 1993 a van was used to carry explosives to the underground parking garage of the World Trade Center. Six people were killed, and more than 1,000 were injured. There were 50,000 people in the building, and the goal of the terrorist was to collapse the building into the adjacent tower.

as shown in **Figure 7-3** and **Figure 7-4.** In Shanksville, Pennsylvania, a hijacked plane crashed into a field. As a result of the impact of the planes and the resulting fire, the 110-story twin towers of the World Trade Center collapsed. The death toll from all three sites was more than 2,800 civilians and 415 emergency responders.

In October 2001, a flood of noncredible anthrax letters hit the country. Mixed in with the thousands of noncredible events were several letters, sent to news media outlets and members of Congress, containing real anthrax. These letters were responsible for five deaths and up to twenty-two other illnesses. Regardless of its origin, the potential for terrorism is here in this country and remains in our thoughts as we respond to any incident. Other incidents are occurring on a regular basis; although they do not fit the exact definition of terrorism, they involve the use of large caliber weapons or a large number of weapons. This chapter looks at terrorism, **HAZMAT crimes,** and other potentially dangerous criminal situations, which are sometimes just as deadly as terrorism. Firefighters have to think outside the box in regard to terrorism. There are many events that occur in this country that place responders at risk. The events discussed here represent terrorism, HAZMAT crimes, murders, and other criminal events, all of which present a risk to firefighters. Someone using ricin to murder another person is committing murder, but the weapon is as deadly to the responders as it is to the intended victim. A booby-trapped drug lab is not terrorism, but it does present significant risks to responders.

The Federal Bureau of Investigation (FBI) in 1997 investigated 72 potential incidents of terrorism, and in 1998 they investigated 146 cases. In 1999 they had investigated more than 256 cases, with a good majority of the cases involving anthrax hoaxes. Other than

hoaxes, most of these crimes are against individuals, attempted murders, or murders for hire. Anthrax hoaxes have lost a lot of their impact and will be replaced by another hoax, most likely by radioactive events. These numbers demonstrate the dramatic increase in these types of events. Many criminals are protecting themselves with body armor and fortified vehicles. Many crimes such as bank robberies involve the use of explosives and sophisticated weaponry.

Street violence, often associated with gangs, is becoming increasingly violent. When people are killed or injured, the fire service is called into action.

Crimes such as assaults and murders are even increasing within our schools. In April 1999 fifteen people (one teacher and fourteen students, including the two attackers) died in a suicide attack by two students on Columbine High School in Littleton, Colorado. Each year there has been an increase in the amount of violence in our schools. In April 2007 a student at Virginia Tech killed 32 people and wounded a number of other students in a violent

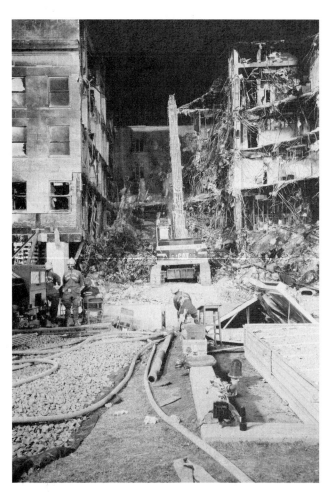

Figures 7-3 and 7-4 On September 11, 2001, two planes hit the World Trade Center complex, one hit the Pentagon, and one crashed in Pennsylvania. The death toll from these incidents was more than 2700 and included hundreds of responders. These photos are from the Pentagon crash site and show the crash and fire damage. *(FEMA photo)*

attack. The EMS and fire services are placed into action in these events and in many cases arrive first. In recent times there have been deadly riots in Los Angeles and in St. Petersburg, Florida, usually after a sporting event.

In Chicago, twenty-one people were killed and fifty were injured in a nightclub after a fight broke out. These deaths may have been caused by a lack of exits and the use of pepper spray by security personnel, which panicked the crowd. Even the workplace is becoming an increasingly dangerous place for emergency responders. In April 1996 a firefighter in Jackson, Mississippi, entered the headquarters fire station intent on killing the fire chief. Several people were killed, and the emotional damage was widespread.

Crimes are becoming more violent, and it is perceived that this trend will continue to intensify. Emergency responders are immediately placed into a dangerous situation and can get caught, literally and figuratively, in the cross fire. As a result of the terrorist attacks on September 11, 2001, the previous bombing of the World Trade Center in 1993, the Alfred E. Murrah Federal Building bombing in Oklahoma City, Oklahoma, and other incidents, our response to some types of incidents will be forever changed. In the late 1990s there were two other bombings that are fairly well known to the fire service, the Atlanta Olympic Park bombing in 1996 and the Atlanta abortion clinic bombing in 1997. That particular abortion clinic bombing is significant because a secondary device was used. It is thought that it was strategically placed with the sole intention of harming the responders. The device was placed near the location where the incident command post was set up. Luckily, only minor injuries were received by the responders who were near the blast. In 1998 there was another bombing at a Birmingham, Alabama, abortion clinic in which an off-duty police officer employed as a security guard was killed by a device that was more accurately placed to target the responders. This secondary device was activated by the person responsible as he watched the victim approach the radio-controlled device.

A brief overview of some of the recent acts of terrorism, HAZMAT crimes, use of terrorism materials, or criminal acts that presented risk to responders follows:

■ A series of bombings in the 1980s targeted primarily at the IRS included an attempt at a chemical bomb using ammonia and bleach. Another attempt included the use of a hot water heater as a very large pipe bomb, but the vehicle carrying the bomb caught fire while the terrorist was driving the vehicle.

■ 1984 Dr. Michael Swango had a long history of suspicious deaths as he moved throughout the United States and abroad. He was stripped of his license to work as a doctor and got a job as a paramedic. Although his crimes do not fit into standard terrorism, he was finally arrested for attempting to poison his paramedic coworkers with chemicals.

■ 1984 The Rajneesh Foundation was responsible for poisoning 715 people with salmonella bacteria. They poisoned the salad bars at ten restaurants. The group also used other biological agents and targeted a water supply for attack. They had previously attempted to poison the water system with raw sewage and dead rodents. They used nursing home patients as test targets for some of the biological attacks. The attacks were an attempt to change a local election.

■ 1985 Members of a militant group developed a plot to poison a water supply. They were going to use 35 gallons of cyanide that was in their possession. Other members of the group were arrested for church arsons and attempting to blow up gas pipelines.

■ 1993 The World Trade Center was damaged by a van bomb; six people were killed.

■ 1995 A man was arrested after manufacturing **ricin,** an extremely deadly toxin, with the intention of killing someone of whom he was jealous.

■ 1995 (April 19) The Alfred P. Murrah Federal Building in Oklahoma City was the target of a truck bomb. Several hundred people were injured and 186 people were killed in the blast.

■ 1995 Members of a militia group known as the Patriots Council were arrested for the manufacture of ricin in the attempted assassination of a U.S. marshal.

■ 1995 Two separate ricin incidents by two doctors provide an overview of typical ricin use. Dr. Deborah Green killed two of her three children by burning her house, and she attempted to kill her husband three times with ricin. In Virginia, Dr. Ray Mettetal attempted to murder his boss with a syringe filled with boric acid and saline. He also possessed ricin and a number of other materials.

■ 1995 (and throughout the 1990s) Aum Shinryko is the cult group that is best known for a 1995 sarin nerve agent attack in the Tokyo subway. For a long period the group carried off chemical and biological weapons attacks and it went to great lengths to develop its weapons program. An examination of this group yields many valuable lessons on chemical and biological attacks. The group had assets of more than $300 million. It was also home to more than 100 scientists whose sole function was to develop chemical and biological weapons. The group used, or explored the use of, clostridium botulinum, bacillus anthracis (anthrax), Q fever, Ebola, and viral hemorrhagic fever. It carried off ten chemical attacks and nine biological attacks, not to mention a large number of small and large full-scale tests.

■ The cult group killed 7 people and injured 200 in a sarin nerve agent attack in Matsumoto in 1994. The subway attack in 1995 killed thirteen, but injured more than 5,000. The attack consisted of placing sarin in three subway lines. A total of eleven sarin bags were pierced with umbrellas. The deaths occurred to those who came in contact with liquid or were in small, confined spaces. Over 4,000 people went on their own to 278 hospitals. The Tokyo Fire Department ambulances transported 688 patients. The breakdown of the injured was: 17 critical and 37 severe. Some 948 patients were suffering from miosis (pinpointed pupils). All of the moderately injured patients were treated and released 6 hours after they arrived at the hospital. Eighty-five percent of the patients did not require any treatment but still flooded the hospitals.

■ 1995 A train was derailed near Hyder, Arizona. A terrorist group claimed responsibility, although no one has been charged in the attack. One person was killed and the derailment seriously injured twelve others.

■ 1995 The Anaheim Fire Department in California was made aware of a potential sarin nerve agent attack at Disneyland. The fire chief was notified at midnight to be at the command post 4 hours later. Police agencies, federal law enforcement, and the military had known about the threat for 5 days, but except for notifying the fire chief and assembling resources prior to the event, no other planning had occurred. The fire department was placed in charge of the incident and quickly developed a plan of action. Disneyland did not close, and 30,000 to 40,000 people visited the park during the threat period. Luckily, the threat never materialized.

■ 1996 A man used his employer's Nuclear Regulatory Commission (NRC) license to order several radioactive sources. The purpose of the purchase is not known, but it is suspected that foul play was being planned.

■ 1996 Members of a paramilitary group who had access to preplanning information obtained through the fire department were arrested when they attempted to blow up a Department of Justice complex in West Virginia.

■ 1996 and 1997 The Atlanta area was besieged with bombings, including one at the Olympic Park that killed one person. Another at an abortion clinic in Birmingham, Alabama, had a booby-trapped device that killed an off-duty police officer and wounded a nurse.

■ 1996 Theodore Kaczynski, known as the Unabomber, was responsible for a string of bombings that lasted 18 years. He sent 16 bombs, which killed 3 and injured 23. After he wrote a manifesto that appeared in the *New York Times,* he was recognized by his brother, who turned him in to the FBI.

■ 1996 A disgruntled firefighter in Jackson, Mississippi, went to his department's headquarters with the intent of killing the fire chief. The fire chief was not injured, but several other fire department personnel were killed.

■ 1996 A lab worker brought in pastries to coworkers that were poisoned with Shigella dysenteria. She made 12 people ill and was found guilty on 5 felony assault charges. Police then learned that she had attempted on more than one occasion to poison a former boyfriend with Shigella and other biological agents.

■ 1997 EMS and police responded to a shooting incident, and the suspect was found to have ricin, *E. coli,* and a mixture of nicotine and DMSO (dimethyl sulfoxide).

■ 1997 In order to create a diversion for an armored car robbery, four members of the Ku Klux Klan plotted to blow up a hydrogen sulfide tank. They were arrested before they could do it.

■ 1998 In Nairobi, Kenya, and Dar es Salaam, Tanzania, bombs exploded at the U.S. embassies, killing 224 people.

■ 1998 Kathryn Schoonover attempted to mail 100 envelopes with cyanide disguised as diet powder all over the United States.

■ 1998 Two men were arrested on the suspicion of **anthrax** possession. They were later released because it was determined that they only possessed a possible anthrax vaccine. One of the men, Larry Wayne Harris, had been arrested in 1995 for possession of plague, a biological toxin that he had ordered through the mail. He had been a member of the Aryan Nation, but was deemed too radical for the group.

■ 1998 In Charlotte, North Carolina, a man with an explosive device held hostages in a government building. The explosive device was also thought to contain some type of chemical agent. It was later determined that the filling agent was harmless, although the explosive was live.

■ 1998 Abortion clinics in South Florida, New Orleans, and Houston were affected by attacks using **butyric acid,** a material with a horrible, irritating odor.

■ 1998 In Lafayette, Indiana, a pickup truck rammed into the courthouse. The bed of the pickup had flammable and combustible materials as well as several explosives.

■ 1998 In Nairobi, Kenya, and Dar Es Salaam, Tanzania, bombers killed 81 people and injured 1,700 others. Followers of Osama bin Laden were convicted for the attack.

■ 1999 Two students, armed with an array of guns and explosives, attacked their own high school. The suicide attack resulted in fifteen deaths and spurred a rash of bomb hoaxes throughout the country.

■ 1999 Early in the year the FBI investigated hundreds of anthrax hoaxes, none of which involved the actual use of the biological agent. Abortion clinics were the targets in most cases.

■ 1999 Ahmed Ressam was arrested in Port Angeles when he crossed over the border into the United States. He was responsible for plotting the Millennium "Border Bomb" and had the components for a large bomb in his car. His target was the Los Angeles Airport (LAX) or the Space Needle in Seattle on New Year's Eve. This plot is the responsibility of an al Qaeda sympathetic group, who have worldwide ties to other terrorist cells.

■ 2000 Police and fire officials scrambled after more than 80 vials filled with an unknown liquid were found throughout Milwaukee and its suburbs.

■ 2000 Dr. Larry Ford's house in Irvine, California, was searched after he attempted to murder his partner. Dr. Ford committed suicide, and his house was found to have held numerous chemicals and chemical agents. The search took several weeks, and chemicals, guns, and biological materials were found.

■ 2000 During a genetics conference in Minneapolis, Minnesota, protestors attacked three restaurants with hydrogen cyanide.

■ 2000 Members of a militia group in Sacramento, California, were arrested for plotting to blow up a 24-million-gallon propane tank.

■ 2001 A Chevron employee in an El Segundo, California, refinery threatened to blow up the facility. The facility produces more than 4 million gallons of jet fuel a day.

■ 2001 Over 2,800 civilians and 415 emergency responders lost their lives at the World Trade Center in New York City, the Pentagon in Arlington, Virginia, and a field in Shanksville, Pennsylvania. Four planes were used in the attack, two hitting the World Trade Center and one the Pentagon. The fourth plane crashed into a field. Members of the al Qaeda group have been held responsible for the attack.

■ 2001 In October, there was a death from an anthrax-laden letter. Eventually, there were a total of seven deaths from anthrax. Emergency services across the country were deluged with calls about white powder. A very small percentage of the calls was investigated by the FBI, but even so, more than 14,000 cases regarding the white powder events were opened. Only five cases involved real anthrax, which was sent to people in Congress, New York City, and Boca Raton, Florida. In December Clayton Lee Waagner was arrested and admitted sending more than 500 noncredible anthrax letters to Women's Reproductive Health Centers. He was a member of the Army of God, a group known for pursuit of the right-to-life cause. As of 2007 no one has been arrested for the anthrax attacks. The FBI is still actively pursuing the perpetrator. They have traveled all over the world looking for clues, and have interviewed more than 9,100 persons.

■ 2001 The Animal Liberation Front (ALF) and the Earth Liberation Front committed a number of terroristic acts throughout the year. ALF admitted 137 illegal actions at a variety of locations. Both groups target buildings and businesses that have any connection to the use of animals or perceived violations of the environment. The acts typically create inconvenience, such as gluing locks, but on occasion they do include arson and other violent destruction of property. For the most part these groups take great care to avoid any potential injuries to humans, but responders could be killed or injured while responding to or handling these acts.

■ 2001 Two Jewish Defense League (JDL) members were arrested for plotting to blow up a mosque and a Congressman's office. Other members of the JDL have been quoted as having a desire to continue the militant work of the two men arrested.

■ 2001 Richard Reid was arrested on a Paris-to-Miami flight for attempting to detonate a PETN explosive that was located in his tennis shoes. PETN is one of the more powerful explosives. He was overpowered by other passengers and the flight crew while trying to light the fuse. He has ties to al Qaeda and is supportive of their cause.

■ 2002 Two different pipebombers created fear in the United States, one in Philadelphia, the other in the Midwest. Preston Lit set pipe bombs off in U.S. Postal Service mail boxes in the Philadelphia area. Lucas Helder, a college student, set off pipe bombs in residential home mail boxes in five states. Helder's intention was to set off bombs so that the explosions when drawn on a map would form a smiley-face. He set a total of eighteen pipe bombs, six of which exploded. His bombings injured six people.

■ 2002 A 15-year-old youngster stole a small plane and flew it into a high-rise building in Tampa, Florida, killing himself. His suicide note stated that he wanted to be just like Tim (McVeigh), Eric (Rudolph), and Osama (bin Laden).

■ 2002 Joseph Konopka, a member of the Realm of Chaos group, stashed potassium and hydrogen cyanide in a Chicago subway tunnel. His group is known for wanting to destroy public utility, water, sewage, and telecommunications systems.

■ 2002 Two militia groups in Kentucky and Pennsylvania had their leaders arrested for possession of weapons and explosives.

■ 2002 The FBI arrested José Padilla, also known as Abdullah al Muhajir, who is associated with al Qaeda in Chicago. He was researching the use of and looking for materials to detonate a radiological dispersion device. He traveled to Pakistan and studied methods to pull off such an attack with al Qaeda operatives. In late 2002 the FBI was still looking for 100 al Qaeda members and at 150 people and groups who may have al Qaeda ties. They made two large arrests in Detroit, Michigan, Buffalo, New York, as well as other arrests throughout the United States, capturing a number of suspects who had al Qaeda ties.

■ 2003 Two bombings occurred in May, one in Riyadh, Saudi Arabia, and the other in Casablanca, Morocco, which killed more than sixty-five people and injured hundreds more. Both bombings were against United States held interests and have been linked to Osama bin Laden.

■ 2003 Mr. Iyman Faris was sentenced to 20 years in prison for providing assistance to al Qaeda and conspiring to commit a terrorist act. Mr. Faris was a U.S. citizen who was born in Kashmir, and last lived in Columbus, Ohio. He was accused of scouting targets, one of which included the Brooklyn bridge. He had traveled to Afghanistan in 2000 to attend a training camp with Bin Laden. His experience as a truck driver and access to cargo planes was of interest to al Qaeda. He performed a number of research tasks for the al Qaeda leadership. In early 2003 he became a double agent for the FBI, providing information on al Qaeda. In 2004 Nuradin Abdi was charged with conspiracy to commit a terrorist act for setting off explosives at a shopping center. When Adbi arrived in Columbus, it was Faris who picked him up.

■ 2004 There was a series of bombings in Madrid, Spain. The attacks that took place on four commuter trains killed 191 persons and injured 2,050. Although it was suspected that the attacks were carried out by al Qaeda, the actual terrorists were local Islamics, and several others who were not Islamic. The terrorists used 13 backpack explosives, of which 10 detonated. The attack was very well coordinated and each of the explosions took place within several minutes of each other in different parts if the city, overwhelming the emergency response system. The attacks did create some attention among other terrorist groups including al Qaeda, as the attack changed the outcome of the national elections. The government immediately placed blame on the Euskadi Ta Askatasuna (ETA). This group has conducted a number of terrorist attacks in Spain, and was a likely suspect group. There were a number of misstatements made by government officials, which did not make the citizens happy. On the days following the attack, thousands of citizens protested in the streets. The demonstrations increased in numbers of protestors and intensity each day new facts were learned that contradicted the government's position. As a result the party that was in power prior to the attack lost the election and government control went to the opposition party.

■ 2004 William Krar was sentenced to 11 years along with Judith Bruey, who was sentenced for just over 4 months in federal prison. They were charged with possession of sodium cyanide and other chemicals that could be used for WMD materials. The chemicals were primarily acids, which could be used to produce sodium cyanide gas if mixed with sodium cyanide. In addition to the chemicals the FBI found hundreds of thousands of rounds of ammunition, a substantial number of pipe bombs, machine guns, and a number of remote-controlled explosive devices disguised as briefcases. Mr. Krar had sent false identification documents, but the package was delivered to the wrong address. The exact motive or intended target of the weapons is not known but Mr. Krar had a history of being antigovernment, and had a number of previous weapons-related arrests.

■ 2004 Demetrius Van Crocker was convicted in a domestic terrorism case. Working through intelligence the FBI was able to work the case with an undercover FBI agent. The FBI agent was posing as a security guard at the U.S. Army Pine Bluff Arsenal, one of the locations that stored military chemical weapons. Van Crocker conspired with the FBI agent to purchase sarin nerve agent and C-4 explosives, which the agent purported that he could steal from the arsenal. Van Crocker was a white supremacist who disliked the government and supported Timothy McVeigh. Although Van Crocker had a low IQ he had studied the process to manufacture military chemical weapons and other toxic industrial chemicals that could be used as effective weapons. Through discussions with the undercover agent he was able to describe several scenarios involving chemical weapons and had more than a general interest in the topic.

■ 2005 The two series of bombings occurred on July 7th and July 21st. Much like the Madrid bombings there were coordinated attacks on the public transportation system in London. Within 1 minute of each other three triacetone triperoxide (TATP)–based bombs exploded in London's subway system known as the Underground. An hour later a fourth backpack bomb exploded on a double-decker bus. The attack killed 52 including four attackers and injured more than 700. The placement and timing of the explosions created hardships for the rescue crews, as the trains had left and were in-between stations. The relationship with al Qaeda is not known, but it is suspected that the terrorists were sympathetic to their cause. The bombings on the 21st were similar but the four explosive devices malfunctioned and did not detonate. Three trains and a bus were targeted but only the detonators exploded, and the main charges failed to detonate. Like the earlier bombs these were made of TATP which is very unstable. Two days later another bomb was found that was similar to the other devices. The explosives in both cases had been homemade and are considered to be very unstable. It is possible that the explosives were made at the same time, and due to the length of time that had passed the explosives used on the 21st had deteriorated and were no longer explosive. In both bombings sixteen people were arrested and a number of others were held and released.

■ 2005 There were a number of explosions in Bali, Indonesia, and two adjacent communities. The targets of the bombings were markets and areas where tourists are known to congregate. In one bombing the SHB walked into a crowded café and set off his backpack bomb. In all twenty-three people were killed which included three SHB. Among the 129 injured there were six Americans injured in the attack. The attacks came before Ramadan, which is a period of significance for Muslims. A number of Australian tourists were in Bali for vacation during this period; four Australians died and nineteen were injured. This was not the first time there had been bombings. In 2004 the Australian embassy in Jakarta was bombed. In August 2003 a SHB detonated a car bomb outside the Marriott hotel. This attack against a U.S. interest killed 12 and injured 150 persons. In 2002 there was another series of well-coordinated attacks in Bali where 202 persons were killed and 209 were injured. The majority of those killed were from other countries, including seven from the United States. Australians suffered the most casualties, as the area is popular with Australian tourists. The attackers used diversionary attacks to move victims to a central location where a larger van bomb was exploded. The SHB in the van did not have to activate the bomb, as it could have been remotely controlled if the bomber changed his mind. In all of the bombings members of the Jemaah Islamiyah, which is an Islamic group, have been arrested or suspected of committing these acts of terrorism.

■ 2006 Police in Ontario, Canada, arrested seventeen persons thought to be part of an Islamic terrorist cell. Allegedly the group had made plans to use truck bombs to target government and other high profile buildings, and to attack events where there were large numbers of civilians. The group had ordered 6,600 pounds of ammonium nitrate which is a common explosive material. As the police had intelligence information and had infiltrated the group, they were able to make the arrests prior to any terrorist actions. In addition to the truck bombs the group had made plans to storm government buildings, take hostages, and execute high level government officials.

■ 2006 The FBI was alerted that Derrick Shareef, also known as Talib, abu Salam Ibn Shareef, had a desire to commit a terrorist act in Rockford, Illinois. He wanted to attack government buildings and commit other crimes to raise money. During discussions with an informant Shareef changed his plan of attack to that of the Rockford

Mall. Shareef discussed with an undercover FBI agent the desire to purchase four hand grenades and two pistols. He traded two stereo speakers for the grenades and one pistol, and as the U.S. Attorney stated he fit the profile of a "wanna be" terrorist. Shareef planned his attack of the mall with the thought to maximize casualties. With the attack of the mall during Christmastime, he hoped to panic the citizens across the United States.

■ 2006 Police in London arrested twenty-four persons for conspiring to hijack and destroy ten airplanes. The planes were departing London and headed to the United States, and the explosives were going to be hidden in the carry-on luggage. As a result of these arrests, liquids were banned from carry-on luggage, as the explosives were allegedly liquid based. The explosives were most likely going to be TATP and hexamethylene triperoxide diamine (HMTD). The use of TATP has attracted the attention of terrorists due to its ease of manufacture using common chemicals.

■ 2007 The American Embassy in Greece was hit with a rocket-propelled grenade (RPG), the second time this type of attack has occurred. The Greek terrorist group November 17 is suspected in this attack. There were no injuries or deaths as a result of this attack, although the Embassy suffered minor damage.

In this chapter we will provide some information about the type of terrorism agents that currently exist along with some possible scenarios. Some concern exists within the emergency services that we should not publish this type of information. Taking this into account, it should be understood that all of this information, plus additional reference sources, is taken from the Internet or is readily available in the public domain, in training programs, and in other accessible material throughout the United States. Most of the exact recipes and how-to instructions for these and many other devices are easily obtained through printed texts and the Internet. We are not providing any information that is not readily accessible elsewhere. Since we are able to acquire this data in such an easy, normal, and legal fashion, rest assured that terrorists already have it. Responders should be aware of the various chemicals and devices that someone may design to kill the first responders and others in your community. The best defense is a broad-based education, with the challenge of staying one step ahead of the terrorists. By staying informed as to how terrorists may operate, you can be more alert to potentially fatal circumstances.

TYPES OF TERRORISM

The types of terrorism are divided into two distinct sources: foreign based and domestic. Until the Murrah Building attack the fear of terrorism was directed at foreign sources. Most Americans believed that any terrorist attack would be from another country. To be deemed foreign based, the motivation or supervision comes from another country. Domestic terrorism originates from within the United States and is not influenced by any foreign party. The perception is that our greatest threat from terrorism is international in nature, but the opposite is proving to be more accurate. From the list provided earlier in this chapter and from other compiled statistics, it is evident that the largest percentage of terrorism has been domestic in origin.

NOTE The FBI defines terrorism as a violent act or an act dangerous to human life, in violation of the criminal laws of the United States or any segment, to intimidate or coerce a government, the civilian population, or any segment thereof, in furtherance of political or social objectives.

The key to this definition is the intimidation of the government or the civilian population. In 1984, a religious group trying to influence the local political process sprayed salmonella on fast-food restaurant salad bars; the attack rendered more than 715 people ill. This attack was originally not considered an act of terrorism, as the health department initially suggested it was food poisoning. Not until more than a month later did the FBI learn that it was an intentional act. This information was discovered while investigating another crime involving the same group.

The Tokyo subway sarin attack was an attempt to destroy a good portion of the police department, which was trying to raid the terrorist compound. Although the main topic of this chapter is terrorism, that label applies to only a minority of the types of incidents to which firefighters respond.

SAFETY Many incidents may not fit the exact terrorism definition, but the hazards from a pipe bomb are the same regardless of the motivation of its builder.

Many responses that could have been routinely handled in the past must now be treated much differently, and responders must always be on their guard for terrorist-style devices or potential terrorism acts.

ASSESSING TERRORIST THREATS

When responding to a possible incident of terrorism involving **weapons of mass destruction (WMD),** there is always the likelihood that the incident could be a hoax. When dealing with a potential explosive incident the statistics show that there are thousands of explosions carried out each year. Although training for a variety of emergencies is crucial to success, responders must determine their most likely event. Most of the explosive incidents involved small devices and are not nearly on the scale of the World Trade Center or Murrah building. No matter the material or the device, several criteria must be considered when determining the validity of a terrorism threat: educational level, availability of raw materials, production equipment, dissemination, and motivation.

The thought process for determining if a threat is credible or not has five elements. If the person known or thought to be responsible is determined to have several of these capabilities, it increases the credibility factor:

1. The first of the five qualifiers involves the potential terrorist's educational ability to make a device or agent that, unlike explosives or ricin, is very difficult to attain. To truly make a biological pathogen agent, in most cases one needs an advanced knowledge of biology. Ricin, a biological toxin, does not require any advanced knowledge compared with anthrax. Some of the threats with letters or packages have misstated the origin of the material, such as calling anthrax a virus, or have misspelled the agent's name. If the terrorists do not know the true origin of the material or cannot spell the material correctly, they probably do not have the education necessary to make the material they claim to have produced. This does not take into account a person who may purchase the material.

2. The next qualifier is a person's ability to obtain the raw materials necessary to make the agents. Many of the materials necessary to make chemical warfare agents are banned for sale. Others appear on hot lists, which means they are only sold to legitimate businesses. This would not preclude someone from buying the raw materials on the black market or stealing them from a legitimate business.

3. The third qualifier is the ability to manufacture the devices or machinery required to make the agent. The manufacture of chemical warfare agents requires the use of a reactor vessel, which requires about a 10-foot by 10-foot (3-meter by 3-meter) space to produce less than a gallon (3.7 liter). There are some agents that could be produced in a bathtub using backyard chemistry, but these are not the high-end agents that attract much attention. Many people who have attempted to make agents in less than ideal conditions have died during the production process. Many criminals do not take the time to follow standard industrial safety precautions.

4. One qualifier that is often overlooked is the ability to disseminate these agents. The military conducted many tests on chemical and biological warfare agents, and although they have some good methods of dissemination, even they lack a 100 percent effective method of dissemination. The Aum Shinrikyo cult in Japan is a perfect example; they were a group with millions of dollars in assets and full chemical and biological lab and production facilities. They employed the services of 235 scientists to develop and manufacture chemical and biological agents. The Aum Shinrikyo abandoned their biological weapons program after a full-scale release of anthrax that failed. They used sarin nerve agent twice, the first time in Matsumoto, Japan, in which seven people were killed and 200 injured. The dissemination method used in the Matsumoto attack was much more effective than the one used in the Tokyo subway attack. If they had used the same dissemination method they would have greatly altered the course of events. They would have been limited by the amount of agent that could have been produced in a short period of time.

5. The last qualifier, which is the most important, is whether the person or group has the motivation to pull off the attack. The intentional killing of one person takes significant motivation, and the intentional killing of hundreds requires a whole lot of motivation. There is always the potential for an attack, but it takes considerable education, raw materials, manufacturing, and dissemination ability to pull off a chemical or biological terrorist attack. On the other hand, explosives are easy to manufacture and do not require much education, only simple tools, and the materials required are easy to assemble. It is for this reason that explosives are used in the majority of cases and are quite successful in completing an attack. In many cases the terrorist can be successful because of the hysteria associated with a potential terrorism incident. A balance must be struck between a cautious approach and one that does not allow the terrorist to win by crippling a community and causing hysteria.

Another consideration, as we have seen in England and Ireland, is the disturbing trend of viewing emergency service personnel as targets. One probable theory regarding the Eric Rudolph bombings is that the second explosion at one of the Atlanta abortion clinic bombings was aimed at the emergency responders.

The terrorist's motivation is to produce fear, which may be aimed at the general public or the government. This goal can be accomplished by large-scale actions such as the bombing of the World Trade Center in 1993 or the bombing of the Alfred P. Murrah Federal Building in Oklahoma City. The rail attacks in Madrid affected the outcome of Spain's national election. It may even be on a greater scale, like the sarin attack in Tokyo. A terrorist can also incite this level of fear by planting the threat of terrorism or by devising a hoax. The latter scenario is the most likely, and can be very difficult to handle from an emergency service perspective.

SAFETY Responders must always be alert and aware of the potential for a secondary device.

Although the use of secondary devices in the United States is not common, the loss of responders' lives would be a high priority for a terrorist. The use of secondary devices has been a common tactic and is still being used overseas. Terrorists, especially those who are patient and study methodologies and tactics worldwide, will use this tactic. It is a diversionary tactic that responders should be aware of, and although responders should not overanalyze an event they should keep secondary devices in the back of their minds. Just as the use of toxic industrial chemicals and suicide/homicide bombers has been predicted and now has come to fruition, the use of secondary devices could significantly endanger responders' lives.

POTENTIAL TARGETS

Potential targets exist throughout every community in the nation. Terrorists may select commercial buildings, high-rise buildings, even residential homes. Although some incidents do not fit the definition of terrorism, the materials used are the same as in a terrorist situation. Whether the objective is terrorism or some other type of crime, the danger to the responder is the same. When considering terrorism, potential targets may be grouped into several categories: public assembly, such as the area shown in **Figure 7-5,** federal, state, and local public buildings; mass transit systems, high economic impact areas; telecommunication facilities; and historical or symbolic locations.

While obviously not an exclusive list, some buildings should be of concern, including FBI, ATF, and IRS offices; military installations; Social Security buildings; transportation areas; city or county buildings, including fire and police stations; abortion

Figure 7-5 Any location is a potential target for a terrorist. Any location where large numbers of people are present, such as a mall or sports event, are prime targets. Many of the arrests that are related to terrorism are in small towns and rural areas. The assumption that terrorism is limited to urban areas is proved incorrect by arrest statistics.

clinics; Planned Parenthood offices; fur stores; laboratories; colleges; cosmetic production/testing facilities; banks; utility buildings; churches; and chemical storage facilities. Transportation facilities such as airports and train, bus, or subway stations are high on the potential list of targets. Given the number of people who may be potential targets and the relative ease of escape, these areas are open for attack. In the southeastern United States, several churches were damaged or destroyed in arson fires in the 1990s, and in some cases explosive devices were used. Many abortion clinics have been the subject of bombings, attacks, and other threats. Any incident in or near one of these facilities should be approached with caution. Responders should know the location of these facilities in their jurisdiction. Preplans for these facilities should be developed by the company officers, but committing these plans to paper is not recommended, to avoid leaving a blueprint for a terrorist. As the battle between groups against abortion and for reproductive choice continues to heat up, incidents at these facilities can only be expected to rise, with emergency responders caught in the cross fire. Family planning clinics are not the only places targeted in this political struggle. Several anti-abortion groups have posted the home addresses of family planning physicians on the Internet, with the alleged intent of promoting violence against these doctors.

Many of the potential targets of some groups have not been buildings at all but events where large numbers of people are present. The Atlanta Olympic Park bombing is an example. Other scenarios involve sports stadiums, public assembly locations, transportation

Figure 7-6 Other than special events, the most common location where large numbers of people are together are sporting events. At this stadium, if an incident were to occur, more than 50,000 people can become part of the incident.

hubs, fairs, and festivals (see **Figure 7-6**). First responders should have some preplans for these types of locations. One possible scenario for a stadium, devised by Captain Richard Brooks (retired) of the Baltimore County Fire Department, describes the first-in medic unit to a stadium where in section 70 there are forty people projectile vomiting. After 5 minutes 200 people are projectile vomiting, and as time goes on, the number increases. What happens to the responders when confronted with a situation of this nature? How many responders would be affected by this massive amount of afflicted people? How many responders would be needed to handle this incident?

This could be accomplished by putting syrup of Ipecac™ in the ketchup container beside one of the hot dog vendors, an easy task. Imagine the hysteria if a note was found stating that a biological agent was distributed in that section. What impact would that have on the remaining 50,000 people in that stadium if the threat got out? Planners and responders involved on the national level to develop response profiles to terrorism are grappling with how to plan for incidents involving 100 people, 1,000 people, 10,000 people, and 50,000 people. Terrorism incidents can very quickly overwhelm the responders and their whole emergency response system.

When dignitaries visit locations, typically there is a lengthy planning process, and the fire department should be involved. When the Pope visited Baltimore in 1997, the planning process took more than 8 months. Planning for such a large event takes the cooperation of local, state, and federal agencies. Even when dignitaries visit locations such as New York City or Chicago, there is some advance planning. Other events such as political conventions also bring the potential for an incident. When one of these events

comes to your community, you need to be prepared for not only the people coming to the event but also the massive federal response that may be prepositioned. For many of the events there are task forces of federal resources hidden away just in case of an incident.

SAFETY The Department of Justice uses OTTO to suggest possible indicators to a terrorism event:
Occupancy and location
Type of event (biological, nuclear, incendiary, chemical, and explosive)
Timing of the event
On-scene warning signs

Certain dates have significance to several militant groups. April 19 is significant as the anniversary of the ATF storming the Waco, Texas, compound that housed the Branch Davidians, a group thought to be a militia group. April 19 was also, not coincidentally, the date of the Oklahoma City bombing. The significance of April 19 goes back to the Revolutionary War. Another symbolic date is the anniversary of Roe vs. Wade[1] on January 22, which harbors the potential for anti-abortion groups to strike. Within the United States there are a number of militia, patriot, or constitutionalist groups; together they are suspected to have members in forty states. Membership counts vary from fifty in some states to several thousand members in other states. Groups of concern include anarchist groups and white supremacy groups such as the Ku Klux Klan, skinheads, and neo-Nazi groups, including the Aryan Nation. Other groups suspected of activity or thought to have terrorism potential include patriots, militia, constitutionalists, and tax protesters. To learn which groups are active, it is relatively simple to use a Web browser and search the Internet for many of these groups; most have Web sites.

INDICATORS OF TERRORISM

An explosion or explosive device is the most common tool of the terrorist, and police across the country have made several arrests of people making and/or storing large quantities of explosives.

CAUTION According to an FBI source, more than 93 percent of terrorist incidents use explosives as the weapon of choice.

Figure 7-7 The most common explosive device is a pipe bomb, and it is very effective. It is a very dangerous device, not only for responders but for the builder as well. The pipe bomb on top is even more dangerous as nails have been added to cause more potential injuries.

The most common device is a pipe bomb, such as the one shown in **Figure 7-7**. In any incident involving an explosion, the safe course is to suspect a terrorist attack.

SAFETY If one explosion has occurred, always assume that a second device is awaiting your arrival.

Any suspicious package should be treated as if it contains explosives, and it should be dealt with by a bomb technician. Never assume you can handle the package or remove it from the area. Handling bombs or suspicious packages is a very specialized field and should be left to people trained in that field.

The presence of chemicals or lab equipment in an unusual location such as a home or apartment is an indication of possible illegal activity. Three possibilities are: drug making, bomb making, or terrorism agent production, either chemical or biological. The most likely, based on statistics, is drug manufacture followed by bomb making. Although very common in some parts of the country, predominately the West Coast and the Pacific Northwest, illicit drug labs are not distributed evenly throughout the United States. Currently the number of drug labs, especially those set up to manufacture methamphetamine, is increasing in the Midwest with law enforcement officers taking down more than twenty a month in some areas. This trend has made a progression from the West and is slowly moving to the East Coast.

In terms of terrorist activity, responders are most likely to encounter suspects making explosive devices. It is possible but highly unlikely that responders would locate a facility where people are attempting to make a terrorism agent such as sarin. The exception to the rarity of biological agents would be for the production of ricin, as the items required to make ricin are easily obtained and the production process is just as easy. Fortunately ricin does not have the easy potential to kill large numbers of people; it presents primarily an injection hazard, even though it is still very toxic through inhalation or ingestion. A responder in a small community or a rural setting is the most likely to run across any of these types of production areas. Several people are arrested each year for the possession of ricin, with the intent to use it for some type of criminal act. Most of the arrests are occurring in small towns throughout the United States. No matter what type of chemical threat responders locate, they should immediately isolate the area and call for assistance. The call should go simultaneously to the police, fire department, the HAZMAT team, and the bomb squad. This request would apply to all three types of labs—drug, bomb, or warfare agent—as any of these labs should be handled as a cooperative effort.

Another indicator of potential terrorism is an intentional release of chemicals into a building or the environment. Finding a chlorine cylinder in a courthouse would be unusual and should put responders on alert to the fact that there is a high probability that a terrorist may be at work. Finding chemical containers such as bottles, bags, or cylinders in unusual locations would also be suspect. In an industrial facility in which chemicals may be common, responders may discover an intentional release of a chemical.

NOTE One of the best indicators of a chemical release by terrorists will be a pattern of unexplained illness or injury. If these indicators show up immediately the probable cause is a release of a chemical. If these illnesses start to appear a few days to a week later the incident is probably biological.

A response to the mall for a seizure patient is not unusual. However, a response to a mall for six people having seizures is very unlikely and is more likely to involve a chemical release that could be from a terrorist attack. Victims twitching or having seizures, tightness in the chest, pinpoint pupils, runny nose, nausea, and vomiting are all signs and symptoms of a warfare agent attack. EMS providers will probably be the first group to identify the use of a chemical agent. Imagine arriving at an explosion, with large amounts of debris and twenty patients. To most, the injuries would appear as blast injuries, as they will in most cases be visible during a quick survey of patients. Other signs and symptoms, primarily the pinpoint pupils, probably will not be identified until patients are given a more thorough exam. Unconscious or dead patients with no obvious outward signs of death, such as trauma, may have been the victims of a chemical attack. This assumption must obviously be put in perspective. Responding to an apartment building in the winter in a Northeastern city in which there are twelve unconscious people points in all probability to carbon monoxide poisoning. However, a call about twelve unconscious people in the mall during the summer would more likely indicate some type of chemical release. (When terrorists distribute a warfare agent via an explosive device to the building or surrounding area, it may not show signs of damage. In some cases the larger the device the more likely the detonation will consume the agent.) As a result, most of the victims will not display signs or symptoms of a blast, although those closest to the device may suffer some of those injuries.

As with other responses, unusual odors or the presence of a vapor cloud may be an indicator that chemicals may have been released. Finding dead birds or other animals should put responders on alert to the chemical hazard and the potential for terrorism. It would also be unusual to respond to an office building or a house and notice unusual security measures of the type that you would not expect in that occupancy. Items such as extra locks, bars on windows, surveillance cameras, fortified doors, guards, and other unusual protection devices may provide a clue to possible illegal activity.

SAFETY The same hazards that exist in a chemical emergency—thermal, radioactive, asphyxiation, chemical, etiological, mechanical, and psychological—also exist at chemical or biological attacks. In some cases the risk would be increased as the incident was intentional and the specific harms may be intended to kill or injure responders.

INCIDENT ACTIONS

A terrorist incident combines four types of emergency response into a large incident. These types of incidents have these four characteristics until proven otherwise.

NOTE Mass casualty EMS incident + Hazardous materials release and/or explosive devices + Crime scene considerations = Incident management challenges.

The first responders can easily be overwhelmed and should commit to basic actions, such as life preservation. The IC will have enormous responsibilities dealing with all of the required actions. All of these components are difficult to handle individually, and now they are combined. The handling of a 100-person **mass casualty** incident is extremely difficult. It has the potential to overload the IMS system, and still it may only be one-fourth of a terrorist incident. Examine your response system, and determine how many patients present a concern. Some systems define a mass casualty as an incident involving five victims, while in other systems it may be ten to twelve. Obviously for a system with one EMS unit, more than one patient begins to cause system problems and may involve delays in treatment and transportation.

Imagine responding to an explosion at a mall, with 100 people injured. Such a mass casualty incident (MCI) will have EMS playing a predominate role. The police department, on the other hand, will be concerned with evidence preservation and crime scene considerations. Responders deal with the combination of these two priorities daily on a small scale. While responding to a crime scene, responders must take some extra time and consideration when treating patients. The possibility exists that the perpetrator(s) also used a chemical or biological weapon, and the explosion was the means of distribution. In such an incident, responders now are dealing with an MCI, crime scene, and potential chemical release with some, if not most, of the patients contaminated. On top of this scenario, add the potential for a secondary explosive device, one aimed at the responders—*us!* Each responding agency has major responsibilities. The HAZMAT team eliminates the chemical, while the bomb squad searches for a secondary device. EMS treats the patients, and the police department handles the criminal element. Identifying whether an incident is a crime, terrorist act, or just an emergency situation can be difficult. In many cases in the first few moments there may be no way to determine a cause. Even a small explosion could be from a number of sources. Differentiating between a chemical or biological attack can sometimes be determined by the immediacy of the symptoms. Chemical agents or toxic industrial chemicals cause immediate reactions and symptoms. Biological agents do not cause immediate effects, and may take days to several weeks for symptoms to show. Part of the issue that responders have to deal with is the public's hysteria regarding biological threat agents. People may be extremely frightened and may exhibit some psychologically driven medical problems. In some cases the symptoms may be real and in other cases they may be suspect.

The other aspect of a suspected terrorism incident is the massive response from the federal government, even if not requested. Later in this chapter more information is provided about available federal resources. In most cases the minimum response to a suspected terrorism incident would come from the FBI, initially from the local agent. There are fifty-six field offices across the country; almost every major city has a field office. Each FBI office has an agent assigned as the WMD coordinator who works with local responders. The FBI agent(s) typically arrive 1 to 2 hours into the incident. If terrorism is suspected, the FBI becomes the lead agency of the incident as is provided by a Presidential Decision Directive known as **PDD 39.** As with hazardous materials incidents, it is important to know all involved players before terrorists attack. Knowing your local FBI agent can be very important, as meeting for the first time in front of an incident is not conducive to effective scene management.

Working out "who's in charge" concerns prior to an incident is crucial. This is important to figure this out with your local police, EMS, and emergency management agency prior to the incident. These and many other responders will be involved and have a variety of responsibilities at an incident, such as that shown in **Figure 7-8.** In general a unified command is recommended, although this does not mean command by consensus. All involved agencies supply input. The agency with the majority of the tasks to be accomplished is generally the group in charge. In the initial stages the fire/EMS authority would be in charge while rescuing victims, but after the victims are removed and evidence recovery is the next priority, then command may switch to law enforcement. The response to assist with the incident can in many cases be the most difficult part of the incident to manage. The IC will be overtaken by a large number of federal agency representatives, who on a regular basis will be replaced by a later-arriving supervisor. Response groups consisting of two to seventy responders may arrive uninvited, all trying to assist with the incident. The one group that will also overwhelm the system will be the media. In most cases the local media will react as normal and, for the most part, hopefully, cooperate.

Figure 7-8 Many local, state, and federal agencies will be involved in responding to a terrorist incident, each trying to accomplish similar goals. The use of an incident management system is essential to a successful outcome.

NOTE In any incident to which federal resources respond, rest assured that the national media will be following close behind.

The national media do not know or follow local protocol and will require information. A lack of accurate information can be disastrous to an incident, as the media may use inaccurate information, which can cause deterioration of the incident and create more hysteria than may be present already. Media relations are critical in these types of events.

A crucial task at an incident perpetrated by terrorists is evidence preservation. As much care as possible should be taken to preserve evidence and ensure that it is handled appropriately. The collection of evidence is primarily one of law enforcement's responsibilities.

CAUTION Unless properly trained, fire service personnel should not collect evidence but should alert the police of its presence.

A whole host of issues goes along with the collection of evidence, including the chain of custody. This chain of custody, or the paper trail, that follows any evidence is crucial to the successful prosecution of the persons responsible. The failure to follow proper procedure or document the travel of evidence can result in a case being dismissed, regardless of any other evidence. In rare cases, such as the Murrah bombing, fire service personnel have been called on to collect evidence. If your agency is asked to collect evidence, it is best to assign a fire investigator or a fire marshal as they typically are trained in evidence collection. Another alternative is to double up and use a firefighter and a police officer to collect evidence.

It will take a cooperative effort to combat a terrorist attack. The primary function areas are rescue/life safety by fire and EMS personnel, hazard identification by the HAZMAT team, identification of possible secondary devices by a bomb technician, and incident management. It is important to communicate hazards to all personnel and to limit the response to essential personnel. Instead of having the whole alarm assignment report to the front of the building, use one or two companies to investigate while staging the other companies, away from dumpsters, mailboxes, or dead-end streets. When dealing with victims, it is essential to isolate them until the cause is identified. The victims can have a large amount of information and should be questioned quickly. Questions to ask include:

■ What did you see, hear, or smell?
■ Is there one area that this was coming from or was it throughout the building?

■ Did you see anything else that may be suspicious?
■ What type of signs and symptoms do you have?

In addition, the police will need to conduct interviews. Documentation and preservation of the evidence will be essential in the successful prosecution of the terrorist.

GENERAL GROUPINGS OF WARFARE AGENTS

Terrorists can use a number of possible destructive warfare agents. These agents can be classified in four broad areas. The first is weapons of mass destruction (WMD), a term commonly used by the military. Some of the regulations that prohibit making, storing, or using terrorism agents are called WMD laws or regulations. Any item that has the potential for causing significant harm or damage to a community or a large group of people is considered a WMD. The other descriptions are nuclear, biological, and chemical (NBC), chemical, biological, radiological, nuclear, incendiary, and explosive (CBRNE). These are descriptions of the types of materials that could be used in a terrorism attack. Although they have some slight variations, they all are used to describe the various types of agents that a terrorist could use. Most of the language differences come from funding legislation or a specific federal agency.

The military has devised a naming system for many of these agents, many of which are provided in **Table 7-1.** Responders should become familiar with these names, which are used in much of the literature and help guides. A military detection device will usually refer to the military name. When dealing with terrorism, we are entering another world that has its own language. The fire service must learn to adopt this new language. The broad areas of terrorism agents are further subdivided into several categories.

Nerve Agents

Nerve agents include tabun, sarin, soman, and V agent, all related to organophosphorus pesticides. They were designed for one purpose—to kill people. Although very toxic, the capacity of these agents to kill large numbers of people requires that the dissemination device function correctly, and a number of other critical factors must be in place for the attack to be effective. Although several gallons of sarin were used in the Tokyo subway attack in 1995, there was no effective method of distribution, and

TABLE 7-1

Military Designations for Agents		
Name	**Military Designation**	**UN/DOT Hazard Class**
Anthrax	n/a	6.2
Biological agents	n/a	6.2
Chlorine	CL	2.3
Cyanogen chloride	CK	2.3
Distilled mustard	HD	6.1
Hydrogen cyanide	AC	6.1
Lewisite	L	6.1
Mace	CN	6.1
Mustard	H	6.1
Nitrogen mustard	HN	6.1
Pepper spray	OC	2.2 and 6.1
Phosgene	CG	2.3
Ricin	n/a	6.2
Sarin	GB	6.1
Soman	GD	6.1
Tabun	GA	6.1
Tear gas	CS	6.1
Thickened soman	TGD	6.1
V agent	VX	6.1

out of the thirteen people who died, it has been reported that the two people killed by sarin itself actually touched the liquid. The chemical and physical properties of these agents hinder their capacity to be effective as stand-alone killers. To produce the toxic effect the agent must touch people in liquid form or be breathed while the material is in aerosol form. The material will not stay in aerosol form for very long. It has a very low vapor pressure, so it will not create vapors as standing liquid. These examples of vapor production are at 68°F, and the major impact on the development of vapor is temperature. When there is an increase in temperature, there will be more vapors produced, thereby increasing the hazard. When there is a temperature decrease, vapor production is minimized. Other factors that can influence the movement and development of vapors are the surface area of the spill, wind speed, humidity, and any possible dissemination devices. All of the military nerve and blister warfare agents have a vapor pressure less than water, which means they do not evaporate quickly. Unless the liquid is touched or placed on the skin, it does not present a large hazard.

Incendiary Agents

For the sake of classification we will place **incendiary agents** into the chemical classification, as chemicals are used in these devices. The most commonly used chemicals are flammable and combustible liquids. The standard Molotov cocktail is an example of an incendiary device that could be used by a terrorist. In some cases arsonists have used a mixture of chemicals, usually oxidizers, to create very fast, high-temperature fires that could be used by a terrorist.

Blister Agents (Vesicants)

Vesicants, or more commonly called **blister agents,** include the chemical compounds mustard, distilled mustard, nitrogen mustard, and lewisite. These materials were never designed to kill, but were meant to incapacitate the enemy so that if one person was affected by one of these agents, several more would be needed to take care of the victim. The effects were to cause irritation and blistering to a severe degree. Although at high concentrations these materials can be toxic, their biggest threat is from skin contact, causing blistering. Their chemical and physical properties

SIGNS AND SYMPTOMS OF NERVE AGENTS

All nerve agents present the same types of signs and symptoms as organophosphorus pesticides, and in reality the difference is minor. Nerve agents are pesticides for humans and are a stronger, more concentrated version of commercially available pesticides. The signs and symptoms can be generally described using the term SLUDGEM, which stands for:

Salivation—victims will be drooling.
Lacrimation—their eyes will be tearing.
Urination—victims may lose control and urinate on themselves.
Defecation—victims may lose control of their bowels (diarrhea).
Gastrointestinal—they may experience nausea and vomiting.

Emesis—vomiting is another symptom.
Miosis—pinpoint pupils are also signs.

SLUDGEM describes all of the symptoms of exposure to a nerve agent, from the minor ones to the extreme signs. A slight exposure to any of the nerve agents will cause pinpoint pupils, runny noses, and difficulty breathing. A person who has come in contact with the liquid will be experiencing all the SLUDGEM signs in addition to convulsions. A person who is in convulsions needs immediate decontamination and medical treatment to survive. This treatment sequence must occur in less than 10 minutes, and sufficient medication must be available on scene to accomplish the treatment. Most paramedic units carry enough medication to treat one or two patients who have severe symptoms.

make them less of a hazard than the nerve agents. One large concern with these agents is the fact that the effects from an exposure are delayed sometimes 15 minutes to several hours. Quick identification of a blister agent by the HAZMAT team is key to keeping the victims safe.

Blood and Choking Agents

In addition to being terrorism agents, the substances in this group are also common industrial chemicals. The first category, **blood agents,** includes hydrogen cyanide and cyanogen chloride. Both of these materials are gases and disrupt the body's ability to use the oxygen within the bloodstream; they are referred to as chemical asphyxiates. **Choking agents** are even more common; chlorine and phosgene are gases commonly used in industry. Chlorine is in almost every town, as it is used for water treatment and

swimming pools. Any community that has a water system or swimming pool has some form of chlorine. Chlorine comes in containers ranging from 150-pound (68-kg) cylinders to 90-ton (82-metric ton) railcars; it also comes in tablet form (HTH(tm)) typically used in residential pools. The release from a 90-ton (82-metric ton) railcar can result in hundreds of thousands of injuries and possibly many deaths as well, especially in an urban area. A small amount of chlorine can be very deadly or, at a minimum, create substantial panic in a community.

Irritants (Riot Control)

The most commonly used materials that are classified as potential terrorism agents are irritants, including Mace, some other form of pepper spray, or tear gas. The response to one of these types of incidents is common and usually impacts a large number of

SIGNS AND SYMPTOMS OF BLISTER AGENTS EXPOSURE

One of the biggest risks with the blister agents is the fact that they may present delayed effects. If not detected early, patients might be released only to have problems later. In general blister agents are not designed to kill; they were designed to incapacitate the enemy, resulting in troops being assigned to assist with the wounded. It is possible to create scenarios in which fatalities could occur, but these would be unusual cases. The effects from blister agents include irritation of the eyes, burning skin, and difficulty in

breathing. The more severe exposure will result in blisters, which may be delayed. The only real street treatment for these signs and symptoms is decontamination and supportive measures. It is important that anyone who has had contact with the liquid blot the liquid off the skin and avoid spreading the agent. Luckily, the chemical and physical properties of these agents make them difficult to disseminate, and coming in contact with the liquid would be the primary means of injury.

SIGNS AND SYMPTOMS OF BLOOD AND CHOKING AGENTS

Many of the blood agents are commonly found in industry and may be found at typical chemical facilities. The signs and symptoms of blood agents for a slight exposure include dizziness, difficulty breathing, nausea, and general weakness. With cyanides the breathing initially will be rapid and deep, followed by respiratory depression, usually leading to death. The two most common choking agents are widespread in industrial use, and chlorine can be found in every community. Signs and symptoms include difficulty breathing, respiratory distress, eye irritation, and, in higher amounts, skin irritation. Phosgene may present delayed effects, while chlorine's effects are immediate.

people. The usual target is a school, mall, or other large place of assembly. The use of one small container can immediately affect large numbers of people and make them immediately symptomatic. Luckily these materials are not extremely toxic, although they are extremely irritating, and the symptoms will usually disappear after 15 to 20 minutes of fresh air. The response to one of these incidents is difficult as responders must treat patients with real medical problems and an unknown source that is difficult to identify.

Biological Agents and Toxins

Other than explosives the most likely agents that could be used in a terrorism scenario are **biological agents** and **toxins.**

Some of the materials in this grouping include anthrax, mycotoxins, plague, tularemia, and ricin. Out of all of the agents for a terrorist to make, this grouping is the easiest, especially ricin. The fatal route of entry differs with each of the agents and may be skin contact, inhalation, or injection. These agents are difficult to distribute effectively, and in some cases exposure to sunlight may neutralize them. Some of the potential indicators of biological terrorism include presence of a powder or liquid material, containers associated with biology, and in some cases a threat letter expressing the use of a biological material. The use of a dissemination device or placement so that the material is disseminated is usually more robust than with chemical agents. The visual sighting of a dust or vapor cloud being released in an unusual fashion could be indicative of a biological threat agent release. A small explosion followed by an uncharacteristic dust or vapor cloud could be a dispersal device. Any unusual packages, equipment, or boxes that are out of place could be holding a biological material.

The Centers for Disease Control (CDC) have three categories of biological agents that could be used for weapons as shown in **Table 7-2.** Category A agents are those that can be easily disseminated or transmitted person to person, have potential to cause a large-scale public health emergency, might cause public panic and social disruption, and may require special action for public health preparedness. Category B agents are those that are moderately easy to disseminate, result in moderate to low fatality rates, and require specific enhancements of CDC's diagnostic capacity and enhanced disease surveillance. Category C agents are those that might be emerging pathogens that can be used for mass dissemination in the future because of their availability, ease of production and dissemination, and potential for high fatalities and major health impact. Specific information about the two most popular agents follows Table 7-2.

The two most commonly used agents in this category are anthrax and ricin.

Anthrax is a naturally occurring bacterial disease that is commonly found on dead sheep. This naturally occurring anthrax is contagious through skin contact or by inhalation of the anthrax spores. Although relatively easy to obtain, it is more difficult to culture and grow the proper grade of anthrax. To produce fatal effects the type of anthrax required is military grade anthrax, which is very difficult to produce.

SIGNS AND SYMPTOMS OF IRRITANTS

The signs and symptoms for a slight exposure to a high dose are the same, with the exception of increasing severity. The signs and symptoms are eye and respiratory irritation. There is no real treatment except removal to fresh air; in 15 to 20 minutes the symptoms will begin to disappear. Supportive care can be provided.

TABLE 7-2

CDC Categories of Biological Agents

Name	Notes
Category A Agents	
Anthrax	Noncontagious bacteria; highly infectious through inhalation
Smallpox	Highly contagious virus, with high potential for fatalities
Botulinum toxin	Produced from clostridium botulinum; one of the worst toxins known
Ebola	Also known as viral hemorrhagic fever; high fatality rates
Plague	Caused by the Yersinia pestis bacteria; easy to manufacture and spread by flea bites; can be inhaled through the spread of pneumonic plague
Marburg	A viral hemorrhagic fever with high fatality potential
Tularemia	Caused by the Francisella tularensis bacteria, also known as rabbit fever; incapacitates as opposed to causing fatalities
Category B Agents	
Brucellosis	
Epsilon toxin	*Clostridium perfringens*
Food safety threats	*Salmonella* species, *E. coli* o157:H7, *Shigella*
Glanders	*Burkholder mallei*
Melioidosis	*Burkholder pseudomallei*
Psittacosis	*Chlamydia psittaci*
Q fever	*Caxiella burnetii*
Ricin toxin	*Ricinus communis*
Staphylococcal enterotoxin B	
Typhus	Rickettsia prowazekii
Viral encephalitis	Alphaviruses, Venezuelan equine encephalitis, Eastern and Western equine encephalitis
Water supply threats	*Vibrio cholerae, Cryptosporidium parvum*
Category C Agents	
Nipah virus	
Hantavirus	
Multi-drug–resistant tuberculosis	

Even if developed, it must be distributed effectively and under the right conditions.

Although ricin is easy to make, the required distribution method makes a large-scale incident unlikely. It must be injected to be truly effective.

Ricin is 10,000 times more toxic than sarin; a small amount, 1 milligram, about the size of a pin head, can be fatal. Death will usually occur several days after the injection. By other routes of entry such as inhalation or ingestion, the most likely consequence is that many people would get sick but would eventually recover. Next to explosives, ricin is the leading choice of domestic terrorists; several times a year someone is arrested for possession of ricin. Some example ricin cases include:

■ 2002 Kenneth Olsen was arrested for possession of ricin that he kept in his office cubicle. While at work at a high tech company in Spokane, Washington, he researched other poisons and explosives, and left this information throughout the office.

■ 2003 A letter containing ricin was mailed to the White House. It was one of two letters from the "fallen angel" who was protesting the change of work hours being mandated by DOT. The second letter was located at an airport postal facility in Greenville, South Carolina.

■ 2003 FBI agents arrested Bertier Ray Riddle of Omaha, Arkansas, for sending a letter supposedly containing ricin to the local FBI office in Little Rock.

■ 2004 Ricin was located in Senator Bill Frist's office in the Dirksen Senate Office building. It is not known what the source of the ricin was as no letter was found. Frist from Tennessee was the Majority Leader at the time. There is a possibility that the ricin is connected to several other ricin letters from the "fallen angel."

■ 2004 Richard Alberg was arrested for possession of ricin after he placed a large order of castor beans. Although no specific attack was determined Alberg fantasized about the use of biological agents to kill.

■ 2005 Six persons were arrested in London for conspiring to commit terrorist acts. Inside their apartment the police found ricin and materials that can be used to manufacture ricin. The Islamic group that was arrested has ties to Chechnya, and to al Qaeda terrorists who plotted the millennium bombing attempt in the United States. The establishment of an al Qaeda and Chechnya relationship is very troublesome. The Chechnyan region of Russia is responsible for well-coordinated terrorist attacks in Russia; their large-scale attacks have killed thousands of Russian citizens. This is the group responsible for the Beslan school attack, the Moscow theater takeover, downing of two planes, and a subway bombing, among many other attacks.

■ 2006 Ricin, along with pipe bombs and blasting caps, were found in the home of William Matthews of Nashville, Tennessee. In addition to problems at work, his wife had sought a protection order due to his substance abuse. It is not known what he intended to do with the ricin, which was kept in a baby food jar.

The most successful attack using a biological material involved the use of salmonella bacteria by the Rajneesh cult in Oregon. In an attempt to influence a local election, members of the cult put the material in the salad dressings at several fast-food restaurants. More than 715 people fell ill from the deliberate contamination.

The FBI is still investigating the anthrax attacks of October 2001 that started in Boca Raton, Florida, at the American Media building. It is suspected that a letter was sent and then opened at this facility, which produces the *National Enquirer.* One person died, and several others in the building fell ill. No letter was ever recovered from the facility.

The later letters that were sent targeted more media and members of Congress. In addition to the American Media building, the NBC studios, the *New York Post,* and ABC all received letters. Two members of Congress, Senator Daschle and Senator Leahy, also received letters. The letters killed five people, and made eighteen others ill. Some of the deaths and illnesses involved people at various post offices. It is thought that the mail handling process enabled anthrax to aerosolize and get into the air.

The letters contained small amounts of anthrax, which has been tested to be the Ames strain of anthrax that was produced in the United States. Much has been learned about anthrax and its ability to be used as a weapon. The theoretical dose of anthrax in each of the recovered letters was estimated to be enough to kill several hundred thousand people. The reality is that each letter typically killed one person. The other major factor in the deaths was a delay in treatment, and in some cases misdiagnoses. People who did not seek quick treatment or were misdiagnosed had difficult or unsuccessful recoveries. Those who sought quick medical treatment, during which the agent was recognized as anthrax, survived.

The CDC reported that there were ten cases of inhalational anthrax and twelve confirmed or suspected cases of cutaneous anthrax. Of the ten inhalational cases, seven occurred to postal workers in New Jersey and Washington, DC, at mail sorting facilities. In the American Media case, one person who received the letter died, and a person sorting the mail fell ill. Six of the ten inhalational cases of anthrax survived after treatment, which is higher than originally anticipated.

The signs and symptoms of inhalational anthrax are a 1- to 4-day period of malaise, fatigue, fever, muscle tenderness, and a nonproductive cough, followed by a rapid onset of respiratory distress, cyanosis, and sweating. The recent cases also had profound, often drenching, sweating, along with nausea and vomiting.

Radioactives

Nuclear agents, unfortunately, have to be put back onto the list of possibilities that a terrorist could use. There are two types of radiation events, **nuclear detonation** and **radiological dispersion devices (RDD).** The use of an actual nuclear detonation device is unlikely and very improbable given the security these materials have. The amount required

RESPONSE TO ANTHRAX HOAXES

Emergency responders should establish a plan for response for noncredible anthrax events. In 2001 and 2002 these events kept HAZMAT teams very busy, as well as other emergency response personnel. From the outset it would be difficult for someone to manufacture and successfully pull off an anthrax threat. In many of the hoax incidents people received letters proclaiming that they had been exposed to anthrax. In some of these cases the only thing present in the envelope was a letter. Anthrax is not invisible, and a quantity of the material must be present if the attack is to succeed. To harm victims, the material must be inhaled, and to be inhaled, it must be distributed into the air. Just the placement of the material in an envelope does not present a high risk to the letter opener. To pose a risk, the solid material must be dispersed into the air, and if no dispersal occurs, the risk to people in a building is very small. The only action required is to double- or triple-bag the envelope, package, or material in bags suitable for evidence collection. Procedures for the collection of evidence should be followed, and the local police and FBI should be consulted.

People who touched the material should be instructed to wash their hands with soap and water. They do not require full body decontamination, nor are special solutions required such as a bleach and water mixture. The military recommends that humans exposed to anthrax rinse thoroughly with water. Anyone who was within several feet of the person opening the package does not require decontamination. The only reason for a full-body wash would be if victims spilled the material on themselves from head to toe, but no immediate action is required. Victims can be taken to a shower and be provided privacy as they wash. People directly exposed to material that may contain anthrax should be entered into the health care system so that they can be advised of the signs and symptoms of exposure and have emergency contact information relayed to them.

There is no need to start prophylactic antibiotics on people just because they opened a letter with a powder in it. There is sufficient time to complete lab analysis on the material before medical treatment is required. The threat from a solid material (no matter how toxic) is from inhalation of the dust or touching the material. Gloves and SCBA are adequate protection for collection of the evidence. The FBI has a number of labs around the country set up to assist with the identification of WMD agents, and the local FBI office should be contacted for assistance.

and the specific type also make use unlikely. There is current speculation that there are some small nuclear devices missing from Russia; but this has never been substantiated.

There is the potential, however, for some nuclear material that could be used in an RDD, much of which is coming from Eastern Europe. An RDD is a device that disperses radiological materials, usually through an explosive device. One example would be the use of a pharmaceutical grade radioactive material attached to a pipe bomb, which could cause an amount of radioactive material to be distributed. The

FBI AND CDC LABORATORY RESPONSE NETWORK

When discussing WMD detection it is important to note the laboratory system that the FBI and CDC have set up. The Laboratory Response Network, known as the LRN, handles the initial analysis of potential WMD materials. The LRNs have been in existence since 1999, but really came to light in 2001 after the anthrax attacks. The FBI partnered with the CDC to ensure that standardized protocols were implemented and that both public health and evidentiary concerns were addressed. The WMD coordinator can assist in getting evidence or samples to a LRN facility. The gold standard for analysis is laboratory testing, which is the desire of the FBI. Without laboratory testing they cannot make their case, and the suspect may be allowed to go free. They do not place much emphasis on street detection methods and generally discourage tampering with the potential evidence. Prior to shipment they want the evidence screened for fire, corrosive, toxic, and radioactive hazards. The FBI has a number of HAZMAT response teams throughout the United States and maintains a highly technical response group known as the **Hazardous Materials Response Unit (HMRU)** who can provide assistance when potential WMD materials are found or suspected.

There are 140 LRN labs set up for biological threat agents throughout the United States, and generally the local or state health departments run the labs. For chemical attacks there are 62 laboratories that can conduct chemical threat agent analyses. These are the labs that most responders would send their samples to, through the FBI WMD coordinator.

strength of the radiation source dictates how serious an RDD would be. In many cases the RDD would present more of an investigative issue than a significant health issue. There are some radiation sources, however, that would present a risk, but these are not in common use. The other limiting factor is the explosive device and its limitations. Technologically, an RDD is viable for a small area, such as a one-half acre or less of land. Attempting to distribute a significant amount of radioactive material takes a significant explosive device, not to mention the radiation source. The larger the RDD, the more danger there is to the terrorist who has to assemble, transport, and detonate the device.

There is an advantage to a noncredible RDD or a small RDD, and that is the public's reaction. The perception of the public, and of many responders, is that this would be a radioactive disaster. Radiation meters would indicate that there were radioactive materials spread around. The reality is that the amount of radiation would not be dangerous, and as time passed, the danger would lessen as the radioactive material becomes less hazardous. Unless it were put in perspective and good public information were used, there would be tremendous concern in the community. Radiation is one of the big unknowns and causes fear because it is unknown, making it a prime weapon for a terrorist. Education on the hazards of radiation and the effective use of radiation monitors can ease this fear and allow responders to make an informed response.

Other Terrorism Agents

While considerable emphasis is placed on warfare agents, in reality many other common industrial or household materials can be just as deadly, if not more so.

NOTE Remember that a terrorist wants to create panic, not necessarily kill or injure massive numbers of people.

The most likely scenario is a pipe bomb, followed by the use of ricin. Other industrial chemicals can also be used, many of which are just as deadly, if not more so than most warfare agents. Responders should not get lulled into a false sense of security if they do not find a warfare agent or if the device is only a small pipe bomb, not a moving truck with ammonium nitrate and fuel oil. Nerve agents, anthrax, or other chemical warfare agents have only been used twice each in Japan and have never been used in the United States, where they are not manufactured by anyone other than the military. The FBI

Bomb Data Center reports that out of the 3,000 bombings that occur each year, they have only two large bombings; however, these "small" bombs kill an average of 32 and injure 277 people a year. Although responders should be on the watch for secondary devices, an 11-year history of bombings shows only four to five secondary devices out of approximately 33,000 bomb incidents.

DETECTION OF TERRORISM AGENTS

The detection of terrorism agents is difficult and, given the potential circumstances, must often be done under severe conditions. The response to terrorism incidents has changed how many HAZMAT teams operate and has increased their capability to handle other situations. The confirmation that a terrorist agent has been used is difficult; as a group they are extremely hard to detect. The exceptions are standard industrial materials, such as chlorine, which are easy to detect and confirm.

Although the detection of terrorism agents should be done by HAZMAT technicians, this section provides an overview of some of the common detection devices.

Test Strips

NOTE Many responders should have a minimum method of detecting warfare agents.

The lowest cost method is to use **M-8** and **M-9** indicating papers, as shown in **Figure 7-9** and

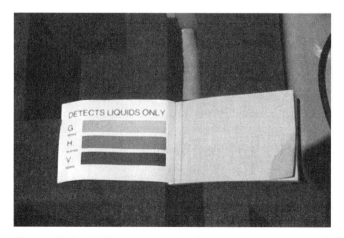

Figure 7-9 M-8 test paper indicates the presence of liquid military nerve and blister agents. The color change indicates the type of agent present.

Figure 7-10 M-9 indicating paper detects the presence of liquid military nerve and blister agents by turning red. The paper is adhesive backed, the same as a roll of tape.

Figure 7-10. These two devices are easy to use and maintain; their only drawback is that they can only be used on liquids. Another major issue with these devices is the fact that they react to a large number of common chemicals and have a tendency to false positive.

M-8 paper is like pH test strips for warfare agents, and it indicates for nerve and blister agents. It provides a separate test for VX, and through a color change indicates which type of agent is present.

M-9 paper is better known as M-9 tape, as it is a roll of tape with an indicating layer on the outside surface. It is commonly placed on vehicles, boots, gloves, and other parts of the responder. M-9 indicates the presence of nerve and blister agents, through a red color change. M-9 does not differentiate between nerve or blister; it only indicates the presence of a warfare agent. M-8 and M-9 are used together as they both react to a differing set of false positives.

A combination test strip, as shown in **Figure 7-11**, developed by Safety Solutions, is a disposable unit that has the ability to detect nerve agents, chlorine, sulfides, oxidizers, fluorides, cyanide, arsenic, and corrosives. The test strip is designed to be worn as a warning badge and changes colors in the presence of the listed materials. It is a low-cost option that is easy to use.

Colorimetric Sampling

NOTE One of the best methods to detect warfare agents without using an electronic device is to use colorimetric sampling tubes, as shown in **Figure 7-12**.

Only a few companies (Drager and MSA) make colorimetric tubes for specific warfare agents, but most companies provide tubes for the hydrolysis products of the agents. There are two primary methods in use for colorimetrics: One is a combination system that samples for a variety of agents simultaneously, and the other samples for individual agents independently of each other. Obviously, with time of the essence, multisampling has the advantage. The individual tests allow for more sampling flexibility and will assist in the final identification, so a combination strategy should be employed.

NOTE One major plus for the colorimetric tubes is the fact that they have the least amount of false positives when dealing with warfare agents. If they indicate the presence of warfare agents, they are most likely present.

Figure 7-11 This is a combination test strip that detects a variety of chemicals, which is indicated by color change.

Figure 7-12 This is a colorimetric tube which is showing a positive reaction, after drawing a sample. The color change on the glass tube provides an indication as to the type and amount of material present.

Handheld Electronic Detection Devices

There are a number of devices in this category, and responders should look at all the devices before buying one for emergency response. These devices use a variety of sensing technologies, with each having some drawbacks. Each of the electronic devices has false positives, which is an issue. The basis for the electronic devices is that they draw in a sample and develop an algorithm. This algorithm is also known as a signature, fingerprint, or footprint. The signature developed is then compared against those in the library. If a match is found, the meter alarms, warning the user.

The **APD-2000** developed by ETG (now Smith's Detection), shown in **Figure 7-13**, uses **ion mobility spectrometry (IMS)** as its detection method. It also has the ability to detect mace and pepper spray, which is an invaluable benefit to this detector. This ability exists for hours after the discharge of the spray, and it can detect small amounts in the air. In addition this detector also has the ability to detect gamma radiation.

The SABRE 2000, also by Smith's Detection, uses ion mobility to detect chemical warfare agents, just like the APD-2000, but it also detects explosives and drugs. The library covers the explosives and illegal drugs that are likely to be found. It is a device that should be used, as it covers the most likely threat—explosives.

There are several devices that use **surface acoustic wave (SAW)** technology to detect terrorism agents. The HAZMAT CAD Plus and the Joint Chemical Agent Detector (JCAD) use fairly unique detection technology. These devices have two or more piezoelectric crystals that are absorbed into the coatings on the sensor surface, much like the metal oxide sensor used in flammable gas indicators. This absorption causes a change in the resonance frequency of the instrument. This resonance outputs an algorithm that is compared to those in the library. As there are two or more sensors, these readings are compared against those stored in the library, and if a close match is found, the unit alarms. Both of these detection devices offer some ability to detect some families of industrial chemicals.

The **AP2C** is a French version of the Chemical Agent Meter (CAM) and uses a hydrogen flame to detect chemical agents. The unit requires the user to refill the hydrogen cylinders, which can be costly. The AP2C only detects nerve and blister agents; it does not have the ability to detect any industrial materials. This unit and all of the others have false positive issues. Combining any two of these detection devices would make detection more reliable. If both indicated a nerve agent, then a nerve agent is likely to be present. If only one indicated, then the alarm is most likely a false positive.

Biological Agent Detection

The detection of biological agents such as anthrax is very difficult; the technologies used have been adapted from lab-based systems and can be complicated. Dramatic improvements have been made, and the systems are getting more reliable. There are two basic methods, **handheld assays (HHA),** shown in **Figure 7-14,** and **polymerase chain reaction (PCR),** shown in **Figure 7-15.** The handheld assays are available from a number of manufacturers, but they do vary in quality. The HHAs are much like home pregnancy tests, although the detection method is a more complicated process. The user places a prepared sample onto the test bed of the assay and waits for a color change. Most of the systems provide an electronic reader, which is highly recommended because it dramatically improves the accuracy of the assays. The PCR technology is new as a response method, but has been used in lab settings for many

Figure 7-13 This detection device indicates the presence of military chemical warfare agents such as nerve and blister agents. It also detects a small number of toxic industrial chemicals.

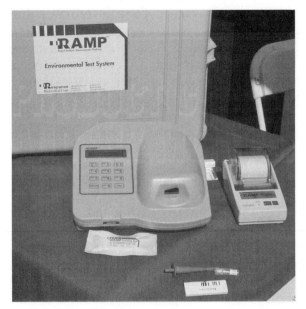

Figure 7-14 This is a handheld bioassay that detects common biological threat agents, including anthrax, smallpox, ricin, and others.

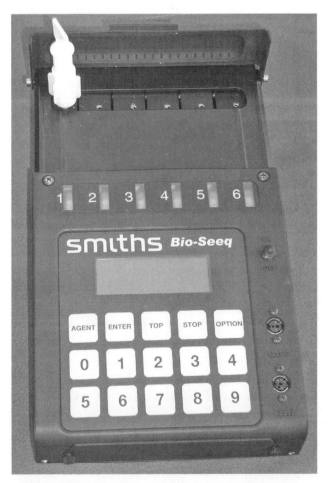

Figure 7-15 This is a polymerase chain reaction (PCR) biological test device. This device replicates the DNA structure of a sample and tries to match the structure to that of known biological threat agents, such as anthrax and several others.

years. The PCR unit replicates the sample's DNA and compares it to DNA stored in the library. If a match is found, the user is alerted to the presence of the material. Proper sample preparation is essential to success with any of these devices. The best methods for the detection and confirmation of biological agent detection are lab-based systems. The FBI and CDC have established a Laboratory Referral Network (LRN) in every state that provides assistance in the identification of biological agents.

FEDERAL ASSISTANCE

After the attacks on September 11, 2001, the federal government created the Department of Homeland Security (DHS) by merging twenty-two previously separate agencies into one department to protect the nation against threats. The department's first priority is to prevent further terrorist attacks. Several key agencies are now under one department. The Federal Emergency Management Agency (FEMA) is under DHS and is now in the emergency preparedness and response section of DHS.

Prior to the formation of DHS the federal government has established roles and responsibilities in case of a terrorist event, and these roles are provided in Presidential Decision Directive 39 (PDD 39). This document names the FBI as the lead agency during the emergency (crisis management) stage of the incident. FEMA becomes the lead when the incident is no longer in the emergency phase (consequence management). The FBI has a Hazardous Materials Response Unit (HMRU), which on a potential terrorist incident provides identification, mitigation assistance, and evidence collection. The HMRU is a multifaceted group that responds not only to terrorism incidents, but also to scenes involving explosives, drug labs/incidents, and environmental crimes. Since its inception, the HMRU has been responding to an ever-increasing number of incidents. The majority of these incidents involve biological threats, although a number of incidents involve other materials. HMRU is also prepositioned at a number of events throughout the nation and works closely with local HAZMAT teams.

A response from FEMA will vary with the incident, and the number of FEMA personnel will resemble a response similar to any other disaster to assist in restoration and recovery issues. The Urban Search and Rescue (USAR) teams fall under FEMA and are activated by following the emergency management chain from local to state to FEMA. At this time there are 25 USAR teams across the country.

They provide expertise in heavy rescue operations such as building collapses. They are a multifaceted team of 70 people and are composed of rescue specialists, dog search teams, medical specialists, Hazmat WMD specialists, communications specialists, and an engineering and rigging component. The greatest majority of the victims are rescued by the first responders and other local specialized resources. But in some cases trapped victims may be alive for many days and may require the expertise and equipment of a USAR team. One unfortunate aspect is that the USAR teams have a delayed response. It will take them several hours to become airborne, not including travel time. If there is a possibility they may be needed, their assistance should be requested early. Like many teams, USAR may send an advance party hours before the arrival of the remainder of the team.

A number of other agencies may be involved in the response to a terrorist incident. The military has a couple units with terrorism response capabilities and responsibilities. The Army has the Technical Escort Unit (TEU) headquartered at Aberdeen Proving Grounds, near Baltimore, Maryland. There are other units at Dugway Proving Grounds, Utah; Pine Bluff Arsenal, Arkansas; and Fort Belvoir, Virginia. The TEU is assigned to provide escort services for warfare agents and be the troubleshooting group in the event of an incident involving military warfare agents or explosives. When an incident involves potential chemical agents, the TEU will respond at the request of the FBI to assist with identification and mitigation of the incident. The TEU handles chemical, biological, and explosive materials and other hazardous materials. This self-contained unit employs the lab resources of the Research Development and Engineering Command (RDEC).

The United States Marines have a unit known as the Chemical and Biological Incident Response Force (CBIRF), based in Indian Head, Maryland. This unit has responsibility for the response to terrorism across the world. Its three main components are decontamination, medical, and security. The CBIRF is a self-contained unit and requires only a water source for long-term operations. It can also provide detection and mitigation. CBIRF responds nationwide upon request to terrorist incidents. All of the federal resources, including CBIRF and TEU, can and have been pre-positioned for certain events, such as the Olympics, political conventions, and visits by dignitaries. These resources integrate within the local system and are immediately available, usually hidden away from the public.

Other federal programs include training under the Nunn-Lugar-Domenici legislation, which was passed in September 1996 (PL 104-201). The law, known as the Domestic Preparedness Training Initiative, mandated that the Department of Defense provide training to the 120 major cities across the United States. It was intended to enhance the capability of the local, state, and federal response to incidents involving NBC materials. Part of this process is an assessment after the training to determine the cities' capabilities. This program has been moved to the DHS and is coordinated with the Department of Justice. It has been expanded beyond the original 120 cities, and there are a number of training initiatives around the country for all first responders. These training programs are coordinated through the emergency management agencies in each state. The training is free, and there are a number of quality courses held throughout the United States.

One other federal asset is the National Guard Civil Support Teams (CST) that are set up like a local hazardous material response team. These teams were formerly known as Rapid Assessment and Initial Detection teams (RAID), which were designed to be placed in each of the FEMA regions. They are comprised of 22 persons and carry detection devices, protective clothing, state of the art communications abilities, and decontamination equipment. There are 55 such teams across the United States, activated by the states, even though they are federally funded (although some states have implemented teams that are not funded by the federal government). Also set up around the country are Metropolitan Medical Response Teams (MMRT), which are teams of 129 people in each area. They are trained and equipped to handle the medical component of a terrorism event. The plan is to have them in the 120 largest metropolitan areas in the country.

BASIC INCIDENT PRIORITIES

Rescue live victims using full protective clothing and SCBA.

■ All terrorist agents are predominantly hazardous through inhalation. Avoid touching any unknown liquids or solids, as most are also toxic through skin contact.

■ If there are live victims in an area, then the atmosphere cannot be that toxic.

Use a quick in/quick out approach.

■ Do not treat victims; remove them from the area.

　■ Swoop and scoop is the order of the day.

　■ Any possible fatalities should be left in place, as they are considered evidence.

- Remove the live victims, and then check the vital signs outside the hazard area. Do not take the time to treat the victims in the hazard area.
- Keep in mind the terrorist may be among the injured.
- Ensure police presence in and around the event.
- Watch for secondary devices.
  - Use bomb-sniffing dogs to search for secondary devices.
  - Avoid staging all equipment together.

Request HAZMAT and bomb squad assistance. The sooner you can eliminate the potential for chemical agents or a secondary device, the better off you will be. The HAZMAT team may have the ability to detect terrorism agents, and most HAZMAT teams are working with bomb squads on a more frequent basis.

Limit personnel operating in the hazard area. Establish multiple staging areas—out of the line of sight.

Notify your emergency management agency so that it can mobilize the state and federal resources.

If a building has collapsed or there is potential for a building collapse, request a tactical rescue team or a USAR team to assist.

Isolate all victims, separating contaminated from clean victims.

Establish a safe triage, treatment, and transport area—away from the impact or hot zone.

Notify all area hospitals of the incident.

Remember that the incident is a crime scene and make provisions to preserve as much evidence as possible.

If you suspect chemical agents, use the DOT ERG or other reference sources, such as the *Management of Chemical Warfare Agent Casualties* guidebook, to guide you through safety precautions and patient treatments.

SUMMARY

Within each community there exist multiple agencies that may respond to a terrorist attack. One of the major issues is the coordination of those resources. The response to a potential terrorism incident can be very challenging, and every responder must be aware of its possibility. To be safe, wear all PPE. Do not

delay in the environment, and once the live victims are out, relocate to a safe area. Be aware of the potential for secondary devices, and request HAZMAT and bomb squad assistance quickly. There is potential for one to thousands of patients who may be injured or dead. Some scenarios involve tremendous loss of life, including a large number of responders. One of the major challenges involves the evidence collection process and collection of potentially hazardous evidence. The possibility exists that we may lose and the terrorist may win, but we can avoid this outcome by training, planning, and preparing for such an incident. There are a number of federal resources that can assist with the response to a terrorist attack, or even a suspected attack. Working with and coordinating this response are important considerations. Special response companies or teams should know who their local contacts are and should develop a relationship with the federal liaisons. As the FBI has the responsibility and is the lead agency to investigate an act of terrorism, responders should know that the FBI has WMD coordinators who are responsible for interacting with local responders to help coordinate the overall response.

KEY TERMS

Advanced Portable Detection 2000   Is a warfare agent detection device that also monitors for mace, pepper spray, gamma radiation, and some toxic industrial chemicals.

Anthrax   Naturally occurring biological material that is severely toxic to humans; common agent cited in hoax incidents.

AP2C   French chemical agent meter (CAM).

APD-2000   See Advanced Portable Detection 2000.

Biological agents   Toxic substances that are living materials or are obtained from living organisms.

Blister agents   See *Vesicants.*

Blood agents   Chemicals that affect the body's ability to use oxygen and in many cases prevent the body from using oxygen, producing fatal results.

Butyric acid   Fairly common lab acid used in many attacks on abortion clinics; although not extremely hazardous, it has a characteristic stench that permeates the entire area where it is spilled.

Choking agents   Chemicals that produce an immediate reaction (coughing, difficulty breathing); terrorism agents that are considered choking agents are chlorine and phosgene, both very toxic gases.

Handheld Assay   Detection device for biological agents, such as anthrax.

HHA   Handheld assay.

IMS   See *Ion mobility spectrometry.*

Incendiary agents Chemicals used to start fires, the most common being a Molotov cocktail.

Ion mobility See *Ion mobility spectrometry.*

Ion mobility spectrometry Is a detection technology that measures the travel time of ionized gases down a specific travel path.

M-8 paper Is a paper detection device like pH paper that detects liquid nerve, blister, and VX agents.

M-9 paper Is a detection device that is used for liquid nerve and blister agents; comes in a tape form.

Mass casualty Incident in which the number of patients exceeds the capability of the EMS system to manage the incident effectively. In some jurisdictions this can be two patients, while in others ten may qualify as a mass casualty.

Nerve agents Very toxic materials used by the military to attack opposing troops; can be equated to high strength pesticides.

Nuclear detonation Explosion caused by a nuclear device.

PCR Polymerase chain reaction, a detection method for biological agents.

PDD 39 Presidential Decision Directive 39, which established the FBI as the lead agency in terrorism incidents; the FBI is responsible for crisis management; and FEMA is the lead for consequence management.

Radiological dispersion device (RDD) Radioactive material that is coupled with an explosive device, which spreads the radioactive material. It is not a nuclear detonation, as the explosion is caused by a standard explosive charge.

Ricin Biological toxin that can be used by a terrorist or other person attempting to kill or injure someone; the easiest terrorist agent to produce and one of the most common.

Toxins Substances that are biological in nature and are obtained from naturally occurring materials.

Vesicants Group of chemical agents that cause blistering and irritation of the skin; commonly referred to as *blister agents.*

Weapons of mass destruction (WMD) One of the terms used to describe military, chemical, or biological weapons systems.

REVIEW QUESTIONS

1. What are four potential targets?
2. Name four indicators of potential terrorist activity.
3. What are the most common agents that are readily available and could be used in a terrorist attack?
4. Who are the three main local/regional agencies or groups that should be notified immediately of a suspected terrorist attack?
5. What is the process for requesting federal assistance?
6. Out of the BNICE agents, which is the most likely to be found at an incident?
7. What is the second most likely agent?
8. Which agent is designed to kill immediately?
9. What are the immediate signs and symptoms of sarin exposure?
10. Who is the lead agency to investigate acts of terrorism?
11. What is significant of category A biological agents?
12. Which biological toxin is used most of the time in the United States?
13. Which is more toxic, ricin or sarin?
14. What are the outward indicator differences between a chemical attack and a biological attack?

ENDNOTES

[1]Roe vs. Wade was the U.S. Supreme Court decision in 1973 that allowed legalized abortions. This case has created much of the turmoil on both sides of the abortion battle.

Suggested Readings

Bevelacqua, Armando S., *Hazardous Materials Chemistry,* 2nd ed., Delmar Cengage Learning, Clifton Park, NY, 2006.

Bevelacqua, Armando S., and Richard Stilp, *Terrorism Handbook for Operational Responders,* 2nd ed., Delmar Cengage Learning, Clifton Park, NY, 2004.

Bevelacqua, Armando S., and Richard H. Stilp, *Hazardous Materials Field Guide,* 2nd ed., Delmar Cengage Learning, Clifton Park, NY, 2007.

Buck, George, *Preparing for Terrorism: An Emergency Services Guide,* Delmar Cengage Learning, Clifton Park, NY, 1998.

Buck, George, *Preparing for Biological Terrorism: An Emergency Services Guide,* Delmar Cengage Learning, Clifton Park, NY, 2002.

Buck, George, Lori Buck, and Barry Mogil, *Preparing for Terrorism: The Public Safety Communicator's Guide,* Delmar Cengage Learning, Clifton Park, NY, 2003.

Hawley, Chris, *Hazardous Materials Incidents,* 2nd ed., Delmar Cengage Learning, Clifton Park, NY, 2004.

Hawley, Chris, Michael Hildebrand, and Gregory Noll, *Special Operations: Response to Terrorism and HazMat Crimes,* Red Hat Publishing, Chester, MD, 2002.

Hawley, Christopher, *Hazardous Material Air Monitoring and Detection Devices,* 2nd ed., Delmar Cengage Learning, Clifton Park, NY, 2006.

Henry, Timothy V., *Decontamination for Hazardous Materials Emergencies,* Delmar Cengage Learning, Clifton Park, NY, 1999.

Medical Management of Biological Casualties, U.S. Army Medical Research Institute of Infectious Diseases, Ft. Dietrick, Fredrick, MD, 1996.

Medical Management of Chemical Casualties Handbook, U.S. Army Chemical Casualty Care Office, Medical Research Institute of Chemical Defense, Aberdeen Proving Ground, MD, Sept. 1995.

Pickett, Mike, *Explosives Identification Guide,* Delmar Cengage Learning, Clifton Park, NY, 1999.

Scnepp, Rob, and Paul W. Gantt, *Hazardous Materials, Regulations, Response & Site Operations,* Delmar Cengage Learning, Clifton Park, NY, 1999.

Smelby, L. Charles, Jr., ed., *Hazardous Materials Response Handbook,* 3rd ed., National Fire Protection Association, Quincy, MA, 1997.

Stilp, Richard, and Armando Bevelacqua, *Emergency Medical Response to Hazardous Materials Incidents,* Delmar Cengage Learning, Clifton Park, NY, 1997.

Stilp, Richard, and Armando Bevelacqua, *Citizen's Guide to Terrorism Preparedness,* Delmar Cengage Learning, Clifton Park, NY, 2003.

HAZMAT AND LAW ENFORCEMENT

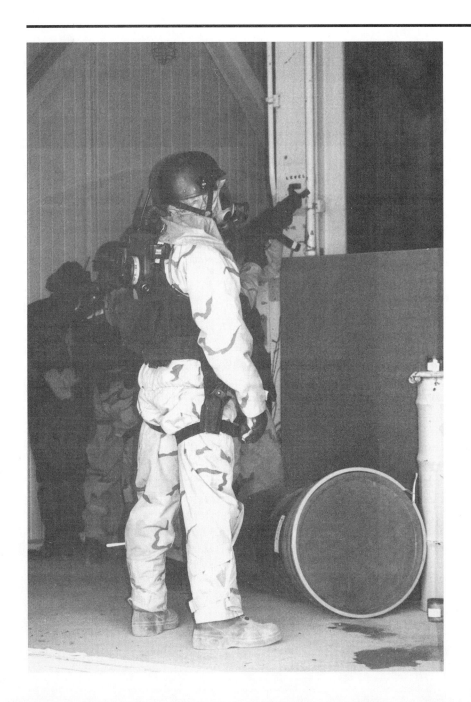

OUTLINE

▌ **Objectives**

▌ **Introduction**

▌ **Common HAZMAT Incidents for Law Enforcement**

▌ **Clandestine Labs**

▌ **Explosives Incidents**

▌ **SWAT Operations**

▌ **Evidence Collection**

▌ **Summary**

▌ **Key Terms**

▌ **Review Questions**

STREET STORY

Two stories are provided with regard to law enforcement officers and hazardous materials, both the result of tunnel vision on the part of the officers. The first involves a direct chemical exposure in which two officers who were staking out a farmer's field were sprayed by a pesticide. Fortunately, after being sprayed the officers had the foresight to seek medical attention. As a general rule, pesticides are designed to kill, although their toxicity to humans varies from compound to compound. These two officers were becoming symptomatic from their exposure and were having a reaction. One of the common reactions is difficulty breathing, followed by the cessation of breathing. Luckily, the medical crew recognized the signs and symptoms and took appropriate action. The actions required to save the officers' lives were to strip them, rinse them off, and scrub them for a period of time, then medicate them and send them to the hospital for further decontamination and treatment.

The other story relates to a traffic stop that was a couple of weeks after the bombing of the Alfred P. Murrah Building in Oklahoma City. At about 3 A.M. a patrol car stopped a rental truck on a residential street just off a main road for erratic driving. The officer approached the vehicle, and when he talked to the driver the officer noted the driver's slurred speech and that there was a chemical odor coming from the truck. Another officer arrived to assist and decided to open the back of the truck. In the back of the truck were about fifteen plastic 55-gallon drums, none of which were marked. Although a lot of details were not known at the time about the bomb used at the Murrah Building, it was known that it was a rental truck with 55-gallon drums in the back of the truck. The officers removed the driver of the truck to a patrol car and called for the fire department. When the fire engine arrived, they noticed the contents of the truck, quickly backed up, and requested that a hazardous materials unit respond. Through radio communication the engine asked that the suspect remain on the scene and that the officers who had been in proximity to the truck remain isolated. The HAZMAT crews arrived and concurred with the engine company that the drums in the back of the truck were a scary combination and a little too close to the McVeigh bombing. Of utmost concern were the contents of the drums. The HAZMAT crews wanted to speak to the driver of the truck both to gain information and to determine if his erratic driving was caused by the fumes in the drums or from the suspected alcohol intake. The HAZMAT crews were unable to locate the driver of the truck as he had been taken to the local precinct, along with the first arriving officers. It was not known at this point if the suspect was contaminated, if he had cross-contaminated the arresting officers, or if the contamination had spread. Other HAZMAT crews had to be requested to respond to the precinct to assess that situation, while the first responding HAZMAT crews had the task of seeing what was in the truck. The contents of the drums were a mixture of waste pesticides and insecticides. The suspect would collect various chemicals and place them into a drum mixed with other chemicals. He was operating a pesticide removal business illegally and was not following any labeling, storage, or transportation regulations, creating a potentially dangerous situation. It was determined that his erratic driving was probably a combination of alcohol intake and chemical exposure, since he was over the DUI legal limit and did need medical treatment for his exposure. The officers involved were also decontaminated and treated at the hospital.

—Christopher Hawley

AUTHOR'S NOTE Although this section is labeled for law enforcement officers, it has applicability for all emergency responders and should be included in all training programs. The material in this section would be considered at the operations-plus level.

OBJECTIVES

After completing this chapter, the reader should be able to identify and explain:

- Types of HAZMAT situations that are common to law enforcement (A)
- Why the use of universal precautions is important (O)
- Types of labs that may be encountered (O)
- Special requirements for SWAT operations (O)
- Considerations at suspected bomb incidents (O)
- Considerations for evidence collection in hazardous environments (O)

INTRODUCTION

Getting shot at is an immediate and very real threat to a law enforcement officer's life. A chemical exposure can have that same type of immediate threat, but more commonly, it may present a long-term hazard that may take years to take a law enforcement officer's life. Just entering a suspect's home to make an arrest places the officer in danger of a chemical exposure that could have long-term effects, as shown in **Figure 8-1.** Law enforcement officers in the United States have always had some involvement with hazardous materials; however, it was usually limited to traffic control or security issues. Throughout the United States they encounter drug labs on a frequent basis, which are very dangerous operations. Once what was a West Coast problem, are now very common in the Midwest and are becoming more frequent in the East. Even without drug labs law enforcement officers are getting more involved with hazardous materials response. In some states the local or state law enforcement agency is the regional HAZMAT team. In other cases law enforcement agencies are training certain components of their departments for some type of HAZMAT response. What most law enforcement officers are going to learn from this chapter is the fact that they have been involved with hazardous materials for many years, but the question that needs to be asked is whether they were properly prepared and protected.

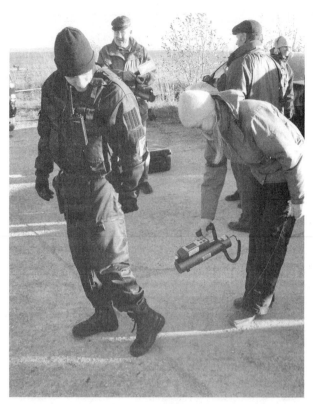

Figure 8-1 A radiation specialist checks a tactical officer for radioactive contamination. Without checking with a radiation monitor the officer could take home lethal amounts of radiation endangering himself, his co-workers, and family.

COMMON HAZMAT INCIDENTS FOR LAW ENFORCEMENT

When law enforcement officers respond to auto accidents, medical calls, fires, or known chemical emergencies, they are at risk for chemical exposure. All of these present some type of risk for the law enforcement officer. In many cases the law enforcement officer is the first arriving responder. The actions provided in Chapter 5 should be followed by this first arriving responder.

Actions taken by the first arriving responders who may not have appropriate PPE should be limited to isolating the event, requesting appropriate resources, and then coordinating the arriving resources.

SAFETY Since many law enforcement officers have limited protective clothing, they should avoid all types of chemical exposures.

SAFETY Going into a chemical situation without respiratory protection is like going into a gun battle without bullets or a bullet-proof vest.

All responders, whether they are fire, EMS, or law enforcement officers, require some form of HAZMAT training. Chapter 1 provides more detailed information on the training levels, but the specialty areas described in this chapter require some additional training. Just entering a building that has had or potentially could have a chemical release requires a minimum of operations-level training. If the officer is going to collect evidence that may involve hazardous materials, then some additional training is required, which may involve the officer becoming a HAZMAT technician. The investigators who are handling the drum dump, which is an environmental crime, as shown in **Figure 8-2,** are required to have HAZMAT

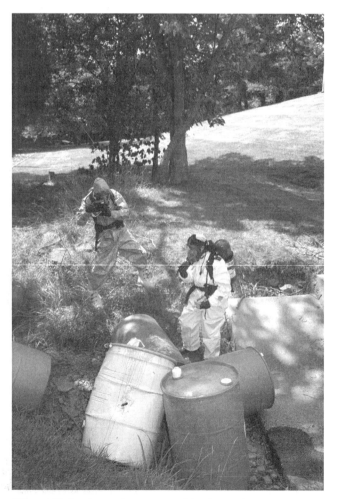

Figure 8-2 Investigators at this drum dump exercise are developing the basis for a sampling plan. They are using air monitors, Polaroid photographs, a site map, and container listings to document the site.

training. SWAT operators who may enter a chemical environment require operations training geared toward their work tasks. OSHA requires that all employees work safely and that job-specific training be provided to the employees. This training also requires annual recertification, which must be provided by the employer.

Blood-Borne Pathogens

Probably the most common exposure to hazardous materials for law enforcement officers is to blood-borne pathogens. The materials in this category include blood, saliva, urine, and other bodily fluids. Hopefully, with the awareness of the dangers of hepatitis A, B, and C, human immunodeficiency virus (HIV), and tuberculosis, law enforcement officers will take universal precautions for these materials. Hepatitis C is now becoming more common, and many emergency response workers have suffered some serious health effects from being exposed to hepatitis C. The issue of tuberculosis is one that is probably not thought of by law enforcement officers. Being in close proximity to a victim who is coughing, and is later determined to have tuberculosis, will place the officer in danger of contracting tuberculosis. Fortunately, the process of protecting oneself against these pathogens is relatively easy, just by using universal precautions.

NOTE In 1991 OSHA issued the Blood-Borne Pathogens Regulation, which requires that workers who may be exposed to these types of materials be trained in the hazards of these materials and be provided protective equipment.

Drugs and Chemical Effects

The next biggest exposure issue for law enforcement officers is most likely drug related. Many drug users use chemicals as part of the process to get high, and the drugs themselves are usually toxic, which may present other hazards. The next section provides information related to drug labs, so we will not discuss them in this section. Drug users may use a combination of things to get themselves high, and many of these items are toxic and flammable. In some cases ether will be used to assist in the heating process of several types of drugs. This material in pure form is extremely flammable, and the container may eventually become a shock-sensitive explosive. Most drugs are in solid form, which means that they present little risk unless eaten or touched with bare hands. In some cases the drugs may be stored or used with flammable and toxic liquids.

UNIVERSAL PRECAUTIONS

In many states EMS providers, law enforcement officers, and firefighters are all covered by this regulation. Part of the regulation requires testing for some materials, exposure reporting programs, and the development of an infection control program. The use of universal precautions is to prevent the transmission of infectious material from the patient to the care provider. Being universal, emergency responders always wear some type of protective clothing, usually gloves, when dealing with all patients. A big part of your protection against these materials involves normal hygiene practices, and simply washing your hands helps prevent problems in the future. In situations where fluids are present added protection is necessary, such as glasses or goggles, and a gown or coverall may also be necessary. OSHA developed specific regulations for workers who may be involved with tuberculosis patients. When dealing with known or potential tuberculosis, patients' respiratory protection is crucial for the responders' protection.

The most common time that law enforcement officers could be directly affected by drug use is when a person is **huffing.** When someone is huffing, they typically use a toxic and/or flammable material. Many common household items such as paints, glues, hair spray, and solvents provide the high that some people desire. The HAZMAT response to this action is that these materials are toxic and flammable and are now in the air. In most cases the user will spray the material into a bag to concentrate the vapors, but when they are done, the vapors will remain in the room of use.

SAFETY Officers must be careful with the evidence from these crimes, as a patrol car is a confined space with limited air movement and the fumes could cause problems for the officers in the car.

CLANDESTINE LABS

There are several types of labs that emergency responders are likely to encounter. The most common would be the drug lab, but other possible labs include **explosives labs, chemical weapons labs,** or **biological weapons labs.** All may be booby-trapped or be set up to harm responders. A key safety note is that a booby trap does not know if the person coming through the door is a police officer, firefighter, or EMT. For the most part a biological weapons lab, like the one shown in **Figure 8-3,** can run unattended without any major concern and may be shut down in any number of ways without major consequence. On the other hand, a drug, explosives, or chemical weapons lab should only be shut down by someone trained to do so, because these labs are especially dangerous.

SAFETY All labs have their inherent dangers to responders, and all are very dangerous locations to occupy.

Drug Labs

There are many types of drug labs; just for methamphetamines there are more than six common methods of production. In 1973 the DEA discovered 41 labs, in 1999 they discovered 2,155 labs, and in 2002 they

Figure 8-3 Much of the equipment at a biological agent lab will resemble the equipment shown here. There will be petri or culture dishes and flasks, and incubators. Some of the culture dishes may have a red gum-like substance known as red agar to facilitate the culture growth. In some cases this lab will be isolated from sunlight and temperature controlled.

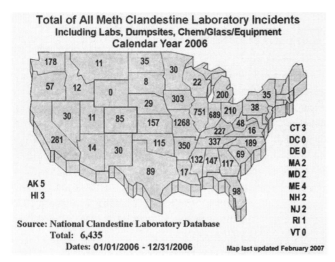

Total of All Meth Clandestine Laboratory Incidents
Including Labs, Dumpsites, Chem/Glass/Equipment
Calendar Year 2006

Source: National Clandestine Laboratory Database
Total: 6,435
Dates: 01/01/2006 - 12/31/2006
Map last updated February 2007

Figure 8-4 Law enforcement officers seized 6,435 drug labs in the United States. Middle America has the largest number of labs, but the East is facing rapidly increasing numbers as well.

raided 12,175 methamphetamine labs. The map shown in **Figure 8-4** shows the location of the 6,435 labs that were seized in 2006, providing proof that the labs are moving eastward at a rapid rate; soon all of the United States will have to deal with this problem. Due to its popularity, meth production is becoming common; however, it does involve a dangerous production process. The production of drugs requires the use of many chemicals.

These chemicals can be purchased outright, stolen, or manufactured using other chemicals. Because many drug-producing chemicals are hot listed or cannot be purchased, the producer must resort to innovative methods to produce the chemicals. The "Nazi" method of meth production involves the use of anhydrous ammonia, which is usually stolen from a chemical facility. If a 150-pound cylinder of anhydrous ammonia developed a leak or catastrophically failed, people downwind for miles could be affected, and those in the immediate area would be in grave danger. Some of the common chemicals used in drug labs are found in **Table 8-1.**

Drug labs can be very complicated to set up and can be found in any number of locations, such as homes, barns, hotels, storage units, and even trucks. Emergency responders routinely encounter these labs inadvertently through other responses. Responders who encounter a drug lab should notify their HAZMAT team and bomb squad and the local office of the DEA. The shutting down of a drug lab is a complicated and very dangerous process. It is this heating and cooling process that indicates the type of lab that may be present. When responders see glassware that is distilling chemicals, in other words evaporating a certain component, and then condensing another portion of the original chemical, this is indicative of possible drug production. The end result of the process is a solid form, usually a powder. There will be a production line approach to the glassware, with some solutions being heated while others are

TABLE 8-1

Common Drug Lab Chemicals

Flammables—Ethers (diethyl ether, petroleum ether), ethanol, isopropyl alcohol, and acetone

Corrosives—Hydrochloric acid, methylamine, hydriodic acid, sulfuric acid, nitric acid, sodium hydroxide, piperidine, hydrogen chloride gas, and anhydrous ammonia

Other chemicals—Mercuric chloride, acetic anhydride, and cyanide

Chemicals used to produce methamphetamines (various methods)

Acetic acid	Hexane
Acetic anhydride	Hydriodic acid
Ammonium acetate	Hydrochloric acid
Anhydrous sodium acetate	Iron filings
Benzaldehyde	Lead diacetate
Butylamine	Mercuric chloride
Ephedrine hydrochloride	Nitroethane
Ethanol	Phenylacetic acid
Ethyl ether	Pyridine
Freon	Red phosphorus

COMMON CHEMICALS USED FOR EXPLOSIVES

Acetone	Graphite	Potassium nitrate
Aluminum powder	Magnesium	Potassium perchlorate
Ammonium nitrate	Nitric acid	Sodium chlorate
Ammonium perchlorate	Phenol	Sodium nitrate
Ethyl ether	Picric acid	Sulfuric acid
Glycerine	Potassium chlorate	Urea (or urea compounds)

cooled. At some point during the process gas cylinders may be used, which allow the gas to mix with some other part of the process. Many of the chemicals involved in the production of drugs are flammable, and although many are also toxic, the predominant hazard is flammability.

For the materials that are toxic, as long as SCBA is worn and materials are not touched or eaten with bare hands, they cannot present harm to the responder. It is highly recommended that personnel from the police and emergency services departments receive training regarding drug lab awareness, with some receiving specialized training in this area.

SAFETY When chemicals are being heated or cooled, there is a chance for a violent reaction.

SAFETY When choosing protective clothing to enter a lab, it is important to protect responders against the predominant hazard.

Explosives Labs

Although not as common, it is possible that law enforcement officers would encounter an explosives lab, which is the predominant weapon of choice for a terrorist. An explosives lab can run from a workbench pipe bomb builder to a full chemical production facility. A person building a pipe bomb does not need much equipment other than some simple hand tools, pipes, caps, and some powder. A person making RDX or other more technical explosives will need some more equipment. Depending on the availability of some materials, the bomb maker may have to make some chemical components, as opposed to purchasing them. It is when materials are made at home that the danger increases for a responder. Many of the processes for making the chemical components for explosives are very dangerous and present a lot of

risk for the builder and responders. An explosives lab differs from a drug lab in the fact that most of the processes do not involve heating, cooling, condensing, or distilling. However, some processes, such as those used to make some of the chemicals, do use some of these procedures, so there are no black-and-white rules for identification. The major work at an explosive lab is usually mixing of materials, in most cases solids and liquids. If gases are used, they are usually coupled with the explosive device and may be used to increase the heat of the explosion, boosting its efficiency. Someone who is making explosives will usually have a large amount of powders in their house and will have a large number of chemicals, such as those listed in the box "Common Chemicals Used for Explosives."

Some of the other indicators of an explosives lab will be the presence of ignition devices, boosting charges, or blasting caps. Many people who play with explosives manufacture are doing so to make fireworks and will have cardboard tubes for the explosives to be placed into. In one case it was originally thought that the explosives maker was just making fireworks, but some other explosive devices were found to have BBs and nails taped and glued to the outside of the explosive. Neither of these has an effect on the display ability of the explosive device and are only designed to kill or maim. When an explosives lab is discovered, the local bomb squad and the local HAZMAT team should be called in to assess the materials. Since the stability of many of the materials cannot be assured, it is usually necessary to destroy many of the found materials in a remote region.

Terrorism Agent Labs

These are the least likely encountered labs, but responders must be aware of their possibilities and of the unique features of these types of labs. The two types of terrorism agent or WMD labs are chemical weapons and biological weapons. Statistically, the most likely lab is a biological toxin lab, which may

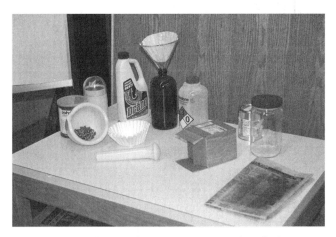

Figure 8-5 Shown above is an example of what a ricin lab may look like, with some representative equipment and chemicals.

be used to manufacture ricin, as shown in **Figure 8-5**. The FBI arrests a few people a year for the possession of ricin, usually accompanied by a threat. Biological labs may be set up to attempt to make other biological materials. Other than ricin, botox, and a few other biological materials, the manufacture of biological weapons is very difficult. Chapter 7 provides a box "Assessing Terrorist Threats" about this subject. The manufacture of ricin is a simple process requiring only a few items, such as castor beans and some readily available chemicals. The process for making some of the more advanced biological agents such as anthrax is more difficult and requires a higher level of education, sophisticated equipment, and access to raw materials. The development of biological agents involves culturing of the material, usually in a petri dish. These petri or culturing dishes may be kept in an incubator or oven-like device to keep the material warm and at a constant temperature. Depending on the type of agent, there may be grinders, dryers, and sieves present to finish off the product. The major route of entry for many of these products is through contact or ingestion. The only inhalation concern would be during the grinding process, but a simple HEPA mask offers more than enough protection for a responder. Some of the materials used as part of the process may be flammable with some toxicity, but the chemicals used to clean glassware and tools are usually highly corrosive and in most cases will be sodium hydroxide (lye) and/or bleach. A bio weapons lab will involve some chemicals, but in some cases it resembles a greenhouse more than a chemical lab. Some biological materials are sensitive to light and will be kept in the dark; therefore, the characteristic distilling and condensing glass apparatus will be missing.

A chemical weapons lab will have two methods of production: the development of new chemical agents through standard production methods or the synthesis of existing materials. The development of new chemical agents is the more difficult of the two processes and is nearly impossible except for someone with a chemistry background and access to some chemistry equipment. The recipes for chemical weapons are fairly sophisticated and require access to many raw materials that are not listed. There have been only two arrests of persons in the United States for attempting to make a chemical weapon such as sarin, a nerve agent. Both were arrested for ordering the raw materials, and it is thought that the two individuals did not have the mental capacity to manufacture the agent. The production of these agents can be very risky to the people involved, and without safety precautions they may end up killing themselves in the process. Some of the off gases from the production of warfare agents are extremely dangerous.

The most probable scenario for the development of a chemical warfare agent is through synthesis of an existing product. The criminal would take existing materials, which are usually in diluted form, and synthesize them or reduce them down to a concentrated product. Luckily, most if not all existing products do not present much risk to humans, as they are strictly engineered to harm only bugs. This would not, however, stop a terrorist from attempting to use one of these products in an illegal manner. There are some pesticides on the market that are applicable to this type of scenario. The standard household pesticide usually has 0.5 or 0.05 percent of pure product mixed with an inert ingredient. A mixture that is used by a farmer may be in the order of 40 to 50 percent pure form, which is then diluted in the farmer's tank. In order to be more effective on humans the pesticide must be concentrated and not diluted. The criminal must devise a method of removing one or more of the inert ingredients. Since the inert ingredient is usually flammable or combustible, this is not that difficult of a task. A mechanism must be devised to off-gas the inert ingredient and then capture the pesticide and collect it.

Pesticides that are already concentrated are known as technical grade pesticides. These technical grade pesticides are in pure form and are not diluted in any way. These materials are not in common use, but can be found in and around the United States. One type, known as phorate, which is made in the United States for use overseas, is a common example. It is frequently shipped in 55-gallon drums in sea containers because it is a commercial equivalent to VX nerve agent. It is almost as toxic and has the same immediate effect on humans as does VX.

EXPLOSIVES INCIDENTS

Bomb Squad and HAZMAT Interface

The tasks of bomb squads and HAZMAT teams are combining, and in many cases these two specialty groups are working hand in hand (**Figure 8-6**). Many bomb technicians are becoming HAZMAT technicians, while many HAZMAT technicians are learning more about the bomb squad duties. One cannot replace the other, but the best approach is a team or coordinated approach to an explosive device that may be chemical in nature. The division of duties is pretty clear: The bomb squads handle anything explosive, and the HAZMAT teams handle the chemicals. The problem comes into play when the device combines both of these hazards. Preplanning and training are key to success at one of these events. A disturbing trend, although slight, is the use of chemicals and explosives. The building of an explosive chemical dispersion device is complicated, and unless done just right, it will not disperse the chemical agent. Just the search of a suspect's home is complicated as both the bomb tech and the HAZMAT tech are looking for two entirely different hazards. When chemicals or explosives are suspected, the best approach is to use people from both of these disciplines to make the first search. The level of protective clothing is determined by the hazards presented, but may involve a bomb suit with a chemical suit underneath and a SCBA for respiratory protection.

There are situations in the HAZMAT world in which the tools used by a bomb technician have some applicability. The most common tool would be the remote control robot, which has live video capability and can be specially outfitted with air monitors that use radio telemetry to communicate with a computer to provide air monitor readings and images of a HAZMAT situation. The pan disrupter, or water cannon, is another tool that may be used to open dangerous containers from a remote location. In some special situations a vent and burn may be necessary, which is a job suited for a bomb technician and other tank specialists.

General Rules

Law enforcement's actions are limited at known explosives events, usually to evacuation, isolation, and crime scene tasks. The bomb squad will be called to handle the actual device. Even if a bomb dog did not hit on a package, only a bomb technician should handle the package. Local protocol will dictate how responders respond to a bomb threat, and as each jurisdiction has differing thoughts on the subject, it is not discussed here. But if a suspicious package or device is found, then responders should evacuate everyone from around the package and isolate the area. The distances used for these types of events are provided in **Table 8-2;** however, always consult with the local bomb squad for its preferences. The entry of a bomb technician to examine a suspicious package is equivalent to that of a HAZMAT technician going in to look at a leaking container. Both are risky, but both are very comparable in setup and mitigation. If one were to follow standard HAZMAT practices for a suspected package, the end result would be a safe and efficient operation. There is the same need for isolation, identification, protective clothing, monitoring, scene setup, mitigation, and termination. Other responders can assist in the handling of explosives events by following the same protocols and making sure the scene is safe.

SAFETY As tempting as it may be for a responder to touch, open, or move a suspicious package, this job should be left to a bomb technician who will most likely open the package remotely.

Special Situations

One of the major reasons that HAZMAT and bomb technicians are coordinating their efforts is the increase of chemical bombs. Another possibility is that criminals may use their knowledge of bomb squad operations to their advantage. Much of what a bomb technician does to a suspicious package may be explosive, or at least provide an ignition source,

Figure 8-6 A bomb technician is assisting law enforcement HAZMAT technicians with a suspicious item in an illegal laboratory.

TABLE 8-2

Explosives Standoff Distances			
Threat Description	Explosives Capacity[1]	Building Evacuation Distance[2]	Outdoor Evacuation Distance[3]
Pipe bomb	5 pounds or 2.3 kilograms	70 feet or 21 meters	850 feet or 259 meters
Briefcase or suitcase bomb	50 pounds or 23 kilograms	150 feet or 46 meters	1,850 feet or 564 meters
Compact sedan	500 pounds or 227 kilograms	320 feet or 98 meters	1,500 feet or 457 meters
Sedan	1,000 pounds or 454 kilograms	400 feet or 122 meters	1,750 feet or 534 meters
Passenger or cargo van	4,000 pounds or 1,814 kilograms	640 feet or 195 meters	2,750 feet or 838 meters
Small moving van or delivery truck	10,000 pounds or 4,536 kilograms	860 feet or 263 meters	3,750 feet or 1,143 meters
Moving van or water truck	30,000 pounds or 13,608 kilograms	1,240 feet or 375 meters	6,500 feet or 1,982 meters
Semi-trailer	60,000 pounds or 27,216 kilograms	1,570 feet or 475 meters	7,000 feet or 2,134 meters

1—Based on maximum volume or weight of explosives (TNT equivalent) that could reasonably fit into the briefcase or vehicle.

2—Governed by the ability of an unstrengthened building to withstand severe damage or collapse.

3—Governed by the greater of fragment throw distances, glass breakage, or falling glass hazard distance. Note that pipe bombs and briefcase bombs assume cased charges which throw fragments farther than vehicle bombs.

CASE STUDY

The bomb squad was requested to a federal facility where a package was thought to contain an explosive device. The federal facility was one where the package was X-rayed as was all of the mail coming into the facility. The foot locker, which had a legitimate return address, had been overnighted to a specific person within the agency. When X-rayed, the security people saw what they thought was an explosive device. The bomb technician looked at the X-rays and determined that in addition to bomb-like items there were containers of chemicals present. The X-rays showed batteries, wires, jars with chemicals, pipes, and compressed gas cylinders. He requested the assistance of the HAZMAT team because this was a combination device. The decision was made to move the package to a remote location where it could be opened. When it was opened, it was determined that junk parts from several motorcycles were in the trunk, which included batteries, wires, mufflers, spray cans, and oil. Following discussions with the security people, it was noted that this was not the first time this person had mailed weird packages to the building. His previous packages included cat and dog body parts, all mailed to his former supervisor. Since no threats were communicated, there was no crime involved in this junk mail case. The concern though is that when the next package arrives, the security personnel might downplay the need for proper safety precautions, and the next package may be the real deal.

which could potentially level a building if the criminal left a flammable material behind. The remote opening of a package may be just the "booster" charge a large explosive or flammable device needs to be detonated. Another possibility is that the bomb technician through the remote opening of the package becomes the dispersal method for a chemical agent that can then hurt or harm all of the investigators. Keep in mind that the criminal may not want to hurt or harm anyone just then; instead he may be looking for the harm to show up years later.

There have been several recent events involving chemicals that were either the initiating device or were the target of an explosive charge. A cylinder can be ruptured and land upward of a mile, and a 55-gallon drum can be ignited and fly several thousand feet. One recent threat involved an explosive device that was to be placed on a liquid propane tank. A household-size propane tank, such as that used for a barbeque grill, if ruptured provides a kill zone from the flame 35 feet in diameter. The target of the propane attack just north of Sacramento was an 8-million-gallon tank of liquid propane. As with the placement of an improvised explosive device (IED) on a cylinder, tank car, or facility tank, there is an optimal location to place the device. There are specific methods to ensure detonation, which require specific devices, placements, and weather.

If the tank in Sacramento had been detonated, the crater would have been several miles in diameter. There was a threat a few years ago by several people who were going to place an IED on a hydrogen sulfide tank to create a diversion so they could rob an armored car. The plot was discovered, and arrests were made prior to the device being placed.

A potentially growing threat will be the use of a radiological dispersion device (RDD), which has seen some use overseas. In this country there have been radiological materials used in crimes against individuals, but none involving terrorism. One thought is that the use or threatened use of an RDD will be on the increase and that it will be used in the place of anthrax hoaxes. The bomb tech and the HAZMAT tech will need to coordinate their efforts to handle this type of threat.

SWAT OPERATIONS

Intel

One important segment to any SWAT or tactical operation is the gathering of intelligence prior to the assault. When chemicals are added to the mixture of hazards, this can further complicate the arrest. It is possible that the chemicals could be used against the arresting officers to thwart the assault. Just being in the proximity of some chemicals may place the SWAT team in harm's way. Hopefully, through intelligence gathering the presence of chemicals can be discovered and research on the chemical hazards can be done before the entry. If the chemicals are only present in the building, then normal SWAT operations can occur to make the arrest. If the chemical could be booby-trapped or used as a protection mechanism, then additional planning is necessary. If the assault is going to take place at an industrial location or one that stores or uses chemicals, then information may be available for planning purposes. Chapter 1 provides some background on the Emergency Planning and Community Right to Know Act (EPCRA), which requires information about chemical storage at facilities. If the facility is required to report under this act, then there is a good bit of information that is available on the facility, the chemicals, and where and how they are stored. If a suspect has stolen chemicals from a facility, then the list of potential materials is also available through this information. Every commercial establishment is required to keep a list of chemicals stored on site, and is supposed to have Material Safety Data Sheets (MSDSs). The type and the quantity of the chemicals determine if this information is distributed elsewhere. If the facility is covered by the OSHA process safety regulation or the EPA Risk Management Program, then chemical information on the facility is available for employee, emergency responder, and, in some cases, the public's use. Although critical information on the facility is not supposed to be in public view, the facility had to develop a worst-case scenario if a major accident occurred. Part of the development of the worst-case scenario in the facility is to identify the specific steps that could be taken to make the worst-case scenario come true. Although not for public viewing, it is available for employees and emergency responders, and is the "how-to" information to destroy the plant and, in some cases, cause havoc in the community for many miles. Either the SWAT intelligence component or the terrorism division of the police department should be aware of all the facilities in their area that could have that kind of impact.

NOTE A 90-ton railcar of chlorine, if attacked in the proper fashion, could create serious effects, including fatalities in a path of about 2,000 feet by 6 miles, and could impact up to 14 miles downwind.

Special Considerations

The planning for a SWAT OPS that involves chemicals will have some additional steps to plan for. The basic assault plan will remain the same, with contingencies in place to deal with the chemicals. Many of the issues with the chemicals will be implemented only if the chemicals leak or are caused to be released. The biggest threat in these operations will always be the two-legged threat, with a bullet being the primary killing agent. When dealing with chemicals, even those set to be released on the SWAT operators, the mission is such that the entry can be made and the subject apprehended, but a quick exit will be necessary to make the operation safe.

There are some SWAT teams across the country practicing in Level A chemical protective clothing to perform their assaults. There is nothing stealthy about wearing a Level A suit. It is big, bulky, noisy, and usually a bright neon color, and the heat stress is tremendous. It is nearly impossible to shoot wearing this type of garment. The practical side of SWAT entries has the SWAT operators wearing military-style protective clothing such as the Saratoga gear. The key to a SWAT operator's protection is respiratory, and the military version has the SWAT operator wearing a **Powered Air Purifying Respirator (PAPR),** which is an APR with an air pump attachment as in **Figure 8-7.** Many SWAT teams use a gas mask (also known as an APR) when making entries. These masks require the wearer to inhale deeply with some effort to get a breath of air, while a PAPR provides a steady stream of cool air across the face, which can be easily inhaled.

When making an entry into a building that may have chemicals, the SWAT operator should have some type of respiratory protection available. When it is likely that the chemicals are going to be used as a protection mechanism, then the SWAT operator should

be wearing military chemical gear and a PAPR. When higher levels of protective clothing are required, it may be best to wait out the suspect or use other means.

The best clue as to the safety of a building is to watch the suspect. If the suspect is walking around in a Level A chemical suit, then you may want to wait prior to entry. If the suspect is not showing any signs or symptoms, then the airborne concentration of the chemical must be low. SWAT operators should have protective clothing in the event that the area is booby-trapped or during the arrest something gets knocked over or spilled. The use of Nomex should be at a minimum, since the fire hazard will probably be the most likely hazard for an operator to encounter.

In many cases entries are made into drug labs or other chemical areas using chemical protective clothing, and the biggest risk to the entry person is the fire risk. Chemical protective clothing and heat do not mix, you could instantly become shrinkwrapped. The determination of protective clothing is based on the type of hazard present. Chapter 4 discusses standard protective clothing, and Chapter 5 discusses rescue operations.

The SWAT operator is performing the same type of operation as a firefighter making a rescue, and the risks are going to be comparable. The biggest change is that the person the SWAT operator is going after has made a conscious decision to be in that environment, and when endangered, the SWAT operator should wait out the suspect. The presence of hostages obviously changes the picture and makes it much more complicated.

SAFETY SWAT operators must not let the chemicals get in the way of their operation; they must remain focused on their primary mission.

SAFETY Training with the gear, especially shooting, is essential for proficiency.

SAFETY Prehydration is very crucial for the crew's ability to survive, and even when training operators must continually hydrate.

Figure 8-7 The use of air purifying respirators adds another element of difficulty to a tactical officer's tasking while arresting high risk suspects. If there is a suspicion of chemicals being present, respiratory protection is essential. The chemical filtering cartridge used with the APR should be matched to the risk, as cartridges protect against different materials.

Setup

When hazardous materials are involved in a SWAT operation, there are a few extra steps that need to be added to the planning process. Several contingencies

PROTECTIVE CLOTHING AND THE SWAT ENVIRONMENT

A SWAT operator should always be able to wear his body armor, and the chemical protection is secondary to gunshot protection. A SWAT operator must be able to see and be able to pull the trigger with a minimum of distraction. The best protection that does offer chemical protection is military gear, known as Mission Oriented Protective Posture (MOPP) gear. Designed for the military environment, which has some crossover into the civilian SWAT world, the military gear offers protection against chemical exposures. The nature of the SWAT entry means that the actual contact time with any chemical will be short. Exceptions to this thought would be a booby trap that covers a SWAT operator or an operator who falls into a chemical release. The military gear is designed for military warfare chemicals and is to be used for long periods of time in a contaminated environment. The SWAT environment is a quick-in/quick-out, hopefully minimizing contact with any chemicals. The standard military issue clothing is the Saratoga Joint Service Lightweight Integrated Systems Technology (JSLIST) suit designed for chemical and biological warfare. It is accompanied by the MCU2P gas mask, which is an air purifying respirator (APR) as described in Chapter 4. The Saratoga comes with a SCALP suit, which is an overgarment used for liquid splashes, and it also comes with chem/bio underwear. The mask, which is called the M40 mask in the military world, can only be used when the oxygen level is above 19.5 percent for non-IDLH atmospheres and nonflammable atmospheres. The cartridge filter is for military chemicals. There are other canisters available for industrial chemicals that may be better suited for a wider range of chemicals. The C2 filter that comes with the mask will protect against military chemical and biological materials, acid gases (hydrogen chloride, chlorine, hydrogen fluoride, sulfur dioxide, sulfuric acid, and nitric acid), fuels, oils, and alpha and beta radiation. It does not protect against ammonia, carbon monoxide or carbon dioxide, nitric oxide, nitrogen dioxide, or metal carbonyls. The basis for the protection for other materials is based on the vapor pressure of the chemical. The gear offers protection for chemicals with a vapor pressure of less than 7.5 mm Hg and offers limited protection for vapor pressures between 7.5 and 75 mm Hg. The higher the concentration, the less time that can be spent in that environment; the lower the concentration the more time that can be spent. Chapter 2 provides more information on vapor pressures. The biggest challenge to this gear would be an ammonia environment; however, a commercial cartridge is readily available to protect against the ammonia. The glove issue is one that is controversial, but the SWAT operator should not be touching any open chemicals. Thus the hand protection should allow the operator to shoot well. A thin nitrile or surgical glove with a Nomex flight glove overcover is adequate protection for a SWAT operator.

If chemical contact or immersion is a real possibility then the hand protection can be increased. As long as there are live people without protective clothing in the environment, no matter what happens, with the exception of an explosion or rapid fire spread, the Saratoga gear along with Nomex jumpsuits would be the preferred gear for SWAT operations. While using this gear you lose vision, communication, stealth ability, and quickness and you gain heat stress.

---

need to be added to the process so that in the event of a release operators are prepared. When chemicals are known to be involved, someone must be assigned to research the materials and learn about their hazards. Storage information, typical containers, basic hazards, and the type of PPE required are all essential information. The operators must be briefed on what to look for, what type of odors may be present, and what conditions would trigger a retreat situation.

The entry into the building may change slightly, especially if explosive breaching is going to be done to gain entry. If flash bangs are going to be authorized, then some additional checks will have to be done. Before detonating a **breach charge** the breacher or someone next in line should check the door area for at least 5 seconds with a multigas monitor as well as a photoionization detector, which is shown in **Figure 8-8.** If the bad guy is sure that breaching or flash bangs are going to be used, then a flammable atmosphere could be created and the SWAT operators provide the ignition source to launch themselves and the building a considerable distance. Most air monitors can be modified to vibrate when in alarm, and the operator should not concentrate on the instrument. He should make the initial check, then fall into place, make the assault, and capture the suspect. If the meter alarms, the operator should evaluate the alarm, determine the risk, and in most cases complete the mission, and then with the suspect evacuate the building. When operating in environments that are likely to have chemical contamination or very toxic materials, the operators should don respiratory protection prior to entry. If they

Figure 8-8 Prior to entering the building a SWAT operator is checking the atmosphere near the door. If conditions are OK, then the green light is given to initiate the entry. In some cases the team may need to withdraw.

choose not to use respiratory protection, then an alarm on the meter will mean that respiratory protection is needed.

Another consideration is decontamination of the operators, hostages, and the suspect. The best group to perform this decontamination is a group of HAZMAT-trained law enforcement officers, who could stand by prior to the assault. The second-best group to perform this task is a HAZMAT team, who could be standing by prior to the assault. Obviously **operational security** is key prior to an assault, so having a good relationship with the local HAZMAT team is crucial. Unfortunately, a large-scale decon setup takes time to set up, which means preparation time is crucial. The local HAZMAT team can provide some guidance based on the chemicals on some decon methods that could be employed while a decon setup is being deployed. Any SWAT operation, whether chemicals are thought to be present or not, should have a contingency for decon. If a suspect throws a jar of sulfuric acid at an officer entering the suspect's bedroom, there should be no hesitation in moving that officer to a location where the officer should be immediately decontaminated. A shower, a hose, or even a simple garden sprayer are all adequate forms of emergency decontamination. The commander must always have a plan to deal with such emergencies, and have personnel assigned to complete those tasks.

Emergency Conditions

The SWAT team enters the door of a house, and as soon as they enter the house, they find a liquid on the floor. If they are wearing the Saratoga gear with respiratory protection, they can continue on to complete the mission. If they are not wearing Saratoga gear or respiratory protection, a decision must be made whether to continue or to evacuate. That decision has to be made with an air monitor to determine if the risk is fire, corrosive, or toxic hazards. No matter the situation, decontamination is crucial because some of the operators will have direct contact with the product. Those who did not come in contact do not need to be decontaminated, but do require a medical evaluation depending on the type of chemical found. There are a number of ways for a SWAT team to be set up and for a criminal to use known entry procedures against the SWAT team. Some criminals may inadvertently have chemicals in their proximity, and the SWAT team through a normal operation ignites them or causes a release, as shown in **Figure 8-9.** Imagine a bank robber who is using acetone to wash the dye pack ink from the money and has the money and acetone in a bucket. The SWAT team enters the building and throws a flash bang near the bucket. Acetone evaporates quickly, is heavier than air, is highly flammable (explosive), and is readily ignited. The resulting

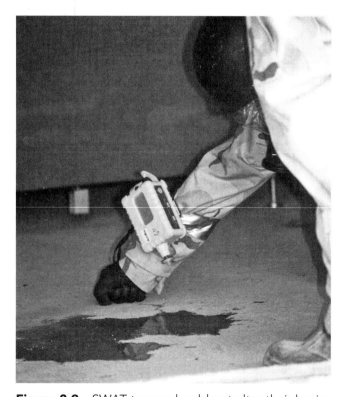

Figure 8-9 SWAT teams should not alter their basic assault plan, just because chemicals may be present. The biggest threat comes from the criminal inside, and the chemicals are a secondary threat. If leaking materials are encountered, then they should be checked for fire and corrosive or toxic hazards after the building has been cleared of the human threat.

explosion could blow the SWAT team across the street and severely damage the building, injuring or killing those in or near the room. The same change of mind-set has to occur with SWAT operators as happened with the bomb technicians.

SAFETY Hazardous materials present some risks in normal handling; however, in the hands of a criminal they may have disastrous effects.

When planning an assault, there should be operators assigned to handle the air monitor(s), operators assigned to check for hazardous materials, as shown in **Figure 8-8** and **Figure 8-9,** along with other normal checks, operators assigned to perform rescues, and some assigned to do decontamination. These operators do not have to stand by while the others make the entry, they can be part of the team. In the event of a contamination, they then switch to these collateral duties as long as the human threat is removed. If intelligence reveals that chemical contamination is possible, then other personnel may need to be added to the assault group, possibly staged away from the event. The local HAZMAT team could be requested to stand by at a nearby location. This can be accomplished without operational security being compromised. Many large-city HAZMAT teams are used as standby details for other search warrants and are provided a time and a location for the standby, with no other details being provided. Knowing the type of chemicals involved and the expected tasks to be performed will speed up the HAZMAT team's response once notified. Liaisons should be established prior to an emergency so that relationships and work tasks are predefined, making the operation safer and more efficient.

EVIDENCE COLLECTION

Considerations

The collection of evidence in a hazardous environment brings a whole host of challenges. Luckily, in most cases the collection of evidence is not time dependent and fingerprinting can often wait until the HAZMAT hazard is reduced or removed. There are cases, though, when time is of the essence and when it needs to be determined if the environment is safe to collect the evidence. In many cases protective clothing must be worn to determine the nature of the hazard and to determine what type of equipment must be used to collect the evidence, much like the **environmental crime** shown in **Figure 8-10.**

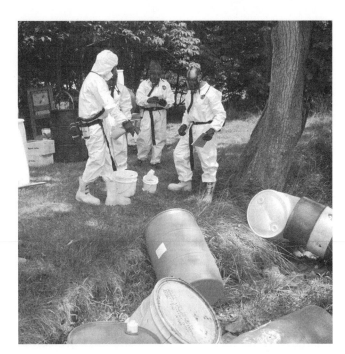

Figure 8-10 An example of a drum dump, in which the investigators are continuing on with the sampling plan that was developed on the previous entry. Drums must be sampled and characterized, and evidence collected.

The considerations for evidence collection are multiple. First and foremost, the persons collecting the evidence must be trained to a level of HAZMAT training. This level is determined by the tasks that are going to be performed by the investigator. If the investigator is collecting normal evidence with chemicals present or will be taking chemical samples from open sources, then the operations level of training is required. If the investigator is going to be specifically collecting chemical evidence or is going to be opening containers, then the technician level is required. Both the operations courses and the technician courses should be geared toward law enforcement tasks, and the operations course specifically should cover standard HAZMAT operations issues and needs to add some specific chemical sampling training to meet OSHA requirements. The tiered approach works well in that operations-level personnel can evaluate the hazards and, if necessary, request additional resources as required. There are not many law enforcement departments that have officers trained to the HAZMAT technician level. Thus, they may need to rely on the local HAZMAT team for this level. The local HAZMAT team can assist and offer guidance, but with some exceptions most HAZMAT teams are not trained in evidence collection. The best approach is to combine the two groups and make a combined entry, with both focusing on their specific tasks. Other possible resources are the bomb technicians, since most are becoming HAZMAT technicians, and the fire investigators, since to investigate a chemical fire they are required to be HAZMAT technicians. The EPA offers evidence collection courses geared toward hazardous materials, and the local EPA contact is a good resource. The local EPA criminal investigator is also a good contact, since most of their investigative work involves the collection of evidence in hazardous atmospheres. The Coast Guard may also have investigators who can provide assistance in this area. Another potential source of assistance is the local or state environmental crime investigators who are also trained in this specialty. Other considerations for evidence collection are the sampling media, tools, and equipment. Each of these materials must be compatible with the chemical so as not to cause a reaction or change in the chemical makeup of the sample. The protective clothing that is used must protect the wearer from the type of hazards that are present. Some form of respiratory protection may be required, which has time limitations and may have other conditions of use, especially when using APRs and PAPRs. The time that personnel can collect evidence is another issue, as crews must be sent in teams of two with a backup team of at least two. If using SCBA, then the time is limited to a maximum of 30 to 40 minutes, if not shorter. Crews must be rotated in because the PPE will wear them down, and rehabilitation and hydration are vital.

Evidence must be collected in compatible containers, and how soon the material needs to get to a lab must be considered. Will the local lab accept this type of material, and are they equipped to store and analyze the material? Labs will have certain requirements for storage containers and may have requirements for certain types of collection media. An important question to ask is how much of a sample is required. Discussions need to take place with the lead investigator and the prosecutor as to the amount of chemicals that are needed for evidence and for prosecution. If the suspect had a gallon jar of a flammable substance, does the whole jar need to be kept or can a photograph and a 40-mL sample suffice for evidentiary purposes?

Sampling and Evidence Collection

One of the major concerns of local law enforcement and the FBI regarding the prosecution of a terrorism case is the purity of the evidence. In order for the evidence to be used in a successful prosecution the evidence must be pure beyond all doubts. If there is any suspicion that the evidence has been tampered with, altered, or contaminated it could affect the outcome

of the prosecution. When responders enter the hazard area there is great potential for evidence to be destroyed. The best way to proceed is to coordinate your efforts with your local FBI WMD coordinator, who can assist you with the preservation of evidence. There are several questions that arise regarding evidence. When are we dealing with a potential crime? When should you concern yourself with evidence? At what point do you follow evidentiary procedures? The easiest answer is to follow the rules for every situation. If a HAZMAT response team has been called to a situation, no matter the circumstances there is a good likelihood that lawyers and the courts may get involved. The case may not be a criminal one, but may be a workman's compensation lawsuit or other civil matter. Responders may be called to testify about their activities or the results of their actions. The responders or their agency may be the target of a lawsuit which may require a defense.

Evidence collection is very process driven, but procedures can be implemented that will help preserve any potential evidence. The gold standard for any courtroom is laboratory analysis, and so evidence must be available to be analyzed by a laboratory. This does not mean that every incident requires that material be sent off to the laboratory, but that steps are taken to preserve some evidence in the event that laboratory testing is required.

All collection tools and containers must be certified clean and/or sterile, and should have lot numbers. They must come from sealed boxes, which have been tracked and have documentation about their certifications. Each new sampling point or test requires a new tool and container. The process must be documented as to who performed which sampling, with dates, times, and specific notes on the actions taken. Each evidence or sampling container must be numbered, with the collector's initials, and date and time included. Collection of evidence uses a process known as clean person–dirty person. The clean person should never come in direct contact with the dirty person or any evidence. The dirty person is the one performing the actual evidence collection, using tools to pick up the evidence. The dirty person takes great care not to cross-contaminate, nor come in direct contact with potential evidence and a contaminated tool.

When there is a large amount of potential evidence the process can be overwhelming. But following a process and a procedure helps eliminate any errors and will preserve evidence. There is great potential to get involved in a court case, and following the proper procedures every time minimizes your risk in the courtroom. The one time that you do not follow evidentiary procedures will be the time that you are grilled in the courtroom for contaminating the evidence. The FBI and local law enforcement have response teams that collect evidence in these situations, and can provide great assistance when trying to process a potential crime scene.

Operations and Entry

Operationally, the setup will be very similar to a HAZMAT accident and will require the same sectoring and operational concerns. The complicating factor in collecting evidence in a HAZMAT situation is that **chain of custody** must be maintained. The actual sectoring of the event will make tracking personnel easier, and access points are usually well controlled at a crime scene and may need to be strengthened when chemicals are involved. Safe work practices need to be employed so that cross-contamination does not occur and that evidence is collected in a manner that cannot be disputed in court.

SAFETY When chemicals are discovered at a crime scene or are known to be present, the first priority is life safety.

Persons in and around the area should be evacuated until the nature of the hazard is determined. This may mean a single apartment or the entire apartment building. The chemical determines the actual size of the isolation zone. Reference books and air monitors can assist in this decision. The ultimate decision needs to be made with an air monitor, as that provides real time on scene conditions. Prior to entry the criminal threat should have been removed and should not be an issue.

Initial entry crews should enter the building in appropriate protective clothing and accompanied by air monitors that protect them against fire, corrosive, and toxic hazards. A minimum backup crew of two is required, and someone outside must be designated as the Incident Commander. If the event is large enough, a separate safety officer must be assigned; if not, the Incident Commander assumes that responsibility. If available, assisting in this entry should be a HAZMAT technician, and if there is any chance that explosives or explosive-making materials may be present, a bomb technician. The hazards should be characterized, and leaking or open containers noted and possibly mitigated. A basic drawing that will be used to develop a **sampling plan** should be completed, and Polaroid photographs should be taken so that other crews can be briefed prior to entry, as is shown in **Figure 8-11.** Although all of this information is discoverable, crews are not yet collecting evidence because this recon is to assist in the sampling plan, which will provide subsequent crews with the

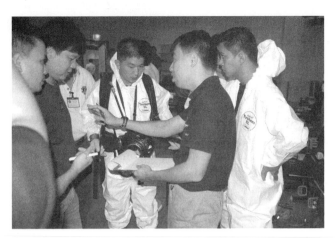

Figure 8-11 A mixed entry crew of HAZMAT technicians and law enforcement officers are briefed on safety concerns, tasks to be completed, and emergency procedures prior to entry.

knowledge of where to begin the evidence collection. Decontamination must be planned for, although unless the initial entry crew steps in something or comes in contact with a solid or a liquid they should not require much, if any, decon. Other crews who will be sampling chemicals will require some decon, but this may be accomplished with dry decon or a bucket to clean their hands. Contingency plans should be developed in the event of a fire, massive release, failure of a container, or gross contamination of an investigator.

An investigator should be assigned to track the entry team, to monitor the time on air or the work time, and to possibly coordinate the evidence coming from the hot sector. There are a lot of logistical concerns, and an investigator should be assigned to logistics to make sure all of the equipment is readily available. If required, the decon crews usually take a beating, since they are dressed in PPE for long periods of time and are usually not switched out as the entry crews are switched. They should be monitored by the safety officer, and someone should be assigned to monitor these crews.

The initial entry crew, in addition to developing the concerns for the development of a sampling plan, should also be determining the need for PPE. If they mitigate some hazards or ventilate, the level of PPE may be able to be downgraded for the next entry. The level of PPE is determined by the type of chemical and the risk that is present. If the chemical released presents a fire risk, then fire-resistive clothing is required, such as a Nomex or PBI jumpsuit or firefighters' turnout clothing. Crews should not be entering a flammable environment with chemical protective clothing. The chemical has to be removed or the building ventilated so as to remove the fire hazard. When dealing with chemicals, respiratory protection is always recommended, but the least amount of protective clothing is better for the entry crews. The other factors such as heat stress, mobility, vision, and hearing also need to be considered in the hazard assessment. Conditions should be reevaluated on a regular basis to make sure conditions have not changed.

SUMMARY

The addition of chemical hazards in the law enforcement field complicates an already complex task. A criminal with bad intentions could wreak havoc with law enforcement officers using chemicals, and officers must be prepared for this possibility. Terrorism has forced HAZMAT teams to become better at their present jobs, and hopefully this trend also moves the law enforcement community to make the same transition. On the federal level several law enforcement groups have been providing HAZMAT training for their investigators, much like the HAZMAT teams across the country. This trend needs to move to the local level for a number of reasons: First, the law requires it, and second, most importantly, it is for your protection.

The unfortunate thing is that some chemicals do not cause noticeable or immediate effects, but instead they may affect your family later or may even kill you 5 or 10 years later.

SAFETY Going into a crime scene ill prepared for chemicals is asking to get shot by a different means.

KEY TERMS

Biological weapons lab   A lab designed to manufacture biological warfare agents.

Breach charge   An explosive charge such as a strip or shape charge that is used to explosively remove a door or window.

Chain of custody   All evidence must be tracked and must have a written description of how it moved from the crime scene to the lab and then to storage. Each time it passes from one person to another it must be documented.

Chemical weapons lab   A lab designed to manufacture chemical warfare agents. This type of lab is either manufacturing the agent from precursor chemicals or is synthesizing commercially available pesticides.

Environmental crime   The spilling, release, or potential release of chemicals into the environment. Littering or discharging chemicals into a sewer, creek, or other waterway is an environmental crime.

Explosives lab   A lab designed to manufacture explosives.

Huffing   Term used to describe the intentional inhalation of paints, solvents, or other chemicals with the intention of getting high.

Mission Oriented Protection Posture (MOPP) gear   Chemical-resistant clothing designed for military operations.

Operational security   Keeping the details of an operation secret and only divulging critical information to those who have a need to know.

Powered Air Purifying Respirator (PAPR)   An air purifying respirator (APR) that has a blower attachment that blows filtered air to a mask.

Sampling plan   A drawing or a map of an incident that indicates all of the containers that need to be sampled. A priority of containers to be sampled and the resources necessary to do that sampling are part of the sampling plan.

Universal precautions   The wearing of PPE designed for blood-borne pathogen protection, such as gloves, gown, and a face mask.

REVIEW QUESTIONS

1. What types of HAZMAT situations are common to law enforcement officers?
2. When should universal precautions be used?
3. What are the three types of labs that may be encountered?
4. Intelligence reveals that chemicals may be used to booby-trap a building, and the SWAT team is developing a plan of action. What special considerations need to be added to this plan?
5. What are the minimum evacuation distances for pipe, car, van, and truck bombs?
6. What are some additional evidentiary considerations for evidence collection in hazardous environments?

Suggested Readings

Bevelacqua, Armando S., *Hazardous Materials Chemistry*, 2nd ed., Delmar Cengage Learning, Clifton Park, NY, 2006.

Bevelacqua, Armando, and Richard Stilp, *A Citizens Guide to Emergency Preparedness*, Delmar Cengage Learning, Clifton Park, NY, 2003.

Bevelacqua, Armando S., and Richard Stilp, *Terrorism Handbook for Operational Responders*, 2nd ed., Delmar Cengage Learning, Clifton Park, NY, 2004.

Bevelacqua, Armando S., and Richard H. Stilp, *Hazardous Materials Field Guide*, 2nd ed., Delmar Cengage Learning, Clifton Park, NY, 2007.

Hawley, Christopher, *Hazardous Material Air Monitoring and Detection Devices,* 2nd ed., Delmar Cengage Learning, Clifton Park, NY, 2006.

Hawley, Chris, Michael Hildebrand, and Gregory Noll, *Special Operations: Response to Terrorism and HazMat Crimes,* Red Hat Publishing, Chester, MD, 2002.

Henry, Timothy V., *Decontamination for Hazardous Materials Emergencies,* Delmar Cengage Learning, Clifton Park, NY, 1999.

Pickett, Mike, *Explosives Identification Guide.* Delmar Cengage Learning, Clifton Park, NY, 1999.

Scnepp, Rob, and Paul W. Gantt, *Hazardous Materials, Regulations, Response & Site Operations,* Delmar Cengage Learning, Clifton Park, NY, 1999.

Smelby L. Charles, Jr., ed., *Hazardous Materials Response Handbook,* 3rd ed., National Fire Protection Association, Quincy, MA, 1997.

Stilp, Richard, and Armando Bevelacqua, *Emergency Medical Response to Hazardous Materials Incidents,* Delmar Cengage Learning, Clifton Park, NY, 1997.

GLOSSARY

8-Step Process© Management system used to organize the response to a chemical incident. The elements are: detect, estimate the harm, choose an objective, identify the action, do the best you can, and evaluate your progress.

Abortifacient A chemical or material that can cause abortions.

Aboveground storage tanks (AST) Tanks that are stored above the ground in a horizontal or vertical position. Smaller quantities of fuels are often stored in this fashion.

Absorbed dose A measure of the amount of radiation transferred to a material.

Absorption Defensive method of controlling a spill by applying a material that absorbs the spilled chemical.

Activity A measure of the number of decays per second that occur with the source.

Acute Quick, one-time exposure to a chemical.

ADR/RID Abbreviations used by the international shipping community and established by a European Agreement. The U.S. DOT is a voting member of this shipping body which governs Regulations Concerning the International Transport of Dangerous Goods by Rail (RID) and Road (ADR).

Advanced Portable Detection 2000 Is a warfare agent detection device that also monitors for mace, pepper spray, gamma radiation, and some toxic industrial chemicals.

Air bill Term used to describe the shipping papers used in air transportation.

Air monitoring devices Used to determine oxygen, explosive, or toxic levels of gases in air.

Air purifying respirators (APR) Respiratory protection that filters contaminants out of the air, using filter cartridges; requires the atmosphere have sufficient oxygen along with other regulatory requirements.

Allergen A material that causes a reaction to the immune system.

Anthrax Naturally occurring biological material that is severely toxic to humans; common agent cited in hoax incidents.

AP2C French chemical agent meter (Cam).

APD 2000 See Advanced Portable Detection 2000.

Awareness level The basic level of training for personnel who during the performance of their job may come across a hazardous materials/WMD release, the basis of which is the ability to recognize a hazardous situation, isolate and deny entry, and call for assistance.

Becquerel (Bq) A measure of radioactivity; the metric version of Curies.

Biological agents Toxic substances that are living materials or are obtained from living organisms.

Biological weapons lab A lab designed to manufacture biological warfare agents.

Biomimetic Form of a gas sensor that is used to determine levels of carbon monoxide. Typically used in residential CO detectors, it closely recreates the body's reaction to CO and activates an alarm.

BLEVE Boiling liquid expanding vapor explosion, which occurs when a pressurized tank is heated and the pressure inside the tank exceeds the capacity of the relief valves to handle the pressure. When pressurized tanks rupture, they can travel significant distances, and if holding flammable materials, they will usually ignite, resulting in a large fireball.

Blister agents See *Vesicants*.

Blood agents Chemicals that affect the body's ability to use oxygen and in many cases prevent the body from using oxygen, producing fatal results.

Boiling point Temperature to which liquids must be heated to turn to a gas.

Breach charge An explosive charge such as a strip or shape charge that is used to explosively remove a door or window.

Building Officials Conference Association (BOCA) A group that establishes minimum building and fire safety standards.

Bulk tank Large transportable tank comparable to but larger than a tote.

Bump test Used to determine if an air monitor is working; will start alarm if a toxic gas is present. This quick check ensures the instrument responds to a sample of gas.

Bung Cap of a drum. The cap varies in size but the most common is 3/4 to 2 1/2 inches.

Butyric acid Fairly common lab acid used in many attacks on abortion clinics; although not extremely hazardous, it has a characteristic stench that permeates the entire area where it is spilled.

Calibration Setting the air monitor to read correctly. To calibrate a monitor, the user exposes it to a known quantity of gas and makes sure it reads the values correctly.

CAM A detection device for nerve and blister agents; there are American, British, and French versions.

Carcinogen Material capable of causing cancer in humans.

Catalytic bead The most common type of combustible gas sensor, which uses two heated beads of metal to detect the presence of flammable gases.

Ceiling level Highest exposure a person can receive without suffering any ill effects; combined with the PEL, TLV, or REL, it establishes a maximum exposure.

Chain of custody   All evidence must be tracked and must have a written description of how it moved from the crime scene to the lab and then to storage. Each time it passes from one person to another it must be documented.

Chemical weapons lab   A lab designed to manufacture chemical warfare agents. This type of lab is either manufacturing the agent from precursor chemicals or is synthesizing commercially available pesticides.

Chemtrec   The Chemical Transportation Emergency Center, which provides technical assistance and guidance in the event of a chemical emergency. This network of chemical manufacturers provides emergency information and response teams if necessary.

Chip measuring system (CMS)   Form of colorimetric air sampling in which the gas sample passes through a tube. If the correct color change occurs, the monitor interprets the amount of change and indicates the gas level on an LCD screen.

Choking agents   Chemicals that produce an immediate reaction (coughing, difficulty breathing); terrorism agents that are considered choking agents are chlorine and phosgene, both very toxic gases.

Chronic   Continual or repeated exposure to a hazardous material.

Clandestine drug labs   Illegal labs set up to manufacture street drugs.

Colorimetric tubes   Used to detect both known and unknown types of gases. A form of detection for toxic gases, which changes color in the presence of a toxic gas.

Combustible gas indicators   Detection devices used to measure flammable gases in air.

Computer Aided Management for Emergency Operations (CAMEO)   Computer program that combines a chemical information database with emergency planning software; commonly used by HAZMAT teams to determine chemical information.

Consist   Shipping papers that list the cargo of a train; the listing is by railcar of all cars.

Containment system   Provided as a backup system in the event of a tank failure. If the tank fails, the containment system is designed to hold the released product.

Convulsant   A chemical that has the ability to cause seizure-like activity.

Cryogenic gas   Any gas that exists as a liquid at a very cold temperature, always below 150°F.

Curies (Ci)   The measure of activity level for radiation sources.

Damming   Stopping a body of water, which at the same time stops the spread of the spilled material.

Dangerous Cargo Manifest (DCM)   Shipping papers for a ship that list the hazardous materials on board.

DECIDE   Management system used to organize the response to a chemical incident. The factors of

DECIDE are: detect, estimate, choose, identify, do the best, and evaluate.

Decontamination   Physical removal of contaminants (chemicals) from people, equipment, or the environment; most often used to describe the process of cleaning to remove chemicals from a person.

Deflagrates   Rapid burning, which can be considered a slow explosion, but travels at a lesser speed than a detonation.

Department of Transportation (DOT)   Federal agency that issues regulations regarding the transportation of hazardous materials.

Diking   Defensive method of stopping a spill; a common dike is constructed of dirt or sand and is used to hold a spilled product. In some cases, in a facility such as around a tank farm, a dike may be preconstructed.

Dilution   Adding a material to a spill to make it less hazardous; in most cases water is used to dilute a spilled material, although other chemicals could be used.

Diverting   Using materials that can move products around or away from an area; for example, several scoops of dirt can be used to divert a running spill around a storm drain.

Emergency decon   Rapid removal of material from a person who requires immediate cleaning. Most emergency decon setups use a single hoseline to perform a quick, gross decon.

Emergency Planning and Community Right-to-Know Act (EPCRA)   The portion of SARA that specifically outlines how industries report chemical inventory to the community.

Emergency response planning (ERP)   Levels that are used for planning purposes and are usually associated with preplanning for evacuation zones.

Encapsulated suit   Chemical suit that covers the responder, including the breathing apparatus; usually associated with gastight and liquid-tight Level A suits. Some Level B styles are fully encapsulated, but not gas or liquid tight.

Endothermic reaction   Chemical reaction in which heat is absorbed and the resulting mixture is cold.

Environmental crime   The spilling, release, or potential release of chemicals into the environment. Littering, discharging chemicals into a sewer, creek, or other waterway, is an environmental crime.

Environmental Protection Agency (EPA)   Federal agency ensuring protection of the environment and the nation's citizens.

Etiological   Hazard that includes biological agents, viruses, and other disease-causing materials.

Evacuation   Movement of people from an area, usually their homes, to another area that is considered to be safe. Persons are evacuated when they are no longer safe in their current area.

Excepted packaging   Used to transport low risk radioactive materials; the packaging is exempted from DOT regulations.

Exothermic reaction Chemical reaction that releases heat, as in when two chemicals are mixed and the resulting mixture is hot.

Explosives lab A lab designed to manufacture explosives.

External floating roof tank Tank in which the roof is exposed to the outside and covers the liquid within the tank. The roof floats on the top of the liquid, which does not allow for vapor build-up.

Extremely hazardous substances (EHS) List of 387 substances that the EPA has determined present an extreme risk to the community if released.

Fine decon More detailed form of decontamination, usually performed at a hospital by staff who are trained and equipped to perform decon procedures.

Firefighters' protective clothing (FFPC) The protective clothing that firefighters use for firefighting, rescue, and other first response activities; also known as turnout or bunker gear.

First responders A group designated by the community as those who may be the first to arrive at a chemical incident. This group is usually comprised of police officers, EMS providers, and firefighters.

Fissile Isotopes that can be induced into fission, which creates a release of energy.

Fit testing Test that ensures that the respiratory protection fits the face and offers maximum protection.

Flammable range The correct mixture of flammable gas and air in which there can be a fire; for most gases there is a certain span (range) that will allow for a fire to ignite.

Flash point Temperature of a liquid, which if an ignition source is present, will ignite only the vapors being produced by the liquid, creating a flash fire.

Formal decon Washing and scrubbing portion of the decontamination process. The process is usually repeated and is performed by a decon crew.

Frangible disk A pressure-relieving device that ruptures to vent the excess pressure. Once opened, the disk remains open; it does not close after the pressure is released.

Freezing point Temperature at which liquids become solids.

Gas State of matter that describes the material in a form that moves freely about and is difficult to control. Steam is an example.

GEDAPER© Management system used to organize the response to a chemical incident. The factors are: gather information, estimate potential, determine goals, assess tactical options, plan, evaluate, and review.

Gross decon Step in the decontamination process that removes the majority of the chemicals through a flushing of the person. The gross washing employs large amounts of water and is usually done by the individual or a partner.

Gross negligence Disregarding training and acting without regard for others.

Half-life The amount of time for a given radiation source to decay, which means it is emitting radioactivity.

Hazard area Term used to describe the dangerous area around a release; also called the isolation area or hot zone.

Hazardous materials (HAZMAT) Those substances that have the ability to harm humans, property, or the environment.

Hazardous Waste Operations and Emergency Response (HAZWOPER) The OSHA regulation that covers safety and health issues at hazardous waste sites and response to chemical incidents.

HHA Hand-held assay.

Hot, warm, and cold zones All zones of isolation used at chemical releases, from the most dangerous to the safest areas.

Huffing Term used to describe the intention inhalation of paints, solvents, or other chemicals with the intention of getting high.

Hyperbaric chamber Chamber used to treat scuba divers who ascend too quickly and need extra oxygen to survive. The chamber, which recreates the high pressure atmosphere of diving and forces oxygen into the body, is also used to treat carbon monoxide poisonings and smoke inhalation.

Ict$_{50}$ Military term for incapacitating level over set time to 50 percent of the exposed population.

Ignition temperature Temperature of a liquid that will ignite on its own without an ignition source; also called *self-accelerating decomposition temperature.*

IMS See *Ion mobility spectrometry.*

Incendiary agents Chemicals used to start fires, the most common being a Molotov cocktail.

Incident Commander HAZMAT training level that encompasses the operations level with the addition of incident command training. Intended to be the person who may command a chemical incident.

Infrared sensor Monitor that uses infrared light to determine the presence of flammable gases. Light is emitted in the sensor housing and the gas passes through the light; if it is flammable, the sensor indicates the presence of the gas.

Intermodal containers Constructed to be transported by highway, rail, or ship. Intermodal containers exist for solids, liquids, and gases.

Internal floating roof tank Tank with a roof that floats on the surface of the stored liquid; this tank also has a cover on its top to protect the top of the floating roof.

Ionization potential (IP) The ability of a gas or vapor to be ionized; most commonly used to determine if a photoionization detector can detect a gas or vapor.

Ion mobility See *Ion mobility spectrometry.*

Ion mobility spectrometry Is a detection technology that measures the travel time of ionized gases down a specific travel path.

Irritant Material that is irritating to humans, but usually does not cause any long-term effects.

Isolation area Area in which citizens and responders are not allowed; may later become the hot zone/sector as the incident evolves. This is the minimum area that should be established at any chemical spill.

Isotope Material that has a different form due to the number of neutrons that are in the nucleus.

Laws Legislation approved by Congress and signed by the president.

LCt$_{50}$ Military term for lethal concentration over set time to 50 percent of the population.

Leaking Underground Storage Tank (LUST) Term to describe this potential health and environmental hazard.

Level A protective clothing Fully encapsulated chemical protective clothing; gastight and liquid-tight gear offers protection against chemical attack.

Level B Level of protective clothing usually associated with splash protection; Level B requires the use of SCBA.

Liability Being held legally responsible for your actions.

Liquid State of matter that implies fluidity, which means a chemical can move as water would. There are varying states of being a liquid from moving very quickly to very slowly. Water is an example.

Local Emergency Planning Committee (LEPC) Group comprised of members of the community, industry, and emergency responders to plan for a chemical incident and to ensure that local resources are adequate to handle an incident.

Low Specific Activity (LSA) Indicates that a material is emitting low levels of radiation.

Lower explosive limit The lower part of the flammable range, or the minimum required to have a fire or explosion.

M-8 paper A paper detection device like pH paper that detects liquid nerve, blister, and VX agents.

M-9 paper A detection device that is used for liquid nerve and blister agents; comes in a tape form.

Mass casualty Incident in which the number of patients exceeds the capability of the EMS system to manage the incident effectively. In some jurisdictions this can be two patients, while in others ten may qualify as a mass casualty.

Material Safety Data Sheets (MSDS) Information sheets for employees that provide specific information about a chemical, with attention to health effects, handling, and emergency procedures.

Melting point Temperature at which solids become liquids.

Metal oxide sensor (MOS) Coiled piece of wire that is heated to detect the presence of flammable gases.

Mission-Oriented Protection Posture (MOPP) gear Chemical-resistant clothing designed for military operations.

Mission-specific competencies The knowledge, skills, and abilities to perform the mission of the organization. Examples are provided in NFPA 472.

Multi-gas detector Air monitor that measures oxygen levels, explosive (flammable) levels, and one or two toxic gases, such as carbon monoxide or hydrogen sulfide.

National Fire Protection Association (NFPA) A group that issues fire and safety standards for industry and emergency responders.

National Response Center (NRC) Location a spiller is required to call to report a spill if it is in excess of the reportable quantity.

Negligence Acting in an irresponsible manner, or different from the way one was taught to act, differing from the standard of care.

North American Emergency Response Guidebook (NAERG) Reference book provided by the DOT to assist first responders in making decisions at a transportation-related chemical incident.

Nuclear Regulatory Agency (NRC) The government regulatory agency responsible for oversight of nuclear materials.

Occupational Safety and Health Administration (OSHA) Federal agency responsible for enacting safety regulations protecting the nation's workers.

Operational security Keeping the details of an operation secret and only divulging critical information to those who have a need to know.

Operations level The next level of training above awareness that provides the foundation that allows for the responder to perform defensive activities at a chemical incident.

Ordinary tank Horizontal or vertical tank that usually contains combustible or other less hazardous chemicals. Flammable materials and other hazardous chemicals may be stored in smaller quantities in these types of tanks.

Overpack Response action that involves placing a leaking drum (or container) into another drum. Some drums are made specifically to be used as overpack drums and are oversized to handle a normal-sized drum.

Paragraph q Section within HAZWOPER that outlines the regulations governing emergency response to chemical incidents.

PCR Polymerase chain reaction, a detection method for biological agents.

PDD 39 Presidential Decision Directive 39, which established the FBI as the lead agency in terrorism incidents; the FBI is responsible for crisis management; and FEMA is the lead for consequence management.

Permeation Movement of chemicals through chemical protective clothing on a molecular level; does not cause visual damage to the clothing.

Permissible exposure limit (PEL) OSHA value that regulates the amount of a chemical that a person can be exposed to during an 8-hour day.

Personal protective equipment (PPE) Equipment and clothing designed to protect a person from a variety of hazards when responding to a HAZMAT incident; PPE ranges from earplugs to a fully enclosed chemical suit.

Photoionization detector (PID) Air monitoring device used by HAZMAT teams to determine the amount of toxic materials in the air.

Polymerization A runaway chain reaction that can be violent if contained. Once started it cannot be stopped until the chemical reaction has run its course.

Powered air purifying respirator (PAPR) An air purifying respirator (APR) that has a blower attachment that blows filtered air to a mask.

Psychological decontamination Used to make patients who are not contaminated feel better, producing psychological cleaning.

Radioactive decay As a material emits radiation, it decays changing its form. Uranium eventually becomes lead after it decays.

Radioisotope Radioactive form of an element, it is unstable due to the number of neutrons.

Radiological dispersion device (RDD) Radioactive material that is coupled with an explosive device, which spreads the radioactive material. It is not a nuclear detonation, as the explosion is caused by a standard explosive charge.

Radon Radioactive gas that is emitted from the earth, sometimes collecting in basement areas.

Recommended exposure level (REL) Exposure value established by NIOSH for a 10-hour day, 40-hour workweek. Is similar to the PEL and TLV.

Regulations Rules developed and issued by a governmental agency, which have the weight of law.

Relief valve Device designed to vent pressure in a tank, so that the tank itself does not rupture due to an increase in pressure. In most cases these devices are spring loaded so that when the pressure decreases, the valve closes to keep the chemical inside the tank.

Remote shutoffs Valves that can be used to shut off the flow of a chemical; the term *remote* is used to denote valves that are away from the spill.

Reportable quantity (RQ) A term used by the EPA and DOT to describe a quantity of chemicals that may require some type of action, such as reporting an inventory or reporting an accident involving a certain amount of chemicals.

Responsible party Person or company responsible for the spill or release, who is ultimately responsible for the cleanup of the incident.

Retention Digging a hole to collect a spill; can be used to contain a running spill or collect a spill from the water.

Ricin Biological toxin that can be used by a terrorist or other person attempting to kill or injure someone; the easiest terrorist agent to produce and one of the most common.

Risk-based response An approach to responding to a chemical incident by categorizing a chemical into a fire, corrosive, or toxic risk. Using a risk-based approach can assist the responder in making tactical, evacuation, and PPE decisions.

Sampling plan A drawing or a map of an incident that indicates all of the containers that need to be sampled. A priority of containers to be sampled, and the resources necessary to do that sampling are part of the sampling plan.

Sea containers Shipping boxes designed to be stacked on a ship and then placed onto a truck or railcar; also called sea boxes.

Sector See *Zone.*

Self-accelerating decomposition temperature (SADT) See *Ignition temperature.*

Self-contained breathing apparatus (SCBA) Protective gear that safeguards responders' respiratory systems; considered essential for responding to chemical spills.

Sensitizer Chemical that after repeated exposures may cause an allergic effect for some people.

Shearing effect Tearing of the molecules, which can be equated to being ripped apart.

Shelter in place Form of isolation that provides a level of protection while leaving people in place, usually in their homes. People are usually sheltered in place when they may be placed in further danger by evacuation.

Short-term exposure limit (STEL) Fifteen-minute exposure to a chemical; requires a 1-hour break between exposures and is only allowed four times a day.

Sick building Term associated with indoor air quality; occupants in a sick building become ill as a result of chemicals in and around the structure.

Sick building chemical Substances that cause health problems for occupants of a structure.

Solid State of matter that describes chemicals that may exist in chunks, blocks, chips, crystals, powders, dusts, and other types. Ice is an example.

Span of control Number of workers that one commander can effectively manage. When the number of workers exceeds the span of control, the supervision must be divided among more commanders.

Specialist level A level of training that provides for a specific type of training, such as railcar specialist; someone who has a higher level of training than a technician.

Specific gravity Value that determines if the liquid will sink or float in water. Water is given a value of 1, and chemicals with a specific gravity of less than 1 will float on water. Those chemicals with a specific gravity of greater than 1 will sink in water.

Specification (spec) plates Label on trucks and tanks that lists the type of vehicle, capacity, construction, and testing information.

Spent nuclear fuel Radioactive fuel that was used in a nuclear reactor.

Standard of care   Combination of training, experience, standards, and regulations that guide a responder to a specific action.

Standard Transportation Commodity Code (STCC)   Number assigned to chemicals that travel by rail.

Standards   Usually developed by consensus groups establishing a recommended practice or standard to follow.

State Emergency Response Committee (SERC)   Group that ensures the state has adequate training and resources to respond to a chemical incident.

States of matter   Term applied to chemicals to describe the forms in which they may exist: solids, liquids, or gases.

Sublimation   Ability of a solid to go to the gas phase without being liquid.

Superfund Amendments and Reauthorization Act (SARA)   Law that regulates a number of environmental issues, predominately for the chemical inventory reporting by industry to the local community.

Supplied air respirators (SAR)   Respiratory protection that provides a face mask, air hose connected to a large air supply, and an escape bottle; typically used for waste sites or confined spaces.

Surface contaminated object (SCO)   Material that may be contaminated with radioactive waste; is usually low hazard.

Technical decontamination   The process used to perform decontamination of persons or equipment.

Technician level   High level of HAZMAT training that allows specific offensive activities to take place to stop or handle a chemical incident.

Threshold limit value (TLV)   Exposure value that is similar to the PEL but is issued by the ACGIH. It is for an 8-hour day.

Tote   Large tank, usually 250 to 500 gallons (946 to 893 liters), constructed to be transported to a facility and dropped for use.

Toxins   Substances that are biological in nature and are obtained from naturally occurring materials.

TRACEM   Acronym for the types of hazards that may exist at a chemical incident: thermal, radiation, asphyxiation, chemical, etiological, and mechanical.

Type A container   Container designed to hold low risk radioactive material, offers moderate protection against accidents.

Type B container   Container designed to hold high risk radioactive materials, is a hardened transport case capable of withstanding severe accidents.

Underground storage tanks (UST)   Tanks, commonly for gasoline and other fuels, that are buried under the ground.

Universal precautions   The wearing of PPE designed for blood-borne pathogen protection, such as gloves, gown, and a face mask.

Upper explosive limit (UEL)   Upper part of the flammable range; above the UEL, fire or an explosion cannot occur as there is too much fuel and not enough oxygen.

Vapor density   Value that determines if the gas will sink or rise in air. Air is given a value of 1, and chemicals with a vapor density less than 1 will rise in air. Those with a density of greater than 1 will sink and stay low to the ground.

Vapor dispersion   Intentional movement of vapors to another area, usually by the use of master streams or hoselines.

Vapor pressure   Amount of force pushing vapors from a liquid; the higher the force, the more vapors (gas) are being put into the air.

Vapor suppression   Defensive strategy to minimize the production of vapors. For example, flammable liquids produce hazardous vapors that could ignite; firefighting foams are used to suppress the vapors and not allow them to move by forming a barrier on top of the spill.

Vesicants   Group of chemical agents that cause blistering and irritation of the skin; commonly referred to as *blister agents.*

Violent tank rupture (VTR)   This concept is the same as a BLEVE, but there is not a characteristic fireball and explosion.

Waybill   Term that may be used in conjunction with consist, but is a description of what is on a specific railcar.

Zone   Area established and identified for a specific reason; typically a hazard exists within the zone. Zones are usually referred to as hot, warm, and cold to provide an indication of the expected hazard in each zone. Some agencies use the term *sector.*

ABBREVIATIONS

ACGIH American Conference of Governmental Industrial Hygienists

ACS American Chemical Society

ALK Alkalis

ALOHA Aerial Location of Hazardous Atmospheres

ANFO Ammonium nitrate and fuel oil mixture

APR Air purifying respirator

AST Aboveground storage tank

ASTM American Society for Testing and Materials

ATF Bureau of Alcohol, Tobacco, and Firearms

BBP Blood-borne pathogens

BLEVE Boiling liquid expanding vapor explosion

BNICE Biological, nuclear, incendiary, chemical, and explosive agents

BOCA Building Officials Conference Association

CAAA Clean Air Act Amendments

CAMEO Computer Aided Management for Emergency Operations

Canutec Canadian Transportation Emergency Center

CAS Chemical Abstract Service

CBIRF Chemical and Biological Incident Response Force (U.S. Marines)

CERCLA Comprehensive Environmental Response, Compensation, and Liability Act

CGI Combustible gas indicator

Chemtrec Chemical Transportation Emergency Center

CMS Chip measurement system

CORR Corrosives

DCM Dangerous Cargo Manifest

DHS Department of Homeland Security

DNR Department of Natural Resources

DOT Department of Transportation

EHS Extremely hazardous substance

EPA Environmental Protection Agency

EPCRA Emergency Planning and Community Right-to-Know Act

ERG North American Emergency Response Guidebook

ERP Emergency response planning

FEMA Federal Emergency Management Agency

FFPC Firefighter Protective Clothing

HAZMAT Hazardous materials

HAZWOPER Hazardous Waste Operations and Emergency Response

HCS Hazard communication standard

HMIG Hazardous Materials Information Guide

HMIS Hazardous Materials Incident Reporting System

HMRU Hazardous Materials Response Unit (FBI)

ICBO International Conference of Building Officials

ICC International Codes Council

IDLH Immediately dangerous to life or health

IED Improvised explosive device

IM Intermodal

IMS Incident management system

IP Ionization potential

LC Lethal concentration

LD Lethal dose

LEL Lower explosive limit

LEPC Local Emergency Planning Committee

LHPC Licensed Health Care Provider

LOX Liquid oxygen

LSA Low specific activity

LUST Leaking underground storage tank

MCI Mass casualty incident

MOS Metal oxide sensor

MOT Materials of trade

MSDS Material Safety Data Sheet

NA North American identification number

NBC Nuclear, biological, and chemical agents

NDPO National Domestic Preparedness Office (FBI)

NFPA National Fire Protection Association

NIMS National Management Incident System

NIOSH National Institute of Occupational Safety and Health

N.O.S. Not Otherwise Specified

NRC National Response Center

ORM Other regulated material

OSHA Occupational Safety and Health Administration

P Polymerization hazard

PASS Personal Alert Safety System

PEL Permissible exposure limit

PG Packing group

PID Photo-ionization detector

PIH Poison inhalation hazard

PPE Personal protective equipment

RAID Rapid Assessment and Initial Detection teams (National Guard)

RDD Radiological dispersion device

REL Recommended exposure level

R&I Recognition and identification

RMP Risk Management Program

RQ Reportable quantity

SADT Self-accelerating decomposition temperature

SAR Supplied air respirators

SARA Superfund Amendments and Reauthorization Act

SBCCOM Soldiers Biological and Chemical Command

SCA Surface-contaminated articles

SCBA Self-contained breathing apparatus

SERC State Emergency Response Committee

SETIQ Mexican Emergency Transportation System

SLUDGEM Salivation, lacrimation, urination, defecation, gastrointestinal, emesis, miosis

STCC Standard Transportation Commodity Code

STEL Short-term exposure limit

TC Transportation Canada

TEU Technical Escort Unit (U.S. Army)

TLV Threshold limit value

TOG Turnout gear

TRACEM Thermal, radiation, asphyxiation, chemical, etiological, mechanical hazard

TWA Time weighted average

UEL Upper explosive limit

UN United Nations identification number

USAR Urban Search and Rescue team (FEMA)

UST Underground storage tank

VTR Violent tank rupture

WMD Weapons of mass destruction

INDEX

Abbreviations, 269–270
ABC (network), 232
Abdi, Nuradin, 218
Abortifacient, 207
Abortion clinic attacks, 215–217, 221–223
Aboveground storage tanks (ASTs), 63, 64
Absorbed dose, 116
Absorption, 183, 191–192
Accidents, 104
Acetic acid, 248
Acetic anhydride, 248
Acetone, 248, 249, 256
  correction factor for, 201
  exposure levels for, 123
  flash point of, 74
  vapor pressure of, 71
Acetylene, 166, 201
ACGIH (American Conference of Governmental Industrial Hygienists), 153
Acrolein, 123
ACS (American Chemical Society), 115
Activity, of source of radioactive material, 76
Acute exposure, 114
Adsorption, 184
Aerial Location of Hazardous Atmospheres (ALOHA), 99, 107, 152
Aerosolized insecticide/pesticide, 84
AFFF (aqueous film-forming foam), 164
AFFF foam, 166
AFFF-AR foam, 166
AFFF-ATC foam, 166
Agar, 247
Agricultural supply stores, 20
Air bills, 106
Air Guard, 144
Air monitoring, 197–209
  carbon monoxide, 207–209
  colorimetric sampling, 205, 206
  configurations for, 198–199
  and exposure levels, 122
  at first responder operations level, 197–199
  flammable gas indicators, 202–203

and foam use, 162
  importance of, 191
  and isolation zones, 150
  maintenance of devices for, 199
  and meter terminology, 199–201
  multi-gas detection device strategies, 204–205
  oxygen monitors, 200–202
  photoionization detectors, 205
  radiation detection, 205, 206
  regulations/standards for, 198
  and SWAT teams, 255–256
  toxic gas monitors, 203–204
Air monitors, purchasing, 198
Air purifying respirators (APRs), 125, 129, 254, 255, 258
Air reactive materials, 3
Aircraft rescue firefighting, 163
Airplane crashes, 144
Airplane hijackings, 213, 214, 217, 220
Airplane suicides, 218
Al Qaeda, 217, 218, 232
ALARA ("As Low As Reasonably Achievable"), 119
Alberg, Richard, 232
Alcohol-resistant AFFF (AR-AFFF), 164
Alcohol-resistant fluoroprotein (AR-FP), 164
Alcohols, 162, 164
ALF (Animal Liberation Front), 217
Alfred P. Murrah Federal Building, 25, 213, 215, 222, 227
Allergens, 116
ALOHA; See Aerial Location of Hazardous Atmospheres
Alpha radiation, 77, 78, 169
ALS providers, 12
Alternative fuels, 159–160
Aluminum drums, 39
Aluminum powder, 249
American Chemical Society (ACS), 115
American Conference of Governmental Industrial Hygienists (ACGIH), 153
American Media building, 232
American Society for Testing and Materials (ASTM), 128
Americium-241, 76, 77, 117

Ames strain, of anthrax, 232
Ammonia, 13, 138, 215
  exposure levels for, 123
  flammability of, 25
  and Saratoga gear, 255
  spills of, 146
Ammonium acetate, 248
Ammonium hydroxide, 197
Ammonium nitrate, 25, 157, 219, 249
Ammonium nitrate and fuel oil (ANFO), 25, 157
Ammonium perchlorate, 249
Anaheim, California, 216
Anarchist groups, 223
ANFO; See Ammonium nitrate and fuel oil
Anhydrous ammonia, 7, 197, 248
Anhydrous ammonia refrigeration system, 138
Anhydrous sodium acetate, 248
Animal Liberation Front (ALF), 217
Animals, dead, 225
Anthrax, 216, 217, 230–233
  Aum Shinryko use of, 221
  CDC categorization of, 231
  detection of, 237
  effects of, 230
  hazard classification of, 228
  manufacture of, 250
  response to hoaxes involving, 233
Anthrax letters, 214, 217
Antigovernment, 219
AP2C, 236
April 19th (significant date), 223
APRs; See Air purifying respirators
Aqueous film-forming foam (AFFF), 164
AR-AFFF (alcohol-resistant AFFF), 164
AR-FP (alcohol-resistant fluoroprotein), 164
Arlington, Virginia, 217
Army of God, 217
Aryan Nation, 216, 223
"As Low As Reasonably Achievable" (ALARA), 119
Asphyxiants, 119
Asphyxiation, 114
ASTM (American Society for Testing and Materials), 128

ASTs; *See* Aboveground storage tanks
ATF; *See* Bureau of Alcohol, Tobacco, and Firearms
Atlanta, Georgia, 215, 216, 221, 222
Atropine, 2
Aum Shinryko, 215–216, 221
Australian tourists, 219
Auto-ignition temperature, 74
Awareness, 10
Azinphos methyl (AZM), 105
AZM (azinphos methyl), 105

B plus suit, 128
Bacillus antracis, 216
Backpack explosives, 218, 219
Backup crew, 141, 259
Backup systems, 192
Bags, 39
Bali, Indonesia, 219
Baltimore, Maryland, 72, 223
Baltimore County, Maryland, 5, 32, 64, 144
Banking, 163
Batteries, 199
BBPs; *See* Blood-borne pathogens
Becquerel (Bq), 76
Benner, Ludwig, Jr., 3
Benzaldehyde, 248
Benzene, 116, 201
Berms, 195
Beslan school attack, 232
Beta radiation, 77, 78, 169
Bhopal, India, 19
Bill of lading, 105
Bin Laden, Osama, 217, 218
Biohazard/infectious substances, 30, 31
Biological agent laboratories, 247, 249–250
Biological agents, 217, 230–233
    Aum Shinryko's use of, 215–216
    delayed reactions to, 225
    detection of, 236–237
    education needed for manufacture of, 221
    hazard classification of, 228
    hysterical reactions to, 226
    LRN labs for, 233
Biomimetic sensing, 208
Birds, dead, 225
Birmingham, Alabama, 215, 216
Bladders, 39
Blasting caps, 249
Bleach, 179, 180, 215, 250

BLEVE; *See* Boiling liquid expanding vapor explosion
Blister agents, 228, 229, 234–236
Blood, 246
Blood agents, 229, 230
Blood-borne pathogens (BBPs), 120, 246
BLS providers, 12
BOCA (Building Officials and Code Administrators), 20
Boca Raton, Florida, 217, 232
Bodily fluids, 246
Body washing, 182
Boiling liquid expanding vapor explosion (BLEVE), 40, 46, 50–53, 159
Boiling point, 22, 69
Bomb making, 224
Bomb squad, 18, 19, 156, 157, 239, 249
    case study, 252
    and drug labs, 248
    explosives and HAZMAT interface with, 251
    and suspicious packages, 251, 253
Bomb technicians, 224, 259
Bomb threats, 156
Bombings:
    "small," 234
    World Trade Center, 213, 215
Bomb-sniffing dogs, 239
Booby traps:
    and chemical agents, 253
    in clandestine labs, 247
    and SWAT team, 255
Booby-trapped devices, 214, 216
Booms, 195
Boosting charges, 249
Boric acid, 215
Bottles, 38–39
Botulinum toxin, 231
Bq (becquerel), 76
Branch Davidians, 223
Branch positions, 141
Breach charge, 255
Breach in container, 153
Breathing difficulties, 229, 230
Bridges, 31, 162, 165
Briefcase bombs, 156, 252
Brooks, Richard, 223
Brucellosis, 231
Bruey, Judith, 218
Bubble B suit, 128
Buddy system, 124
Buffalo, New York, 218
Building collapse, 239

Building Officials and Code Administrators (BOCA), 20
Bulk containers, 21, 24, 36
Bulk storage tanks, 62–67
Bulk tanks, 40–42
Bump gas cylinders, 199
Bump test, 199
Bung, 155
Bureau of Alcohol, Tobacco, and Firearms (ATF), 155, 156, 222, 223
Bus accidents, 145
Butane, 201
Butylamine, 248
Butyric acid, 217

C2 filter, 255
C-4 explosives, 219
CAAA; *See* Clean Air Act Amendments of 1990
CAD Plus, 236
Calcium carbide, 166
Calibration, 199–200
Calibration gas, 199
CAM (Chemical Agent Meter), 236
CAMEO; *See* Computer Aided Management for Emergency Operations
Canada, 20
Canadian pest control number, 36
Canadian Transportation Emergency Center (Canutec), 34
"Candy-striper" railcars, 62
Canutec (Canadian Transportation Emergency Center), 34
Car bombs, 156, 252
Carbon dioxide, 160, 161, 202
Carbon monoxide:
    as chemical asphyxiant, 120
    detection of, 197, 198, 204, 205
    exposure levels for, 123
    incidents involving, 207–209
    monitoring, 207–209
    sensing technology for, 208
    sources of, 209
Carbon monoxide detectors, 199, 207
Carboys, 38
Carcinogens, 9, 115
Cardboard boxes, 38
Cardboard tubes, 249
Cargo tank specialist, 11
Carry-on luggage, 220
Cartridge filters, 255

CAS; *See* Chemical Abstract Service
CAS registry numbers, 100
Casablanca, Morocco, 218
Castor beans, 250
Catalytic bead sensors, 203, 204
Category A agents, 230, 231
Category B agents, 230, 231
Category C agents, 230, 231
CBIRF (Chemical and Biological Incident Response Force), 238
CBRNE (chemical, biological, radiological, nuclear, incendiary, and explosive) weapons, 227
CDC; *See* Centers for Disease Control
Ceiling levels (PEL-C or TLV-C), 122
Centers for Disease Control (CDC), 230–233
Central nervous system (CNS), 117
CERCLA (Comprehensive Environmental Response, Compensation, and Liability Act), 13
Certification, 3
Certified suits, 161
Cesium-137, 76, 77, 117
CF; *See* Correction factor
CFR (Code of Federal Regulations), 8
Chain of custody, 227, 259
Charlotte, North Carolina, 216
Chechnya, 232
Chemical, biological, radiological, nuclear, incendiary, and explosive (CBRNE) weapons, 227
Chemical Abstract Service (CAS), 100, 115
Chemical agent laboratories, 249, 250
Chemical Agent Meter (CAM), 236
Chemical agents, 217
    Aum Shinryko's use of, 215–216
    immediate reactions to, 225
    intelligence gathering about, 253
    LRN labs for, 233
    manufacture of, 221
Chemical and Biological Incident Response Force (CBIRF), 238
Chemical and physical properties, 67–75
    Auto-ignition temperature, 74
    case study, 72

and containers, 75
    corrosivity, 73
    flammable range, 74, 75
    flash point, 74
    reactivity, chemical, 74
    specific gravity, 72
    states of matter, 68–69
    vapor density, 70–72
    vapor pressure, 69–71
Chemical asphyxiants, 119, 120
Chemical asphyxiates, 229
Chemical bombs, 215, 251, 253
Chemical degradation, 184
Chemical haulers, 45–49
Chemical Hazards Risk Information System (CHRIS), 108
Chemical inventory reporting, 6–9
Chemical protective clothing, 11, 14, 112, 126–130
    fire hazard with, 254
    importance of, 113
    Level A ensemble, 126, 127
    Level B ensemble, 127–129
    Level C ensemble, 129–130
Chemical Protective Fabric (CPF), 128
Chemical reactivity, 74
Chemical releases, 224
Chemical storage facilities, 253
Chemical stress, 153
Chemical Transportation Emergency Center (Chemtrec), 34, 107
Chemicals:
    alphabetical listing of DOT-regulated, 93, 94
    in drug labs, 248
    for explosives, 248
    frequently spilled, 146
    as hazard category, 114
    law enforcement and dangers of, 246, 247
    names of, 95
    numerical listing of DOT-regulated, 89, 91–93
    and SWAT teams, 253–257
Chemtrec; *See* Chemical Transportation Emergency Center
Chevron, 217
Chicago, Illinois, 207, 215, 218
Children, 207
Chip measurement system (CMS), 206
Chlorine, 7, 212, 224, 229, 230
    detection of, 198, 204

exposure levels for, 123
    Graniteville, South Carolina case study, 151
    isolation distance for, 150
    military designation of, 228
    spills of, 146
    vapor dispersion of, 197
    vapor pressure of, 71
Chloropyriphos, 169
Choking agents, 229, 230
CHRIS (Chemical Hazards Risk Information System), 108
Chronic exposure, 114
Churches, 222
Ci (curie), 76
Civil Support Teams (CSTs), 238
Clandestine lab incidents, 247–250
    drug labs, 247–249
    explosives labs, 249
    terrorism agent labs, 249–250
Class 1 explosives, 24–25
Class 2 gases, 25, 26
Class 3 flammable liquids, 26
Class 4 flammable solids, 27
Class 5 oxidizers/organic peroxides, 27
Class 6 poisonous materials, 27, 28
Class 7 radioactive materials, 28
Class 8 corrosives, 29
Class 9 miscellaneous hazardous materials, 29
Class A foam, 163
Class B foam, 163, 164
Clean Air Act Amendments (CAAA) of 1990, 13, 153
Clean person—dirty person, 258
Cleanup contractors, 140
Clean Water Act, 12, 58, 161
Clostridium botulinum, 216
Cloud-shaped vapor cloud, 152
CMS (chip measurement system), 206
CNG (compressed natural gas), 159
CNS (central nervous system), 117
Cobalt-60, 76, 77, 117
Code development organizations, 20
Code of Federal Regulations (CFR), 8
Cold stress, 133
Cold temperature clothing, 131
Cold zones, 148–150
Colonial pipeline, 35
Colorimetric sampling, 205, 206
Colorimetric sampling tubes, 205, 206, 235

Columbine High School (Littleton, Colorado), 214
Columbus, Ohio, 218
Combination test strips, 235
Combustibility, 9, 25, 94
Combustible gas detectors, 197, 198
Combustible liquids, 26, 228
    protective actions for, 161, 163–164
    spills of, 146
Combustion, toxic products of, 78, 79
Communications, 138
    of hazards, 227
    and protective clothing, 133–134
    and safety, 145
Compatibility charts, 25
Competence of Responders to Hazardous Materials/Weapons of Mass Destruction Incidents Standard (NFPA 472), 3, 11–12, 198
Competencies, 11
Competencies for EMS Personnel Responding to Hazardous Materials/Weapons of Mass Destruction Incidents (NFPA 473), 12
Composite cylinders, 40
Comprehensive Environmental Response, Compensation, and Liability Act (CERCLA), 13
Compressed gases, 9
Compressed gas in solution, 25
Compressed natural gas (CNG), 159
Computer Aided Management for Emergency Operations (CAMEO), 99, 107, 152
Computer resources, 106–107
Confined space regulation, 199, 204
Confinement, 191
Consequence management, 237
Consist (shipping papers), 105
Constitutionalist groups, 223
Consumer Product Safety Commission, 207
Container markings, 36
Containers, 37–67
    bags, 39
    bottles, 38–39
    breaches of, 153–154
    bulk storage tanks, 62–67
    cardboard boxes, 38
    case study, 53
    and chemical/physical properties, 75

cylinders, 40
drums, 39, 40
ERG guide to fires and, 97
highway-transportation, 43–50
intermodal tanks, 56–58
pipelines, 41
radioactive-material, 42–43
rail-transportation, 58–62
and response to potential BLEVE situations, 50–52
and righting of overturned trucks, 58
specialized tank trucks, 51, 52, 54, 55
suspicions about, 224
temperatures and pressurized, 54
totes/bulk tanks, 40–42
Containment, 154–155, 191
Containment measures, 10
Containment system, 154
Contamination, 173
    of evidence, 258
    from radiation, 245
Contingency plans, 259
Convulsant chemicals, 116
Convulsions, 229
Cooling process (drug manufacture), 248, 249
Correction factor (CF), 200, 201, 204
Corrosive materials, 9, 248
    as Class 8 hazard, 29
    exposure risk of, 121
    placard indicating, 22, 24
    protective actions for, 170–172
Corrosive railcars, 60
Corrosive tankers, 45, 48, 49
Corrosivity, 73
CPF (Chemical Protective Fabric), 128
Cramps, heat, 133
Crash trucks, 163
Credible threats, 221
Crime scene, 225, 239
Criminal investigators, 258
Crisis management, 237
Cross-contamination, 259
Crosstech vapor barrier, 124
Cryogenic gas, 25
Cryogenics, 131
Cryogenic tankers, 51, 52, 54
CSTs (National Guard Civil Support Teams), 238
Cult groups, 215–216
Curie (Ci), 76
Cutaneous anthrax, 232

Cyanide, 215, 216, 230, 248
Cyanogen chloride, 228, 229
Cylinders, 40

Damming, 163, 192–194
Dangerous Cargo Manifest (DCM), 106
"Dangerous goods," 20
"Dangerous" placard, 29
"Dangerous when wet," 21, 27
Dar es Salaam, Tanzania, 216, 217
Daschle, Tom, 232
Dates, significant, 223
DCM (Dangerous Cargo Manifest), 106
Dead animals, 225
DECIDE process, 142, 143
Decontamination, 8, 173–185
    by absorption, 183
    by adsorption, 184
    by chemical degradation, 184
    by dilution, 184
    emergency, 173–174
    by evaporization, 184
    after fighting fires, 120
    fine, 175
    and hazardous materials management, 146
    by isolation/disposal, 184
    mass, 175–180
    methods of, 183
    by neutralization, 184
    personnel assigned to, 141
    planning for, 259
    process of, 180–183
    by solidification, 184
    by sterilization, 184
    and SWAT teams, 256
    technical, 174–175
    types of, 173
    by vacuuming, 184–185
    by washing, 185
Decontamination corridor, 13
Dedicated railcars, 61
Defecation, 229
Defensive operations, 191–197
    absorption, 191–192
    diking/damming, 192–194
    dilution, 196
    diverting, 194–196
    remote shutoffs, 197
    retention, 196
    vapor dispersion, 196–197
    vapor suppression, 197

Deflagration, 27
Dehydration, 133
Delayed health hazards, 9, 229
Density, vapor, 70–72
Department of Natural Resources
    (DNR), 140
Dermal toxicity, 23
Desired evacuation distance, 156
Detection of terrorism agents,
    234–237
Detroit, Michigan, 218
Diazinon, 169
Diesel fuel, 7
    handling, 93, 94
    placard indicating, 24
    vapor pressure of, 71
Diethyl ether, 248
Diffusion mode, 200
Dignitaries, visiting, 223
Diking, 163, 192–194
Dilution, 184, 196
Dimethyl sulfoxide (DMSO), 216
Direct contact, 173
Disease-causing microorganisms or
    toxins (Division 6.2), 28
Disneyland, 216
Dispersal devices, 230
Dispersion, vapor, 196–197
Disposal, 184
Dissemination, of terrorism
    agents, 221
Distilled mustard, 228
Diverting, 194–196
Division number, 24
Dizziness, 230
DMSO (dimethyl sulfoxide), 216
DNA, 237
DNR (Department of Natural
    Resources), 140
Documentation, of evidence
    collection, 258
Domestic Preparedness Training
    Initiative, 238
Domestic terrorism, 212, 220
Dose levels, 205
Dose rate, 206
Dosimeters, 205
DOT book; *See Emergency
    Response Guidebook*
DOT hazardous material (defini-
    tion), 4
DOT identification number, 24
DOT system of, 20–24
DOT Table 1, 21
DOT Table 2, 21, 22
DOT-306 trucks, 44

DOT-331 pressurized tankers, 45,
    49, 50
DOT-406 gasoline tank trucks,
    44–46
DOT-407 chemical haulers, 45–49
DOT-412 corrosive tankers, 45,
    48, 49
Drager, 235
Drain cleaner, 73
Drills, 6
Drug labs, 214, 245, 247–249
Drug making, 224
Drugs, 236, 246, 247
Drum dumps, 246, 257
Drums, 39, 40, 148
Dry bulk tanks, 55
DuPont, 132
Dursban, 169
Dust clouds, 230

E. coli, 216
Earth Liberation Front, 217
Earthen dams, 193
Eastern Europe, 233
Ebola, 216, 231
Educational ability, of terrorist, 221
EHSs; *See* Extremely hazardous
    substances
8-Step Process, 142, 143
El Segundo, California, 217
Elderly people, 207
Elections, 215, 218, 222, 232
Electricity, 158, 204
Electrochemical sensors, 208
Electrolyte solution, 202
Emergency decontamination,
    173–174
Emergency management agency, 239
Emergency management office, 5
*Emergency Medical Response to
    Hazardous Materials
    Incidents,* 12
Emergency planning, 5–8
Emergency Planning and
    Community Right to Know Act
    (EPCRA), 5, 19, 253
Emergency plans, 8
Emergency responders, 13, 221
Emergency response, 10
Emergency Response Guidebook
    (ERG), 85–100, 148
    alphabetical chemical listing
    (blue section), 93, 94
    hazard class system, 87
    instructions for using, 86

intermodal container markings,
    89, 91
isolation/evacuation information
    (green section), 95–99
numerical chemical listing
    (yellow section), 89, 91–93
placards, 87–89
potential hazards/public
    safety/emergency response
    (orange section), 93–96
railcar silhouettes, 90
sections of, 85
truck silhouettes, 89
water-reactive listings, 99–100
Emergency response planning
    (ERP), 153
Emergency response plans, 5, 6, 13
Emergency services, 10
Emergency shutoff valves, 46,
    47, 197
Emesis, 229
Employers:
    general duty clause for, 4, 11
    MSDS provided by, 13
EMS Level I training, 12
EMS Level II training, 12
Encapsulated suits, 126–129
Endothermic reactions, 74
Entry crew, 141
Environmental crimes, 257
Environmental Protection Agency
    (EPA), 104–105, 258
EPA; *See* Environmental Protection
    Agency
EPA hazard categories, 9
EPA hazardous substances
    (definition), 4
EPA registration number, 36
EPCRA; *See* Emergency Planning
    and Community Right to
    Know Act
Ephedrine hydrochloride, 248
Epsilon toxin, 231
ERG; *See Emergency Response
    Guidebook*
ERP (emergency response plan-
    ning), 153
Escort services, 25, 34, 238
ETA (Euskadi Ta Askatasuna), 218
ETG, 236
Ethane, 201
Ethanol, 201, 248
Ether, 157, 246, 248
Ethion, 71, 123
Ethyl ether, 71, 157, 248, 249
Ethyl isocyanate, 98

Ethylene, 201
Etiological hazards, 115
Euskadi Ta Askatasuna (ETA), 218
Evacuation, 259
  and carbon monoxide
    incidents, 208
  for explosives, 156
  sheltering in place vs., 150–153
  stand-off distances for bomb, 252
  and suspicious packages, 251
Evaporization, 184
Events, as targets of terrorism,
  222–223
Evidence collection, 12, 257–260
  with anthrax hoaxes, 233
  considerations for, 257–258
  of drug use, 247
  and operations/entry, 259–260
  and sampling, 258–259
Evidence preservation, 143, 225,
  227, 239
Excepted packaging, 42
Exhaustion, heat, 133
Explosive, 9
Explosive capacity, 156
Explosive devices, 18, 216–217
Explosives:
  chemicals used for, 249
  as Class 1 hazard, 24–25
  detection of, 236
  DOT recommendations for, 155
  exposure risk of, 121
  and fire, 25, 155
  placards indicating, 21, 22
  protective actions for, 155–157
  standoff distances for, 156, 252
  as terrorists' weapon of choice, 223
Explosives incidents, 251–253
  bomb squad/HAZMAT interface
    on, 251
  case study, 252
  general rules for, 251, 252
  special situations of, 251, 253
Explosives labs, 249
Exposure (to hazards):
  levels of, 121–124
  routes of, 119–121
  types of, 114
External floating roof tanks, 65
Extremely hazardous substances
  (EHSs), 7, 106
Eye irritation, 229, 230

Facility documents, 106
"Fallen angel," 232

Family planning clinics, 222
Faris, Iyman, 218
Farms, 19
FBI; *See* Federal Bureau of
  Investigation
FBI Bomb Data Center, 234
Fear, 222
Federal Bureau of Investigation
  (FBI):
  field offices of, 226
  and Laboratory Response
    Network, 233
  as target of terrorism, 222
  terrorism incidents investigated
    by, 214, 217–220
  and terrorist attacks, 237
Federal Emergency Management
  Agency (FEMA), 237
Federal government, 226, 227,
  237–238
Federal Register, 8
Federal regulations, 8
FEMA (Federal Emergency
  Management Agency), 237
Fencing, 192
Fertilizers, 168–169
Fetal hemoglobin, 207
FFFP (fluoroprotein film forming
  foam), 166
FFPC; *See* Firefighter's protective
  clothing
FGIs; *See* Flammable gas indicators
Fiberboard drums, 39
Field test, 199
Filter fences, 192
Finance officer, 140, 141
Fine decontamination, 175
Firefighter safety, 13–14
Firefighter's Handbook, 140, 197
Firefighter's protective clothing
  (FFPC), 113, 120, 124,
  144, 145
Firefighters' turnout clothing,
  113, 259
Firefighting foam, 162–164, 166
Fire hazard, 9
  as Division 1.3 category, 24
  ERG guide to transport containers
    and, 97
  and explosives, 25
  NFPA 704 marking system for, 33
  for SWAT teams, 254
Firehouses, violence in, 215, 216
Fire investigators, 227
Fire marshals, 227
Firescope, 139

Fireworks, 249
First aid, 98
First responders, 4
Fissile material, 28
Fit testing, 125
Flame-ionization detectors, 206
Flammability, 9
  combustibility vs., 25, 94
  of drug labs, 249
  NFPA 704 marking system for, 33
  placards indicating, 22
Flammable gases:
  detection of, 197
  Division 2.1—ignitable at 14.7
    psi, 25, 26
  EPA monitoring guidelines
    for, 202
  LELs of, 200
  placards indicating, 22
  protective actions for, 157–158
Flammable gas indicators (FGIs),
  198, 202–204, 208
Flammable liquids, 228
  as Class 3 hazard, 26
  exposure risk of, 121
  protective actions for, 161
  spills of, 146
Flammable metals, 166, 167
Flammable range, 74, 75
Flammables, 248
Flammable solids:
  as Class 4 hazard, 27
  exposure risk of, 121
  placards indicating, 22
  protective actions for, 161, 166
Flares, 160, 166
Flash bang, 255, 256
Flash fires, 161
Flash point, 22, 74
Flash suit overgarments, 127
Flashpoint, 25
Floating absorbent booms, 195
Floating roof tanks, 65, 66
Floating skimmers, 195
Florida, 217
Fluoroprotein film forming foam
  (FFFP), 166
Fluoroprotein foam, 166
Fluoroprotein foam (FPF), 164
Flushing, 196
Foam, 165
Foam concentrates, 163
Foam eductors, 163
Foam engines, 163
Foam nozzles, 163
Foam proportioners, 163

Foams, 162, 166, 197
Food safety threats, 231
Ford, Larry, 217
Foreign-based terrorism, 220
Formal decontamination, 181
Four-digit ID number, 24
FPF (fluoroprotein foam), 164
Frangible disks, 40, 154
Freezing point, 68, 69
Freon, 248
Frist, Bill, 232
Fuel oil, 24, 25, 157
Full-body washing, 233
"Fumigated" marker, 31
"Fumigation" label, 31
Functional test, 200

Gamma radiation, 78, 169,
    206, 236
Gangs, 214
Gases; *See also* Flammable gases
    as Class 2 hazard, 25, 26
    exposure risk of, 121
    protective actions for, 157–161
    rising, 71
Gas leaks, 172–173
Gas line valves, 158
Gas masks, 254
Gasoline, 7
    flash point of, 74
    handling, 93, 94
    placard indicating, 24
    protective actions for, 161
    spills of, 146
    and trapped victims, 146
    vapor pressure of, 71
Gasoline spills, 20
Gasoline stations, 7, 64
Gasoline tank trucks, 44–46, 165
Gastrointestinal symptoms, 229
GEDAPER process, 142, 143
General duty clause, 4, 11, 14
Glanders, 231
Glasses, protective, 247
Glossary, 263–268
Gloves, 255
Glycerine, 249
Goggles, protective, 247
GORE CHEMPAK, 131, 132
Gore, W. L., 124
Government buildings, 216–217, 219
Graniteville, South Carolina, 151
Graphite, 249
Gray (Gy), 116
Greece, 220

Green, Deborah, 215
Greenville, South Carolina, 232
Grenades, 220
Gross decontamination, 174–175,
    181
Gross negligence, 12
Ground spills, 193
Gy (gray), 116

Half-life, 76
Hand protection, 255
Hand washing, 247
Hand-aspirated pumps, 200
Handheld assays (HHAs), 236, 237
Handheld electronic detectors, 236
Hantavirus, 231
Harbor booms, 195
Hard sleeves, 194
Harris, Larry Wayne, 216
Hazard area, 148
Hazard classes, 21, 24–31
    Class 1 explosives, 24–25
    Class 2 gases, 25, 26
    Class 3 flammable liquids, 26
    Class 4 flammable solids, 27
    Class 5 oxidizers/organic
        peroxides, 27
    Class 6 poisonous materials,
        27, 28
    Class 7 radioactive materials, 28
    Class 8 corrosives, 29
    Class 9 miscellaneous hazardous
        materials, 29
Hazard classification system, 87
Hazard communication, 13
Hazard communication standard
    (HCS), 100
Hazard identification codes, 89, 91
Hazard zone A, 25, 28
Hazard zone B, 25, 28
Hazard zone C, 25
Hazard zone D, 25
Hazardous materials (HAZMAT),
    3–4
Hazardous materials branch
    officer, 11
Hazardous materials branch safety
    officer, 11
Hazardous Materials Information
    System (HMIS), 33–34
Hazardous materials management
    processes, 142–147
    case study, 144
    isolation/protection, 142, 143, 145
    rescue, 143–147

Hazardous Materials Response Unit
    (HMRU), 233, 237
Hazardous waste cleanup
    contractors, 195
Hazardous Waste Operations and
    Emergency Response
    (HAZWOPER), 8, 142
HAZMAT; *See* Hazardous materials
HAZMAT box, 138
HAZMAT branch management, 141
HAZMAT branch safety, 141
HAZMAT crimes, 214
HAZMAT team, 4, 225
    and drug labs, 248
    and evidence collection, 257–259
    explosives and bomb-squad
        interface with, 251
    and explosives labs, 249
    and SWAT teams, 256, 257
    and terrorist attacks, 239
HAZWOPER; *See* Hazardous Waste
    Operations and Emergency
    Response
HCS (hazard communication
    standard), 100
Health hazards, 113–121
    categories of, 115–116
    and exposure levels, 121–124
    and exposure types, 114
    NFPA 704 marking system for, 33
    organs/systems affected by, 117
    radiation, 116–119
    routes of exposure to, 119–121
    and toxicology, 114
    types of, 114–115
Heat cramps, 133
Heat exhaustion, 133
Heat of explosion, 249
Heat stress, 131–133
Heat stroke, 133
Heating process (drug manufacture),
    248, 249
Helder, Lucas, 218
Hemispheric vapor cloud, 152
HEPA filters; *See* High efficiency
    particulate accumulator filters
Hepatitis, 246
Hepatitis C, 246
Hepatoxins, 117
Herbicides, 168–169
Hexamethylene triperoxide diamine
    (HMTD), 220
Hexane, 201, 248
HHAs; *See* Handheld assays
High efficiency particulate accumu-
    lator (HEPA) filters, 184, 250

High expansion foams, 163
High temperature clothing, 130
Highly toxic, 9
Highway transportation containers, 43–50
HIV (human immunodeficiency virus), 246
HMIS; *See* Hazardous Materials Information System
HMRU; *See* Hazardous Materials Response Unit
HMTD (hexamethylene triperoxide diamine), 220
Hoaxes, 214, 217, 221, 222, 233
Hog casings, 170
Hospitals, 239
Hostage, 216
Hot lists, 221, 248
Hot materials tankers, 55–56
"HOT" placard, 29, 30
Hot water heaters, 215
Hot zone, 148, 149
Household Hazardous Waste Days, 5
Household quantities, 13
Houston, Texas, 217
Huffing, 247
Human immunodeficiency virus (HIV), 246
Humidity, 204
HVAC system, 172
Hyder, Arizona, 216
Hydration, 133
Hydriodic acid, 248
Hydrocarbons, 162, 205
Hydrochloric acid, 72, 146, 171, 248
Hydrogen, 25, 160, 201
Hydrogen chloride gas, 204, 248
Hydrogen cyanide, 204, 217, 218, 228, 229
Hydrogen flame, 236
Hydrogen sulfide, 123, 198, 199, 205, 216, 253
Hygiene, 247
Hyperbaric chamber, 207
Hysteria, 221, 226, 227

IAEA (International Atomic Energy Association), 37
IC; *See* Incident commander
ICBO (International Conference of Building Officials), 20
ICC (International Codes Council), 20

ICS; *See* Incident command system
ICt$_{50}$ (incapacitating concentration), 122
IDLH (immediately dangerous to life or health), 10
IDLH situations, 13
IDR (instant detection and response), 208
IEDs (improvised explosive devices), 253
Ignition devices, 249
Ignition sources, 149, 158
Ignition temperature, 74
Illegal activity, 224
Illness, unexplained, 224
IM 102 tank, 58
Immediate health hazard, 9
Immediately dangerous to life or health (IDLH), 10
IMO101 tank, 56, 58
Improvised explosive devices (IEDs), 253
IMS (ion mobility spectrometry), 236
IMSs; *See* Incident management systems
Incapacitating agents, 116
Incapacitating concentration (ICt$_{50}$), 122
Incendiary agents, 228
Incident command system (ICS), 10, 139
Incident commander (IC), 10, 259
and air monitoring, 198
legal responsibilities of, 142
tasks of, 139–140
Incident levels, 6
Incident management systems (IMSs), 139–142, 226
Industrial chemicals, 234, 236
Industrialized communities, 20
Industrial packaging, 43
Industrial technical assistance, 108–109
Inert ingredients, 250
Infectious materials, 121
"Infectious Substances" placard, 30, 31
Information resources, 83–109
Chemical Abstract Service, 100
Chemtrec, 107
computer resources, 106–107
Emergency Response Guidebook; See Emergency Response Guidebook
facility documents, 106

industrial technical assistance, 108–109
MSDS, 100–105
reference/information texts, 107–108
shipping papers, 103, 105–106
street story, 84
Information texts, 107–108
Information/research officer, 141
Infrared sensors, 203
Inhalation:
by dusts/mists, 23
poisonous by, 23
Inhalational anthrax, 232, 233
Inhalation hazards, 238
with biological agents, 250
placard for, 29
Inhalation toxicity, 23
Initial entry crew, 259–260
Injection, 120
Injury, unexplained, 224
Insecticides, 84, 169
Instability hazard, 33
Instant detection and response (IDR), 208
Insulated tanks, 46–49
Insurance adjusters, 140
Intelligence gathering, 253
Intermodal container markings, 89, 91
Intermodal containers, 44
Intermodal specialist, 11
Intermodal tanks, 56–58
Internal floating roof tanks, 65, 66
Internal Revenue Service (IRS), 215, 222
International Atomic Energy Association (IAEA), 37
International Codes Council (ICC), 20
International Conference of Building Officials (ICBO), 20
International Safety Equipment Association (ISEA), 199, 200
Interviews, victim, 227
Intimidation, 220
Iodine-129, 77
Iodine-131, 77
Ion mobility spectrometry (IMS), 236
Ionization potentials (IPs), 108
Ionizing radiation, 169
Ipecac, 223
IPs (ionization potentials), 108
Iraq, 212
Iron filings, 248

Irritants, 9, 116, 229, 230
IRS; *See* Internal Revenue Service
Irvine, California, 217
ISEA (International Safety
    Equipment Association), 199
Isolation:
  of contaminants, 184
  importance of, 145, 150
  methods of, 142, 143
  of victims, 239
Isolation distances, 95–98, 148
Isolation zones, 148–150
Isopropanol, 201
Isopropyl alcohol, 74, 248
Isotopes, 76

Jackson, Mississippi, 215, 216
Jakarta, Indonesia, 219
Japan, 216, 221, 234
JCAD (Joint Chemical Agent
    Detector), 236
JDL (Jewish Defense League), 217
Jemaah Islamiyah, 219
Jet fuel, 217
Jewish Defense League (JDL), 217
Joint Chemical Agent Detector
    (JCAD), 236
Joint Service Lightweight Integrated
    Systems Technology (JSLIST)
    suit, 255
JSLIST (Joint Service Lightweight
    Integrated Systems
    Technology) suit, 255
Junk mail, 252

Kaczynski, Theodore, 216
Kansas City, Missouri, 25
Kappler Protective Fabrics, 128
"Keep away from food," 22
Kentucky, 218
Konopka, Joseph, 218
Krar, William, 218–219
Ku Klux Klan, 216, 223

Labeling:
  radiation-source, 37
  specifics of, 32
Laboratories, 258
Laboratory analysis, 258
Laboratory incidents; *See*
    Clandestine lab incidents
Laboratory Referral Network
    (LRN), 237

Laboratory Response Network
    (LRN), 233
Laboratory testing, 233
Lacrimation, 229
Lafayette, Indiana, 217
Lag time, 200
Law enforcement, 243–260
  and blood-borne pathogens, 246
  and clandestine lab incidents,
    247–250
  competencies of, 12
  and drug/chemical effects,
    246, 247
  evidence collection by, 257–260
  and explosives incidents, 251–253
  HAZMAT risks faced by, 245–246
  street story, 244
  SWAT operations by, 253–257
  and universal precautions, 247
Laws, 4
Lawsuits, 150
LAX (Los Angeles Airport), 217
LC; *See* Lethal concentration
LD; *See* Lethal dose
Lead diacetate, 248
Lead wool technology, 202
Lead-214, 76, 117
Leahy, Patrick, 232
Leaking underground storage tanks
    (LUSTs), 64
Leaks, 97, 99, 155
LELs; *See* Lower explosive limits
LEL sensors, 199, 200, 204
LEPCs (Local Emergency Planning
    Committees), 5
Lethal air blast range, 156
Lethal concentration (LC), 22, 23,
    93, 122
Lethal dose (LD), 22, 23, 93, 122
Letters, anthrax-hoax, 233
Level 1 incident, 6
Level 2 incident, 6
Level 3 incident, 6
Level A clothing, 254
Level A ensemble, 126, 127
Level B ensemble, 127–129
Level C ensemble, 129–130
Level D suits, 130
Lewisite, 228
LHCPs (Licensed Health Care
    Providers), 13
Liability, 11, 12
Liaison officer, 140, 141
Licensed Health Care Providers
    (LHCPs), 13
Lightning, 65, 66

Light water, 163
Liquefied oxygen (LOX), 168
Liquefied Petroleum Gas code
    (NFPA 58), 40
Liquid-based explosives, 220
"List of Lists, The" (EPA), 7
Lit, Preston, 218
Little Rock, Arkansas, 232
Local Emergency Planning
    Committees (LEPCs), 5
Local emergency response plans, 5, 6
Lock out, tag out, 158
Logistics officer, 140
London, 219, 220, 232
Los Angeles, California, 215
Los Angeles Airport (LAX), 217
Low birth weight, 207
Low expansion foams, 163
Low specific activity (LSA), 42, 91
Lower explosive limits (LELs), 74,
    75, 198, 200, 202–204
LOX (liquefied oxygen), 168
LRN (Laboratory Referral
    Network), 237
LRN (Laboratory Response
    Network), 233
LSA; *See* Low specific activity
LUSTs (leaking underground
    storage tanks), 64

M-8 test paper, 234, 235
M-9 tape, 235
M40 mask, 255
Mace, 228, 236
Madrid, Spain, 218, 222
Magnesium, 166, 167, 249
Mail bombs, 216
Mail boxes, 218
Mail handling, 232
Mail poisons, 216
Mall incidents, 145, 150, 220,
    224–225
*Management of Chemical Warfare
    Casualties,* 239
Mandatory evacuation distance, 156
Manholes, 204
Manufacture, of terrorism
    agents, 221
Marburg, Germany, 231
Marine pollutant materials, 168
Marine Pollutant placard, 29, 30
Markings:
  container, 36
  pesticide-container, 36–37
  specifics of, 32

Masks, 255

Mass casualty incidents (MCIs), 225

Mass decontamination, 175–180

Mass explosion hazard (Division 1.1), 24

Mass transit systems, 222

Material Safety Data Sheets (MSDSs), 8, 13, 100–105
 elements of, 101
 EPA alert on use of, 104–105
 hazards listed on, 25
 and HMIS, 33–34
 intelligence gathering with, 253
 sample, 102–103

Materials of Trade (MOT), 30

Matsumoto, Japan, 216, 221

Matter, states of, 68–69

Matthews, William, 232

MC (motor carrier) numbers, 45

MC-306 gasoline tank trucks, 45, 46

MC-307 chemical haulers, 45–49

MC-312 corrosive tankers, 45, 48, 49

MC-331 pressurized tankers, 45, 49, 50

MC-338 cryogenic tankers, 51, 52, 54

MCIs (mass casualty incidents), 225

MCU2P gas mask, 255

McVeigh, Timothy, 213, 218, 219

Mechanical hazards, 115

Mechanical stress, 153

Media, 140, 141, 226, 227

Medical monitoring, 10, 11

Medical survey questionnaire, 11, 13

Melioidosis, 231

Melting point, 68, 69

Mercuric chloride, 248

Metal oxide sensors (MOSs), 203, 204, 208

Meter terminology, 199–201
 bump test, 199
 calibration, 199–200
 reaction time, 200
 recovery time, 200
 relative response, 200, 201

Methamphetamines, 224, 247, 248

Methane:
 correction factor for, 201
 detection of, 204
 flammable gas indicators for, 202, 203

Methanol, 201

Methyl alcohol, 71

Methyl isocyanate, 19

Methylamine, 248

Metropolitan Medical Response Teams (MMRTs), 238

Mettetal, Ray, 215

Mexico, 34

mg (milligrams), 23

mg/L (milligrams per liter), 23

mg/M^3 (milligrams per cubic meter), 23

Microbiological oil-eating bugs, 192

Microrem (μR), 116

Microsieverts (μSv), 116

Mid-expansion foams, 163

Militant groups, 223

Military, terrorism response capabilities of, 238

Military designations for warfare agents, 227, 228

Military grade anthrax, 230

Military placarding system, 34, 35

Militia groups, 215, 217, 218, 223

Millennium Border Bomb, 217

Milligrams (mg), 23

Milligrams per cubic meter (mg/M^3), 23

Milligrams per liter (mg/L), 23

Millirem (mrem), 116

Millirem per hour (mR/hr), 29

Millisieverts (mSv), 116

Millisieverts per hour (mSv/hr), 29

Milwaukee, Wisconsin, 217

Minneapolis, Minnesota, 217

Minor blast hazard (Division 1.3), 24

Minor explosion hazard (Division 1.4), 24

Miosis, 216, 229

Miscellaneous hazardous materials (Class 9 hazard), 29

Mission Oriented Protective Posture (MOPP) gear, 255

Mission Specific Competency, 3

Mission-specific competencies, 12, 197, 198

MMRTs (Metropolitan Medical Response Teams), 238

Mobility, 133

Mode of transportation, 105–106

Molotov cocktails, 228

MOPP (Mission Oriented Protective Posture) gear, 255

Moscow theater attack, 232

MOSs; *See* Metal oxide sensors

MOT (Materials of Trade), 30

Motivation, for terrorism attacks, 221, 222

Motor carrier (MC) numbers, 45

mrem (millirem), 116

mR/hr (millirem per hour), 29

MSA, 235

MSDSs; *See* Material Safety Data Sheets

mSv (millisieverts), 116

mSv/hr (millisieverts per hour), 29

Muhajir, Abdullah al, 218

Multi-drug-resistant tuberculosis, 231

Multigas detectors, 198
 strategies for using, 204–205
 and SWAT teams, 255
 tactical use of, 204–205

Multisampling, 235

Murrah Building; *See* Alfred P. Murrah Federal Building

Mustard, 123, 228

Mutagens, 117

Mycotoxins, 230

NA (North America), 89

Nairobi, Kenya, 216, 217

Nashville, Tennessee, 232

National Enquirer, 232

National Fire Protection Association (NFPA), 3, 11–12

National Guard Civil Support Teams (CSTs), 238

National Incident Management System (NIMS), 142

National Institute of Occupational Safety and Health (NIOSH), 108

National Priority List (NPL) sites, 13

National Response Center (NRC), 103

National Toxicology Program, 115

Natural foam, 163

Natural gas:
 protective actions for, 157–159
 spills of, 146

Natural gas leaks:
 flammable gas indicators for, 202
 indicators of, 208
 vapor dispersion of, 196, 197

Nausea, 230

"Nazi" method (of meth production), 248

NBC studios, 232

NBC weapons; *See* Nuclear, biological, and chemical weapons

Negligence, 12
Neo-Nazi groups, 223
Nephrotoxins, 117
Nerve agents, 227–229
    detection of, 234–236
    and protective clothing, 145
Neutralization, 184
Neutron radiation, 169, 170, 206
New Jersey, 232
New Orleans, Louisiana, 217
New York City, 213, 217
New York Post, 232
NFPA; *See* National Fire Protection
    Association
NFPA 58 (Liquefied Petroleum Gas
    code), 40
NFPA 472 Committee, 197
NFPA 704 marking system, 20
NFPA 704 System, 32, 33
NFPA 1500; *See Standard on Fire
    Department Occupational
    Safety and Health Program*
NFPA 1994, 14
NFPA protective clothing standards,
    127
NFPA Standard 1001, 3
NFPA Standard 1001 (Standard for
    Firefighter Professional
    Qualifications), 3
Nichols, Terry, 213
Nicotine, 216
Nightclubs, 215
NIMS (National Incident
    Management System), 142
NIOSH (National Institute of
    Occupational Safety and
    Health), 108
NIOSH Pocket Guide, 70
*NIOSH Pocket Guide to Chemical
    Hazards,* 108
Nipah virus, 231
Nitric acid, 112, 171, 248, 249
Nitroethane, 248
Nitrogen dioxide, 204
Nitrogen mustard, 228
Nomex, 132, 254, 255, 259
Noncredible anthrax events, 233
Noncredible RDDs, 234
Nonflammable, nonpoisonous, com-
    pressed gas (Division 2.2), 25
Nonflammable gases placards, 22
Nonionizing radiation, 169
Non-pressure railcars, 59, 60
North America (NA), 89
N.O.S. (not otherwise specified), 91
Not otherwise specified (N.O.S.), 91

November 17 terrorist group, 220
NPL (National Priority List)
    sites, 13
NRC; *See* Nuclear Regulatory
    Commission
NRC (National Response
    Center), 103
Nuclear, biological, and chemical
    (NBC) weapons, 227, 238
Nuclear agents, 232–234
Nuclear detonation, 232, 233
Nuclear Regulatory Commission
    (NRC), 42, 216
Nunn-Lugar-Domenici legislation,
    238
Nursing home patients, 215

Occupancies, 19–20
Occupational Safety & Health
    Administration (OSHA), 3,
    104, 121
Odors:
    in high rise buildings, 172
    lack of, 67
    and sick building calls, 161
    unusual, 115–116, 205, 208, 225
Off gases, 250
Oil-eating bugs, 192
Oklahoma City bombing, 213,
    215, 223
Oleum, 171
Olsen, Kenneth, 231
Olympic Park bombing, 215,
    216, 222
Ontario, Canada, 219
Operation Respond, 106
Operational security, 256
Operations, 10
Operations officer, 140
OPPs; *See* Organophosphate
    pesticides
Oral toxicity, 23
Orange book; *See Emergency
    Response Guidebook*
Ordinary tanks, 65
Oregon, 232
Organic materials, 204
Organic peroxides, 9
    as Class 5 hazard, 27
    Division 5.2 polymerizable, 27
    exposure risk of, 121
    as hazard class, 21, 22
    protective actions for, 167, 168
Organophosphate pesticides (OPPs),
    84, 116, 227

Organophosphate poisoning, 2
Organophosphates, 84
O.R.M. (other regulated material),
    93
ORM-D (Other Regulated
    Material–Class D) placard,
    29, 30, 32
OSHA; *See* Occupational Safety &
    Health Administration
OSHA hazard categories, 9
OSHA hazardous chemicals
    (definition), 4
OSHA process safety, 253
Other regulated material
    (O.R.M), 93
OTTO (occupancy and location,
    type of event, timing of event,
    on-scene warning signs), 223
Outdoor air quality, 207
Overflow dams, 193–194
Overpacked, 147
Overturned trucks, 58, 59
Oxidizers, 9, 228
    as Class 5 hazard, 27
    exposure risk of, 121
    placard indicating, 22
    protective actions for, 167, 168
Oxygen levels, 199
    detection of, 197, 198
    EPA monitoring guidelines
        for, 202
    in normal air, 201
Oxygen monitors, 200–202
Oxygen saturation monitor, 207
Oxygen-producing materials
    (Division 5.1), 27
Ozone, 202, 204

P (polymerization hazard), 91
Packages requiring special
    handling, 28
Packing groups (PGs), 22–23,
    93, 103
Padilla, José, 218
Pads, absorbent, 191, 192
Pagers, 205
Pan disrupters, 251
Panic, 234
PAPRs; *See* Powered air purifying
    respirators
Paragraph q, 10
Paramilitary groups, 216
Parts per billion (ppb), 205
Parts per million (ppm), 23,
    198, 205

Patriot groups, 223
Patriots Council, 215
PBI jumpsuit, 259
PBI/KEVLAR, 130
PCR; *See* Polymerase chain reaction
PDD 39; *See* Presidential Decision Directive 39
PEL (permissible exposure limit), 122
Pennsylvania, 218
Pentagon, 213, 214, 217
Pentane, 201, 202
Pepper spray, 145, 150, 215
    detection of, 236
    military designation of, 228
Peripheral nervous system (PNS), 117
Permeation, 126, 128
Permissible exposure limit (PEL), 122
Peroxide crystals, 157
Personal protective equipment (PPE), 2, 111–134
    chemical protective clothing, 126–130
    cold temperature clothing, 131
    and communications, 133–134
    decontaminating, 175, 181
    determining need for, 259–260
    ERG listing of, 96
    and exposure levels, 121–124
    and health hazards, 113–124
    and heat stress, 131–133
    high temperature clothing, 130
    Level A ensemble, 126, 127
    Level B ensemble, 127–129
    Level C ensemble, 129–130
    Level D suits, 130
    limitations of, 131–134
    and mobility, 133
    NFPA clothing standards, 127
    ratings for, 128
    research on, 144, 145
    SCBA, 124–126
    selection of, 132
    street story, 112
    types of, 124
    and visibility, 133
Personnel, limiting number of, 239
Pesticide container markings, 36–37
Pesticide exposures, 2
Pesticides, 84, 168, 244, 250
PETN explosive, 217
Petri dishes, 247, 250
Petroleum ether, 248
PG III placard, 29

PGs; *See* Packing groups
PH detection, 197, 206
PH of common materials, 73
Phenol, 249
Phenylacetic acid, 248
Philadelphia, Pennsylvania, 218
Phorate, 250
Phosgene, 228–230
Photoionization detectors (PIDs), 164, 197, 205, 255, 256
Physical examination, 11, 13
Physical properties; *See* Chemical and physical properties
Physicians, as targets of terrorism, 222
Picric acid, 157, 249
PIDs; *See* Photoionization detectors
Piezoelectric crystals, 236
PIH; *See* Poison Inhalation Hazard
Pinpointed pupils, 216, 224, 225, 229
PIO; *See* Public information officer
Pipe bombs, 156, 218, 220, 234
    illustration of, 224
    manufacture of, 249
    radioactive material in, 233
    stand-off distances for, 252
Pipeline markings, 34–36
Pipelines, 41
Piperidine, 248
Pipes:
    in dams, 194
    for storage tanks, 63
Pistols, 220
Placards, 20–31
    biohazard/infectious substances, 30, 31
    case study, 25
    Class 1 explosives, 24–25
    Class 2 gases, 25, 26
    Class 3 flammable liquids, 26
    Class 4 flammable solids, 27
    Class 5 oxidizers/organic peroxides, 27
    Class 6 poisonous materials, 27, 28
    Class 7 radioactive materials, 28
    Class 8 corrosives, 29
    Class 9 miscellaneous hazardous materials, 29
    "dangerous," 29
    DOT system of, 20–24
    in ERG, 87–89
    fumigated, 31
    "HOT," 29, 30
    inhalation hazard, 29

marine pollutant, 29, 30
    for Materials of Trade, 30
    materials requiring, 21, 22
    ORM-D, 29, 30
    and packing groups, 22–23
    PG III, 29
    primary vs. secondary, 32
    problems with, 31
    and quantity of hazardous material, 20
    and toxicity measurement, 23
    white square, 31
Plague, 230–231
Plane crashes, 144
Planning, for special events, 223
Planning officer, 140
Plastic drums, 39
Plastic sheeting, 196
Plumes, 41
Plutonium, 77
Plutonium-239, 76, 117
PNS (peripheral nervous system), 117
Poison:
    placard indicating, 22
    protective actions for, 168–169
Poisoning, 215
Poison Inhalation Hazard (PIH), 21, 25, 32, 91
Poisonous by inhalation, 23
Poisonous gases (Division 2.3), 21, 23, 25
Poisonous materials, 27, 28
Polar substances, 162, 164
Police department, 225
Political conventions, 223
Polonium-214, 76, 117
Polymerase chain reaction (PCR), 236, 237
Polymerization, 27, 28
Polymerization hazard (P), 91
Pooling vapor cloud, 152
Pope, 223
Postal facilities, 232
Port Angeles, Washington, 217
Post offices, 232
Post-entry physicals, 11
Post-traumatic stress syndrome (PTSS), 115
Potassium, 218
Potassium chlorate, 249
Potassium hydroxide, 73
Potassium nitrate, 249
Potassium perchlorate, 249
Powders, 249

Powered air purifying respirators (PAPRs), 125, 254, 258
ppb (parts per billion), 205
PPE; *See* Personal protective equipment
ppm; *See* Parts per million
Pre-entry physicals, 11
Pregnant women, 207
Pre-hydration, 127, 132, 254
Presidential Decision Directive (PDD) 39, 226, 237
Pressure:
  and BLEVE situations, 50–52
  and cylinders, 40
  vapor, 69–71, 150
Pressure conditions, 9
Pressurized containers, 54
Pressurized railcars, 59, 60
Pressurized tankers, 45, 49, 50
Prince Georges County, Maryland, 164
Private sector specialist employee, 11
Product control, 189–209
  absorption, 191–192
  air monitoring; *See* Air monitoring
  defensive operations of, 191–197
  diking/damming, 192–194
  dilution, 196
  diverting, 194–196
  importance of, 191
  remote shutoffs, 197
  retention, 196
  street story, 190
  vapor dispersion, 196–197
  vapor suppression, 197
Projectile hazard (Division 1.2), 24
Propane, 208
  correction factor for, 200, 201
  protective actions for, 157–160
  spills of, 146
Propane leaks, 173, 197
Propane tanks, 50, 217, 253
Propeller label, 37
Properties, chemical and physical; *See* Chemical and physical properties
Protection methods, 142, 143
Protective action distances, 95–98
Protective actions, 137–185
  for corrosives, 170–172
  decontamination, 173–185
  for explosives, 155–157
  extinguishment, 162–166

for flammable solids/water reactives/spontaneously combustible, 161, 166–167
for flammable/combustible liquids, 161
for gases, 157–161
general steps of, 171–173
hazardous materials management processes, 142–147
and IC's legal responsibilities, 142
incident management systems, 139–142
for oxidizers/organic peroxides, 167, 168
for poisons, 168–169
for radioactive materials, 169, 170
site management, 146–153
street story, 138
and types of releases, 153–155
unusual incidents, 170, 171
Protective clothing:
  for drug-lab encounters, 249
  and evidence collection, 258
  exposure ratings for, 33
  NFPA standards for, 127
  ratings for, 128
  and SWAT environment, 255
  and universal precautions, 247
Protein foams, 164, 166
Proximity gear, 130
Psittacosis, 231
Psychological decontamination, 175
Psychological hazards, 115
Psychologically driven medical problems, 226
PTSS (post-traumatic stress syndrome), 115
Public buildings, 222
Public Health Service, 115
Public information officer (PIO), 139–141
Public perception, 13
Public reaction, 234
Public transportation systems, 219
Pulse rate, 120
Pulse-Ox monitor, 207
Purcell, Bill, 144
Purity, of evidence, 258
Pyridine, 248
Pyrophoric, 3, 9

Q fever, 216, 231
Quick in/quick out approach, 238–239

R (roentgen), 116
Radiation, 76–78
  action levels/doses of, 118
  common radionuclides, 77
  common sources of, 76, 77
  contamination from, 245
  as hazard category, 114
  hazards associated with, 116
  health effects of, 118
  protection from, 119
  sources of, 117
  types of, 77, 78
Radiation detection, 205, 206
Radiation monitors, 197, 206
Radiation pagers, 205, 206
Radiation source labeling, 37
Radioactive decay, 76
Radioactive I label, 28
Radioactive II label, 28
Radioactive III label, 28
Radioactive isotopes, 206
Radioactive material (RAM), 43, 216, 232–234
  as Class 7 hazard, 28
  exposure risk of, 121
  placards indicating, 21
  protective actions for, 169, 170
Radioactive material containers, 42–43
Radio-controlled devices, 215
Radioisotopes, 76
Radiological dispersion devices (RDDs), 218, 232–234, 253
Radionuclides, 77
Radium, 77
Radon, 76, 77
Radon-222, 76, 117
Rads, 116
RAID (Rapid Assessment and Initial Detection) teams, 238
Railcars, 58–62
  chlorine in, 253
  markings on, 61–62
  silhouettes of, 90
  specialized, 61
  specialized totes on, 42
  specifications for, 60
Rajneesh Foundation, 215, 232
RAM; *See* Radioactive material
Rapid Assessment and Initial Detection (RAID) teams, 238
Raw materials, obtaining, 221
RDDs; *See* Radiological dispersion devices
RDEC; *See* Research Development and Engineering Command

RDX, 249
Reaction time, 200
Reactivity:
    chemical, 74
    NFPA 704 marking system for, 33
Reactor vessels, 221
Readily combustible solids
    (Division 4.1), 27
Realm of Chaos, 218
Rechargeable batteries, 199
Recognition and identification
    (R&I), 17–79
    and chemical/physical properties;
        See Chemical and physical
        properties
    of container markings, 36
    of containers; See Containers
    HMIS of, 33–34
    importance of, 19
    location/occupancy, 19–20
    military warning system of,
        34, 35
    NFPA 704 System of, 32, 33
    of pesticide container markings,
        36–37
    of pipeline markings, 34–36
    of placards/labels/markings,
        20–32
    and radiation, 76–78
    of radiation source labeling, 37
    senses for, 67
    street story, 18
    and toxic products of combustion,
        79
Recommended exposure levels
    (REL), 122
Reconnaissance team, 141
Record keeping, 13, 180
Recovery time, 200
Red phosphorous, 248
Reefers, 61
Reference texts, 107–108
Refineries, 217
Regulated medical waste
    (RMW), 30
Regulating entities, 13
Regulations, 4
    for air monitoring, 198
    for blood-borne pathogens, 246
    federal, 8
Rehydration, 182
Reid, Richard, 217–218
Reinforced drums, 37
REL (recommended exposure
    levels), 122
Relative gas density (RgasD), 71

Relative response, 200, 201
Release within containment
    system, 153
Relief valves, 40, 51, 52, 54,
    154, 159
Remote control robots, 251
Remote opening, of packages,
    251, 253
Remote shutoffs, 197
Reportable quantity (RQ), 32, 103
Reporting facility, 6
Reporting threshold, 6, 7
Reproductive toxins, 117
Rescue, 143–147
Research Development and
    Engineering Command
    (RDEC), 144, 238
Resource team, 141
Respirators, 129
Respiratory depression, 230
Respiratory distress, 230
Respiratory exposure, 119
Respiratory protection, 13
    and evidence collection, 258
    for SWAT teams, 254–256
    and tuberculosis, 247
Respiratory Protection Regulation
    (1910.134), 10, 11
Response guide, 93–96
Ressam, Ahmed, 217
Reston, Virginia, 35
Retention, 196
RgasD (relative gas density), 71
R&I; See Recognition and
    identification
Ricin, 214–216, 221, 224,
    230–232, 234
    CDC categorization of, 231
    detection of, 237
    effectiveness of, 231
    incident examples of, 231–232
    labs for manufacture of, 250
    military designation of, 228
Riddle, Bertier Ray, 232
Righting of overturned trucks,
    58, 59
Right-to-life groups, 217
Riots, 215
Risk Management Program
    (RMP), 13
Risk-based response theory, 114,
    146, 147
Risk/benefit analysis, 146
Riyadh, Saudi Arabia, 218
RMP (Risk Management
    Program), 13

RMW (regulated medical waste), 30
Robots, 251
Rocket-propelled grenades
    (RPGs), 220
Rockford, Illinois, 219–220
Rocks, 193
Roe vs. Wade, 223
Roentgen (R), 116
Rollovers, truck, 46, 160
Rotten egg odor, 205
Rotting organic materials, 204
RPGs (rocket-propelled grenades),
    220
RQ; See Reportable quantity
Rudolph, Eric, 218, 221
Rule of three, 193
Runny noses, 229
Runoff, 162, 174, 180
Russia, 232, 233

SABRE 2000, 236
Sacramento, California, 217, 253
SADT (self-accelerating
    decomposition temperature), 74
Safety officer, 10, 139, 259
Safety Solutions, 235
St. Petersburg, Florida, 215
Salad bars, 215, 220, 232
Saline, 215
Saliva, 246
Salivation, 229
Salmonella bacteria, 215, 220, 232
Sampling, 258–259
Sampling plan, 259
Sampling tubing, 200
Sand, 167, 193
SARA (Superfund Amendments
    and Reauthorization), 5
SARA facilities, 8
SARA Title III, 66, 106, 151
Saratoga gear, 254–256
Sarin, 216, 219, 220, 224, 227,
    228, 231
    Aum Shinryko use of, 221, 222
    exposure levels for, 123
    manufacture of, 250
    military designation of, 228
    vapor pressure of, 71
SARs (supplied air respirators),
    125
SAW (surface acoustic wave), 236
*Sax's Dangerous Properties of
    Industrial Materials,* 108
SBCCI (Southern Building Code
    Congress International), 20

SBCCOM (Soldiers Biological and Chemical Command), 144
SBIMAP (South Baltimore Industrial Mutual Aid Plan), 109
SCA (surface contaminated articles), 91
SCALP suit, 255
SCBA; See Self-contained breathing apparatus
School violence, 214, 215, 217
Schoonover, Kathryn, 216
SCOs (Surface Contaminated Objects), 42
Screen fences, 192
Sea containers, 44
Seattle, Washington, 217
Secondary contamination, 173
Secondary devices, 215, 222, 224, 225, 234, 239
Security measures, unusual, 225
Self-accelerating decomposition temperature (SADT), 74
Self-contained breathing apparatus (SCBA), 6, 112, 113, 124–126, 138
  and carbon monoxide, 207, 208
  decontaminating, 181
  for drug-lab encounters, 249
  and evidence collection, 258
  problems with, 126
  research on, 144, 145
Self-reactive materials (Division 4.1), 27
Senses, hazard detection with, 67
Sensitizer, 9, 116
September 11, 2001, terrorist attacks, 212–214, 217
SERCs (State Emergency Response Committees), 5
SETIQ, 107
Sewers, 150, 195, 204
Shanksville, Pennsylvania, 214, 217
Shareef, Derrick, 219–220
SHBs; See Suicide/homicide bombers
Shearing effect, 157
Sheeting, plastic, 196
Sheltering in place, 150–153
Shigella dysenteria, 216
Shipping papers, 45, 85, 86, 103, 105–106, 147
Ships, 106
"Shock treatment," 171
"Shoe bomber," 217–218
Shutoffs, 46, 47, 197
SI (System International), 116

Sick building calls, 161, 172, 205, 208
Sieverts (Sv), 116
Simple asphyxiants, 119
Site management, 146–153
  case study, 147, 151
  evacuations/sheltering in place, 150–153
  zone establishment, 148–150
Skimmers, floating, 195
Skin absorption, 120
Skin contact, 228, 238, 250
Skin irritation, 229, 230
SLUDGEM (salivation, lacrimation, urination, defecation, gastrointestinal, emesis, miosis), 229
Smallpox, 231, 237
Smith's Detection, 236
Sodium chlorate, 249
Sodium cyanide, 218
Sodium hydroxide, 72, 167, 248, 250
  isolation distance for, 150
  spills of, 146
  vapor pressure of, 71
Sodium nitrate, 249
Soldiers Biological and Chemical Command (SBCCOM), 144
Solid booms, 195
Solidification, 184
Solid polymer electrolyte (SPE) sensors, 202
Solidification agents, 192
Soman, 227, 228
South Baltimore Industrial Mutual Aid Plan (SBIMAP), 109
South Florida, 217
Southern Building Code Congress International (SBCCI), 20
Space Needle, Seattle, 217
Span of control, 139
SPE (solid polymer electrolyte) sensors, 202
Special hazards, marking system for, 33
Specialists, 10, 11
Specialized tank trucks, 51, 52, 54, 55
Specialized tanks, 66–67
Specific gravity, 72
Specification plates, 45
Spills, 97, 99, 155
Splash protective suits, 14
Spokane, Washington, 231

Spontaneously combustible materials, 22, 166–167
Sporting events, 215, 222, 223
Stadiums, 223
Stainless steel drums, 39
Standard for Firefighter Professional Qualifications (NFPA Standard 1001), 3
Standard of care, 12
Standard on Fire Department Occupational Safety and Health Program (NFPA 1500), 13–14
Standard on Protective Ensembles for Chemical/Biological Terrorism Incident, 14
Standards, 4, 11–12
  for air monitoring, 198
  development of, 4
Standard Transportation Commodity Code (STCC), 106
Standoff distances, 156
Staphylococcal enterotoxin B, 231
State Emergency Response Committees (SERCs), 5
States of matter, 68–69
Static electricity, 158, 161
STCC (Standard Transportation Commodity Code), 106
Stealth F117A fighter crash, 144
Steel drums, 39
Sterilization, 184
Stokes baskets, 145
Stomach acid, 73
Storage types, 9
Storm drains, 164, 194, 195
"Stow Away from Foodstuffs" placard, 16, 29
Streaming vapor cloud, 152
Street violence, 214
Stress:
  chemical, 153
  on container vs. chemical, 153
  heat, 131, 132
  mechanical, 153
  thermal, 153
Stroke, heat, 133
Strontium-90, 77
Structural fire situations, 13
Styrene, 201
Sublimation, 69
Subway attacks, 216, 218–221, 227, 228, 232
Sudden release of pressure reactive, 9

Suicide/homicide bombers (SHBs), 212, 219
Suitcase bombs, 156, 252
Sulfur dioxide, 7, 197, 198, 204
Sulfuric acid, 7, 248, 249, 256
  packaging of, 38
  protective actions for, 171
  spills of, 146
  vapor pressure of, 71
Sulfur trioxide, 171
Superfund Act, 13
Superfund Amendments and Reauthorization (SARA), 5
Superfund sites, 13
Super sacks, 39
Supplied air respirators (SARs), 125
Support garments, 14
Suppression, vapor, 197
Surface acoustic wave (SAW), 236
Surface contaminated articles (SCA), 91
Surface Contaminated Objects (SCOs), 42
Surfactants, 163
Suspicious packages, 251, 253
Sv (sieverts), 116
Swango, Michael, 215
SWAT operations, 253–257
  and emergency conditions, 256–257
  intelligence gathering for, 253
  and protective clothing, 255
  setup for, 254–256
  special considerations for, 253–254
SWAT teams, 255–256
Swipe samples, 185
Swoop and scoop, 145, 238
Synthetic chemical agents, 250
Synthetic foam, 163
Syringes, 215
System International (SI), 116

Tabun, 227–228
Tactical rescue team, 239
Tampa, Florida, 218
Tank car specialist, 11
Tank trucks:
  burning, 164
  load capacities of, 44
  righting of overturned, 58, 59
  silhouettes of, 89
  types of, 45
  valve shutoffs on, 197

Targets, of terrorism, 222–223
TATP; See Triacetone triperoxide
Tear gas, 228
Tearing (of eyes), 229
Technetium-99, 77
Technical assistance, 108–109
Technical decontamination, 174–175
Technical Escort Unit (TEU), 238
Technical grade pesticides, 183, 250
Technicians, 10, 12, 198, 259
Temperature, 9
  and BLEVEs, 159
  exertion and body, 120
  and liquefied gas releases, 115
  and oxygen monitors, 202
  and pressurized containers, 54
  and sarin, 228
Tentative interim amendments (TIAs), 12
10,000-pound limit, 7
Teratogen(s), 117, 207
Terrorism, 211–239
  agents of, 227–234
  assessing threats of, 221
  definition of, 220
  indicators of, 223–225
  potential targets of, 222–223
  recent incidents of, 215–220
  response to, 12
  types of, 220, 222
Terrorism agent detectors, 234–237
  biological-agent, 236–237
  colorimetric sampling, 235
  handheld-electronic, 236
  test strips, 234–235
Terrorism agent labs, 249–250
Terrorism agents, 227–234
  biological/toxic, 230–233
  blister/vesicant, 228, 229
  blood/choking, 229, 230
  detection of, 234–237
  incendiary, 228
  irritant, 229, 230
  and Laboratory Response Network, 233
  military designations for, 228
  nerve, 227–229
  pipebomb, 234
  production of, 224
  radioactive, 232–234
  and secondary devices, 234
Terrorist incidents:
  federal assistance with, 237–238
  priorities for, 238–239
  recent examples of, 215–220
  responses to, 225–227

Test strips, 234–235
TEU (Technical Escort Unit), 238
Thermal, radiation, asphyxiation, chemical etiological, mechanical, and psychological (TRACEMP) hazards, 114–115
Thermal hazards, 114
Thermal stress, 153
Thickened soman, 228
Thorium, 77
Threshold limit value (TLV), 122
TIAs (tentative interim amendments), 12
Tier 2 Chemical Inventory Report, 7–9
"Time, distance, and shielding," 169
Tin dioxide sensors, 203
TLV (threshold limit value), 122
TOFC (trailers on flatcars), 61
Tokyo, Japan, 216, 220–222, 227, 228
Toluene, 201
Tool drop, 181
Total vapor survey instrument, 205
Totes, 40–42
Tourists, 219
Toxic, 9
Toxic by inhalation, 168
Toxic gases, 197
Toxic gas monitors, 198, 203–204
Toxicity, 22, 23
Toxic materials, 121
Toxic materials with transportation risk (Division 6.1), 27
Toxic products of combustion, 79
Toxic sensors, 204
Toxicity measurement, 23
Toxicology, 114
Toxins, 230–233, 249–250
TRACEMP hazards; See Thermal, radiation, asphyxiation, chemical etiological, mechanical, and psychological hazards
Tractor-trailers, 44
Trailer silhouettes, 89
Trailers on flatcars (TOFC), 61
Train bombings, 218
Train derailments, 36, 151, 216
Training, 3, 8
  in domestic preparedness, 238
  in evidence collection, 257–258
  for law enforcement, 246
  levels of, 10
  in protective clothing, 254
Transportation corridors, 20, 150
Transportation facilities, 222
Transportation mode, 105–106

Trapped victims, 146, 147
Trenches, 195
Triacetone triperoxide (TATP), 219, 220
Triage, 239
Tritium, 77
Truck bombs, 156, 213, 215, 217, 219, 252
Truck rollovers, 46, 160, 162, 163
Trucks:
    righting of overturned, 58, 59
    silhouettes of, 89
Tube trailers, 54, 55, 160
Tuberculosis, 231, 246, 247
Tularemia, 230–231
Tunnel vision, 115–116
Tunnels, 31
Two-in-two-out rule, 10, 13, 14
Type A containers, 43
Type A organic peroxides, 27
Type B containers, 43
Type B organic peroxides, 27
Type C organic peroxides, 27
Type D organic peroxides, 27
Type E organic peroxides, 27
Type F organic peroxides, 27
Type G organic peroxides, 27
Typhus, 231

UELs; See Upper explosive limits
Underflow dams, 193–194
Underground storage tanks (USTs), 62–64
Underground utilities, 196
Unified command, 226
United Nations (UN), 89
United Nations/North America (UN/NA) identification number, 24, 92
United Parcel Service (UPS), 105
U.S. Army, 238
U.S. Coast Guard, 108, 258
U.S. Congress, 217, 232
U.S. Department of Defense, 238
U.S. Department of Homeland Security, 142, 143, 237, 238
U.S. Department of Justice, 216, 223, 238
U.S. Department of Transportation (DOT), 20, 23, 85
U.S. embassy attacks, 216, 220
U.S. Health and Human Services Administration, 115
U.S. Marines, 238
U.S. Pine Bluff Arsenal, 219

U.S. Postal Service, 218
U.S. Senate, 232
Universal precautions, 246, 247
UN/NA identification number; See United Nations/North America identification number
Unstable reactive, 9
Unusual packages, 230, 251, 253
Uphill position, 148
Upper explosive limits (UELs), 74, 75
UPS (United Parcel Service), 105
Upwind position, 148, 149
Uranium, 77
Uranium-238, 76, 117
Urban Search and Rescue (USAR), 131, 237–239
Urea, 249
Urea nitrate, 25
Urination, 229
Urine, 246
USAR; See Urban Search and Rescue
USTs; See Underground storage tanks
Utilities, underground, 196

V agent, 227, 228
Vacuum trucks, 195
Vacuuming, 184–185
Valves, 158
Valve shutoffs, 197
Van bombs, 156, 213, 215, 219, 252
Van Crocker, Demetrius, 219
Vapor barrier, 124
Vapor clouds, 41, 99, 225, 230
Vapor concentration and toxicity, 23
Vapor density, 70–72, 158
Vapor dispersion, 196–197
Vapor plume shapes, 152
Vapor pressure, 69–71, 150, 255
Vapor production, 228
Vapors, 65
Vapor suppression, 197
Vehicle exhaust, 205
Vehicles, 159–160
Vent and burn technique, 251
Very insensitive explosives with mass explosion potential (Division 1.5), 24
Very insensitive explosives with no mass explosion potential (Division 1.6), 24
Vesicants, 228, 229
Victims:
    questioning, 227
    rescuing/treating live, 238, 239
    trapped, 146, 147

Violent tank rupture (VTR), 51, 154, 159
Viral encephalitis, 231
Viral hemorrhagic fever, 216
Virginia Tech, 214, 215
Visibility, 133, 138
Volatility, 70
Volunteer employees, 8
Vomiting, 229
VTR; See Violent tank rupture
VX, test for, 235
VX nerve agent, 250

Waagner, Clayton Lee, 217
Waco, Texas, 223
"Wanna be" terrorists, 220
Warm zones, 148–150
Warren, Gary, 144
Washing, 185
Washington, DC, 164, 232
Water:
    pH of, 73
    vapor pressure of, 71
Water cannons, 251
Water reactive, 9
Water-reactive materials, 99–100, 166
Water solubility:
    and absorption, 191
    and damming, 193
    and dilution, 196
    and protection of waterways, 72
    and vapor dispersion, 196
Water spray, 97, 98, 196, 197
Water supply, 215, 231
Waterway spills, 193
    damming, 194
    dilution of, 196
    diversion of, 195
Waybill, 105
Weakness, 230
Weapons of mass destruction (WMDs), 227
    assessing threats from, 221
    monitoring competencies for, 198
Weather conditions, 149, 158
Weather inversions, 207
West Virginia, 216
Wetted explosives (Division 4.1), 27
Wetting agents, 163
Wheatstone bridge sensors, 203
White House, 232
White phosphorus, 166
White powder events, 217

White square, 31
White supremacists, 219, 223
"Who's in charge" concerns, 226
Wind direction, 148, 149
WMDs; *See* Weapons of mass destruction
WMD detection devices, 200

Women, 207
Women's Reproductive Health Centers, 217
World Trade Center, 212–214, 217
World Trade Center bombing, 25, 213, 215, 222
Worst-case scenario, 253

X-ray radiation, 78, 206
Xylene, 200, 201

Zones, 146, 148–150
Zytron 300, 128
Zytron 500, 128